SIGNALING THROUGH CELL ADHESION MOLECULES

SIGNALING THROUGH CELL ADHESION MOLECULES

Edited by

Jun-Lin Guan

Department of Molecular Medicine
College of Veterinary Medicine
Cornell University
Ithaca, New York

CRC Press

Boca Raton London New York Washington, D.C.

Library of Congress Cataloging-in-Publication Data

Signaling through cell adhesion molecules / edited by Jun-Lin Guan.
p. cm. -- (Methods in signal transduction)
Includes bibliographical references and index.
ISBN 0-8493-3385-7 (alk. paper)
1. Cell adhesion molecules--Research--Methodology. 2. Integrins--Research--Methodology. 3. Cellular signal transduction--Research--Methodology. I. Guan, Jun-Lin. II. Series.
QP552.C42S57 1999
571.6—dc21 99-17149
CIP

International Standard Book Number 0-8493-3385-7
Library of Congress Card Number 99-17149
Printed in the United States of America 1 2 3 4 5 6 7 8 9 0
Printed on acid-free paper

CRC METHODS IN SIGNAL TRANSDUCTION SERIES

Joseph Eichberg, Jr., Series Editor

SERIES PREFACE

In the past few years, the field of signal transduction has expanded at an enormous rate and has become a major force in the biological and biomedical sciences. Indeed, the importance of this area can hardly be exaggerated and still continues to grow. The knowledge gained has vastly increased and in some instances revolutionized our perceptions of the vast panoply of signaling pathways and molecular mechanisms through which healthy cells respond to both extracellular and intracellular cues, and how these responses can malfunction in a variety of disease states.

The increase in our understanding of signal transduction mechanisms has relied upon the development and refinement of new and existing methods. Successful investigations in this field now often require an integrated approach which utilizes techniques drawn from molecular biology, cell biology, biochemistry, genetics and immunology. The overall aim of this series is to bring together the wealth of methodology now available for research in many aspects of signal transduction. Since this is such a fast-moving field, emphasis will be placed wherever possible on state-of-the-art techniques. Each volume is assembled by one or more editors who are expert in their particular topic and leaders in research relevant to it. Their guiding principle is to recruit authors who will present detailed procedures and protocols in a critical yet reader-friendly format which will be of practical value to a broad audience, including students, seasoned investigators and researchers who are new to the field.

Cell signaling by adhesion molecules, the theme of this volume, is one of the newest, most exciting and most swiftly advancing areas within signal transduction. Our understanding of the interplay between the extracellular matrix and intracellular events which is mediated by integrins and related molecules is still very much in its infancy. Nonetheless, methodologies which are either entirely new or have undergone adaptation, are being rapidly introduced by the growing cadre of investigators who are actively studying this role of adhesion molecules. This volume presents descriptions of some of the most recently developed of these techniques.

Joseph Eichberg, Ph.D.
Advisory Editor for the Series

Preface

The field of signal transduction research is one of the fastest growing in all of biomedical research in recent years. Signaling through cell adhesion molecules represents one of the newest frontiers within the field. Cell adhesion molecules have long been of interest because of their importance in embryonic development, homeostasis, immune responses, wound healing, and malignant transformation. However, it is only recently recognized that cell adhesion molecules are capable of transducing biochemical signals across the plasma membrane to regulate cellular functions. Like other areas of signal transduction research, studies of signaling by cell adhesion molecules have benefited tremendously from methodologies derived from a variety of traditional disciplines including molecular biology, biochemistry, cell biology, and immunology. However, over the last several years, many novel approaches and methods have also been developed to address the specific features of signaling by cell adhesion molecules.

The aim of *Signaling Through Cell Adhesion Molecules* is to bring together these novel and current methodologies in one volume as presented by leading experts whose research has shaped this nascent area of signal transduction research. One of the unique features of this method book is its topic-oriented format, which allows the experts with "hands-on" experiences to present various techniques in the context of particular biological questions. Many chapters not only contain the recipes on experimental protocols, but also discuss the rationale behind the development of novel method or novel applications of an established method. We hope these discussions will stimulate the continued innovations and development of novel approaches in this exciting field as well as to provide a guide for students and other investigators who wish to enter the field.

Like the field itself, this book covers a wide range of topics and methodology. They represent state-of-the-art technologies in the field. Signal transduction by the integrin family cell adhesion receptor has been one of the most extensively studied among cell adhesion molecules. Many of the chapters in this book describe various methods using integrins as model systems. However, all of the methods described can be easily adapted by researchers working on other cell adhesion molecules. Like other cell surface receptors, the cytoplasmic domains of integrins and their interaction with other cellular proteins play important roles in triggering the intercellular signal transduction cascades. Section I describes diverse methods employed in studying protein-protein interactions as well as some of the key players in integrin signalings. Recent studies demonstrated that integrin signaling is a critical factor in a variety of cellular and biological processes. Section II describes various methods used in these studies. Section III describes various strategies developed recently to study inside-out signaling by integrins. Such ability to mediate bi-directional signaling is one of the unique features of integrins and perhaps also other cell adhesion molecules. Section IV describes some of the general methods used in the study of signal transduction by other cell adhesion receptors. In addition, to be highly useful for study of the specific adhesion molecules as described in these chapters, this section also illustrates the multidisciplinary nature and diverse approaches of the signal transduction field.

Besides the broad range of coverage, another equally important feature of the book is its attention to details and emphasis on a "user-friendly" format. Like any other experimental science, the successful use of a particular technique involves much more than just the recipes. It often involves years of experiences in using and trying to improve the methods. Many of the chapters include comments and tips on these details in a user-friendly manner by the expert contributors who have used (or in some cases developed) the methods successfully to make important discoveries in the field. These tips and comments should be extremely helpful for the readers to master the methods described.

I wish to thank all the authors for their outstanding contributions; they share my hope that this volume will assist the further development of the field as well as help students and other investigators to enter the field. In addition, the excellent secretarial assistance of Cindy Westmiller is gratefully acknowledged.

The Editor

Dr. Jun-Lin Guan is an associate professor in the Department of Molecular Medicine at Cornell University College of Veterinary Medicine. He received his B.S. degree in biology from the Chinese University of Science and Technology in 1982. As the top graduate in the biology class, he was awarded the prestigious Kuo Mu-Rao prize from the University. In 1982 Dr. Guan was selected as one of the CUSBEA (China–United States Biochemistry Examination and Admission) fellows to pursue doctoral studies in the U.S. He performed his graduate training on mechanisms of protein trafficking in mammalian cells with Dr. Jack Rose at the Salk Institute. He received the Ph. D. degree in Biology at the University of California at San Diego in 1987. That same year, he was awarded the Anna Fuller Cancer Fund Postdoctoral Fellowship to pursue his postdoctoral training with Dr. Richard Hynes at the Massachusetts Institute of Technology from 1987 to 1991. Dr. Guan joined the faculty at Cornell in 1991 and was promoted to associate professor in 1997. He received the SmithKline Beecham Award for Research Excellence at Cornell University College of Veterinary Medicine in 1993 and an American Heart Association established investigator award in 1997.

While finishing his postdoctoral work at MIT, Dr. Guan discovered that integrin-mediated cell adhesion to FN induced tyrosine phosphorylation of a 120 kDa protein (pp120). Soon after establishing his laboratory at Cornell, he identified pp120 as a novel tyrosine kinase FAK, and found that FAK activation and phosphorylation are regulated by both cell adhesion and transformation by v-Src. Along with contributions from other laboratories, these studies established that integrins are capable of transducing biochemical signals across the plasma membrane and helped to open up the new research field on signal transduction by integrins. Today, signal transduction by the integrin family cell adhesion receptor is one of the most extensively studied among cell adhesion molecules and remains one of the most active areas of biomedical research on signal transduction mechanisms.

Dr. Guan's laboratory has made major contributions to the development of the field of signal transduction by cell adhesion molecules, in particular the role of protein tyrosine kinases in integrin signaling. Dr. Guan and his colleagues have published significant papers in leading journals, which helped to illustrate the intracellular signaling pathways initiated by integrins and to identify the cellular functions regulated by these signaling pathways. Current research in the laboratory is aimed to further define the mechanisms of signal transduction by integrins, as well as to investigate the role of the integrin signaling pathways in various diseases including cancer and cardiovascular disorders.

Contributors

Andrew E. Aplin
Department of Pharmacology
University of North Carolina at Chapel Hill
Chapel Hill, North Carolina

Richard K. Assoian
Department of Pharmacology
School of Medicine
University of Pennsylvania
Philadelphia, Pennsylvania

Allison L. Berrier
Department of Physiology and Cell Biology
Albany Medical College
Albany, New York

Amy L. Bodeau
Department of Physiology and Cell Biology
Albany Medical College
Albany, New York

Michael C. Brown
SUNY Health Science Center Syracuse
Syracuse, New York

Keith Burridge
Department of Cell Biology and Anatomy
and Lineberger Comprehensive Cancer Center
University of North Carolina at Chapel Hill
Chapel Hill, North Carolina

Leslie Cary
Cancer Biology Laboratories
Department of Molecular Medicine
College of Veterinary Medicine
Cornell University
Ithaca, New York

Qiming Chen
Department of Pharmacology
University of North Carolina at Chapel Hill
Chapel Hill, North Carolina

Magdalena Chrzanowska-Wodnicka
Becton Dickinson Technologies
Research Triangle Park, North Carolina

Gabriela Davey
Department of Pharmacology
School of Medicine
University of Pennsylvania
Philadelphia, Pennsylvania

Shoukat Dedhar
Biochemistry and Molecular Biology
Unversity of British Columbia
B.C. Cancer Research Centre/VHHSC
Jack Bell Research Centre
Vancouver, British Columbia, Canada

Fabrizio Dolfi
LaJolla Cancer Research Center
The Burnham Institute
LaJolla, California

François Fagotto
Department of Cell Biology
Max-Planck Institute for
Developmental Biology
Tübingen, Germany

Csilla A. Fenczik
Department of Vascular Biology
The Scripps Research Institute
La Jolla, California

Steven M. Frisch
The Burnham Institute
La Jolla, California

Miguel Garcia-Guzman
LaJolla Cancer Research Center
The Burnham Institute
LaJolla, California

Filippo G. Giancotti
Cellular Biochemistry and Biophysics
Program
Memorial Sloan-Kettering Cancer Center
New York, New York

Mark H. Ginsberg
Department of Vascular Biology
The Scripps Research Institute
La Jolla, California

Jun-Lin Guan
Cancer Biology Laboratories
Department of Molecular Medicine
College of Veterinary Medicine
Cornell University
Ithaca, New York

Gregory E. Hannigan
Department of Laboratory Medicine
and Pathobiology
University of Toronto
Cancer & Blood Program Research
Institute and
Department of Paediatric Laboratory
Medicine
Hospital for Sick Children
Toronto, Ontario, Canada

Takaaki Hato
Department of Vascular Biology
The Scripps Research Institute
La Jolla, California
and Ehime University School of
Medicine
Ehime, Japan

Suzanne M. Homan
Department of Physiology and
Cell Biology
Albany Medical College
Albany, New York

Alan F. Horwitz
Department of Cell and
Structural Biology
University of Illinois
Urbana, Illinois

Alan Howe
Department of Pharmacology
University of North Carolina at
Chapel Hill
Chapel Hill, North Carolina

Rudy L. Juliano
Department of Pharmacology
University of North Carolina at
Chapel Hill
Chapel Hill, North Carolina

Wendy J. Kivens
Department of Laboratory Medicine
and Pathology
Center for Immunology and Cancer
Center
University of Minnesota Medical
School
Minneapolis, Minnesota

Judith Lacoste
Department of Microbiology
Health Sciences Center
University of Virginia
Charlottesville, Virginia

Susan E. LaFlamme
Department of Physiology and Cell Biology
Albany Medical College
Albany, New York

Margot Lakonishok
Department of Cell and Structural Biology
University of Illinois
Urbana, Illinois

Tsung H. Lin
Department of Pharmacology
University of North Carolina at Chapel Hill
Chapel Hill, North Carolina

Betty P. Liu
Department of Cell Biology and Anatomy
University of North Carolina at Chapel Hill
Chapel Hill, North Carolina

Agnese Mariotti
Cellular Biochemistry and Biophysics Program
Memorial Sloan-Kettering Cancer Center
New York, New York

Petra Maschberger
Department of Vascular Biology
The Scripps Research Institute
La Jolla, California

Anthony M. Mastrangelo
Department of Physiology and Cell Biology
Albany Medical College
Albany, New York

Frederick R. Maxfield
Department of Biochemistry
Cornell University
Weill Medical College
New York, New York

W. James Nelson
Department of Molecular and Cellular Physiology
Beckman Center B121
School of Medicine
Stanford Univeristy
Stanford, California

Carol A. Otey
Department of Cell and Molecular Physiology
University of North Carolina at Chapel Hill
Chapel Hill, North Carolina

J. Thomas Parsons
Department of Microbiology
Health Sciences Center
University of Virginia
Charlottesville, Virginia

Peter A. Piepenhagen
Department of Molecular and Cellular Physiology
Beckman Center B121
School of Medicine
Stanford Univeristy
Stanford, California

Lynda M. Pierini
Department of Biochemistry
Cornell University
Weill Medical College
New York, New York

Joe W. Ramos
Department of Vascular Biology
The Scripps Research Institute
La Jolla, California

Mary C. Riedy
SUNY Health Science Center Syracuse
Syracuse, New York

Kristin Roovers
Department of Pharmacology
School of Medicine
University of Pennsylvania
Philadelphia, Pennsylvania

Sarita K. Sastry
Department of Cell Biology and Anatomy
University of North Carolina at Chapel Hill
Chapel Hill, North Carolina

Sanford J. Shattil
Departments of Vascular Biology and Molecular and Experimental Medicine
The Scripps Research Institute
La Jolla, California

Yoji Shimizu
Department of Laboratory Medicine and Pathology
Center for Immunology and Cancer Center
University of Minnesota Medical School
Minneapolis, Minnesota

Jill K. Slack
Department of Microbiology
Health Sciences Center
University of Virginia
Charlottesville, Virginia

Tho Q. Truong
Department of Cell and Structural Biology
University of Illinois
Urbana, Illinois

Christopher E. Turner
SUNY Health Science Center Syracuse
Syracuse, New York

Kristiina Vuori
LaJolla Cancer Research Center
The Burnham Institute
LaJolla, California

Kishore K. Wary
Cellular Biochemistry and Biophysics Program
Memorial Sloan-Kettering Cancer Center
New York, New York

Chuanyue Wu
Department of Cell Biology and The Cell Adhesion and Matrix Research Center
University of Alabama at Birmingham
Birmingham, Alabama

Xiaoyun Zhu
Department of Pharmacology
School of Medicine
University of Pennsylvania
Philadelphia, Pennsylvania

Contents

Section I: Protein-Protein Interactions and Early Events in Integrin Signaling

Chapter 1
The Use of Chimeric Receptors in the Study of Integrin Signaling........................3
Anthony M. Mastrangelo, Amy L. Bodeau, Suzanne M. Homan, Allison L. Berrier, and Susan E. LaFlamme

Chapter 2
Using Synthetic Peptides to Mimic Integrins: Probing Cytoskeletal Interactions and Signaling Pathways........................19
Carol A. Otey

Chapter 3
Use of the Yeast Two-Hybrid System for the Characterization of Integrin Signal Transduction Pathways: Identification of the Integrin-Linked Kinase, $p59^{ILK}$........................33
Gregory E. Hannigan and Shoukat Dedhar

Chapter 4
Focal Adhesion Kinase and Its Associated Proteins........................49
Jill K. Slack, Judith Lacoste, and J. Thomas Parsons

Chapter 5
Analysis of Paxillin as a Multi-Domain Scaffolding Protein........................63
Mary C. Riedy, Michael C. Brown, and Christopher E. Turner

Chapter 6
$P130^{Cas}$ in Integrin Signaling........................81
Fabrizio Dolfi, Miguel Garcia-Guzman, and Kristiina Vuori

Chapter 7
Specificity of Integrin Signaling 101
Kishore K. Wary, Agnese Mariotti, and Filippo G. Giancotti

Section II: Late Events and Biological Functions of Integrin Signaling

Chapter 8
Methods for Study of Integrin Regulation of MAP Kinase Cascades 117
Rudy L. Juliano, Andrew E. Aplin, Alan Howe, Tsung H. Lin, and Qiming Chen

Chapter 9
Methods for Analysis of Adhesion-Dependent Cell Cycle Progression 129
Xiaoyun Zhu, Kristin Roovers, Gabriela Davey, and Richard K. Assoian

Chapter 10
Integrin Modulation of Mitogenic Pathways Involved in Muscle Differentiation 141
Tho Q. Truong, Sarita K. Sastry, Margot Lakonishok, and Alan F. Horwitz

Chapter 11
Methods for Studying Anoikis 161
Steven M. Frisch

Chapter 12
Functional Analysis of FAK and Associated Molecules in Cell Migration 167
Leslie Cary and Jun-Lin Guan

Chapter 13
Integrin Signaling in Pericellular Matrix Assembly 183
Chuanyue Wu

Section III: Inside-Out Signaling by Integrins

Chapter 14
Studies of Integrin Signaling Through Platelet $\alpha_{IIb}\beta_3$ 203
Takaaki Hato, Petra Maschberger, and Sanford J. Shattil

Chapter 15
Tracking Integrin-Mediated Adhesion Using Green Fluorescent Protein and Flow Cytometry 217
Wendy J. Kivens and Yoji Shimizu

Chapter 16
Expression Cloning of Proteins that Modify Integrin Activation.......................... 235
Csilla A. Fenczik, Joe W. Ramos, and Mark H. Ginsberg

Chapter 17
Regulation of Cell Contractility by RhoA: Stress Fiber and Focal Adhesion Assembly 245
Betty P. Liu, Magdalena Chrzanowska-Wodnicka, and Keith Burridge

Section IV: General Methods for Signaling Studies of Cell Adhesion Molecules

Chapter 18
Intra- and Intercellular Localization of Proteins in Tissue *In Situ*........................ 263
Peter A. Piepenhagen and W. James Nelson

Chapter 19
Optical Microscopy Studies of $[Ca2+]_i$ Signaling... 279
Lynda M. Pierini and Frederick R. Maxfield

Chapter 20
Wnt Signaling in *Xenopus* Embryos.. 303
François Fagotto

Index.. 357

SECTION I

Protein-Protein Interactions and Early Events in Integrin Signaling

Chapter 1

The Use of Chimeric Receptors in the Study of Integrin Signaling

Anthony M. Mastrangelo, Amy L. Bodeau, Suzanne M. Homan, Allison L. Berrier, and Susan E. LaFlamme

Contents

I. Introduction 4
II. Activation of Tyrosine Phosphorylation 5
  A. Overview 5
  B. Protocols 5
  C. Materials 8
  D. Buffers 8
III. Inhibition of Cell Attachment 9
  A. Overview 9
  B. Protocols 11
  C. Materials 12
IV. Inhibition of Cell Spreading 12
  A. Overview 12
  B. Protocols 13
  C. Materials 14
  D. Buffers 14
V. Concluding Remarks 14
Acknowledgments 15
References 15

0-8493-3385-7/99/$0.00+$.50

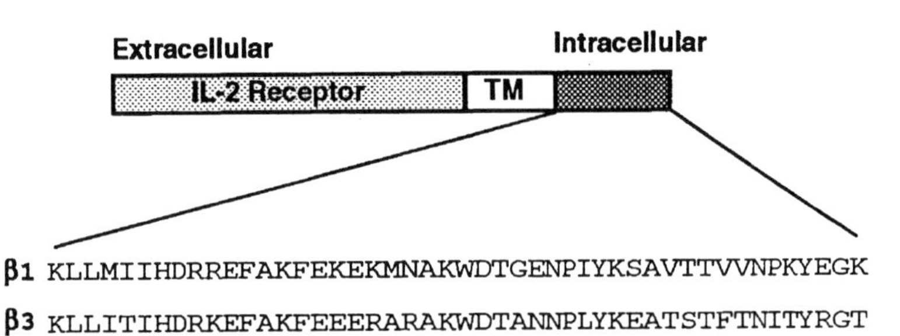

C K

FIGURE 1
Chimeric receptors. Chimeric receptors consist of the extracellular and transmembrane (TM) domains of the tac subunit of the IL-2 receptor[32] connected to various integrin β cytoplasmic domains. The amino acid sequences of the wild-type β1 and β3 cytoplasmic domains are shown. The control receptor (C) consists of the extracellular and transmembrane domains of the IL-2 receptor connected to an intracellular lysine (K) residue.

I. Introduction

Integrins mediate the bidirectional transfer of signals across the plasma membrane. These signals regulate cell adhesion as well as adhesion-dependent aspects of cell behavior including cell proliferation, survival, and differentiation.[1-3] Integrin β cytoplasmic domains function in all steps of the adhesion process, including cell attachment, cell spreading, the formation of focal adhesions, and cell migration.[4,5] They are thought to function in these processes by interacting with specific cytoplasmic proteins, thereby connecting integrins with the cell's cytoskeletal and signal transduction systems. Although several cytoplasmic proteins have been demonstrated to bind to β cytoplasmic domains, their roles in regulating integrin function have not yet been clearly defined.[2,5]

To study the function of β cytoplasmic tails and the molecular mechanisms involved, we constructed chimeric receptors containing wild-type and mutant integrin β subunit cytoplasmic tails connected to the extracellular and transmembrane domain of the human interleukin-2 (IL-2) receptor which functions merely as a reporter domain (Figure 1). Clustering these chimeric receptors on the cell surface can activate signaling pathways by mechanisms similar to endogenous integrins.[6] Therefore, these chimeric receptors can be used to identify and analyze signaling events triggered by integrin β cytoplasmic domains. Additionally, these chimeric receptors can function as dominant inhibitors of endogenous integrin function in a variety of processes including cell attachment and spreading, fibronectin matrix assembly, fibronectin-mediated phagocytosis, and high-affinity ligand binding.[7-11] Identifying the mechanisms and protein interactions involved in these dominant negative effects will provide important insights into the role of β cytoplasmic domains in regulating integrin function.

Recently, we have focused our attention on defining the molecular mechanisms by which these chimeric receptors, when clustered on the cell surface, induce the tyrosine phosphorylation of specific signaling proteins, as well as the mechanisms by which these chimeric receptors inhibit cell attachment and cell spreading. In this chapter, we describe experimental protocols for these studies and provide suggestions on how these protocols can be further utilized to define mechanisms that regulate integrin function.

II. Activation of Tyrosine Phosphorylation

A. Overview

One of the first intracellular signals to be observed upon integrin clustering or integrin-mediated cell adhesion is an increase in the tyrosine phosphorylation of the focal adhesion kinase (FAK), which is a cytoplasmic tyrosine kinase.[12,13] Chimeric receptors have been useful in defining a role for integrin β cytoplasmic domains in triggering FAK phosphorylation, as well as in identifying amino acid motifs within β cytoplasmic domains required to activate FAK phosphorylation.[6,8,14]

The ability of wild-type and mutant β cytoplasmic domains to activate FAK phosphorylation is assayed by transiently transfecting normal human fibroblasts* by electroporation with plasmid DNAs encoding chimeric receptors whose expression is driven by the strong cytomegalovirus promoter. Approximately 15 to 36 h after transfection, the cells are removed from the tissue culture dishes and the chimeric receptors are clustered on the cell surface using magnetic beads coated with antibodies to the IL-2 receptor. The positively expressing cells are recovered using a magnet and then lysed. The tyrosine phosphorylation of FAK in the lysates is then assayed by Western blotting for phosphotyrosine or by immunoprecipitation of FAK, followed by Western blotting for tyrosine phosphorylation. This experimental approach is likely to be useful in testing the ability of β cytoplasmic domains to activate other integrin-triggered events. We have recently observed that the integrin β cytoplasmic domain is also sufficient to induce the tyrosine phosphorylation of p130Cas.** An example of the ability of chimeric receptors to trigger tyrosine phosphorylation is provided in Figure 2.

B. Protocols

1. **Transient transfection** — The transient transfection protocol for normal human fibroblasts is a three-day procedure adapted from a previously published protocol.[15,16] On day 1, confluent cultures are split 1:5. On day 2, in the late afternoon, thymidine is added to a final concentration of 5.6 mM, and the cells are incubated for 10 to 18 h at 37°C to arrest them at the G1/S boundary. The morning of day 3, the thymidine block

* Normal fibroblasts have been used in our studies; however, this experimental approach is likely to be useful with other cell types.

** A. Bodeau and S. LaFlamme, unpublished results.

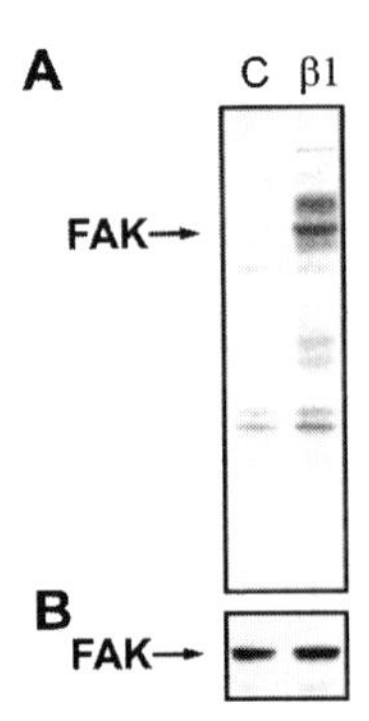

FIGURE 2
Chimeric receptors trigger tyrosine phosphorylation of cytoplasmic proteins. Panel A. Cell lysates were generated from cells transiently transfected with either the control receptor lacking an intracellular domain (C) or chimeric receptors containing the β1 cytoplasmic domain. The lysates were separated by SDS PAGE, transferred to nitrocellulose and then probed with antibodies to phosphotyrosine. Panel B. The filter was stripped and then reprobed with antibodies to FAK. The position of FAK is indicated with an arrow.

is removed by aspirating the culture medium and rinsing the cells once with phosphate-buffered saline, pH 7.2 (PBS), and then adding fresh culture medium. The cells are incubated at 37°C for approximately 8.5 h.* The cells are then removed from the tissue culture dishes by incubation with 0.05% trypsin and 0.5 mM EDTA in Hanks balanced salt solution without Ca^{2+} and Mg^{2+}, washed once with PBS and once with electroporation buffer. The cells are then resuspended at 1.5×10^6 cells per 0.5 ml of electroporation buffer. Plasmid DNA** (20 to 30 μg) is added to the cells, and the mixture is transferred to electroporation cuvettes and electroporated at 170 mV and 960 μF.*** The electroporated cells are then incubated in the cuvettes for 5 min at ambient temperature to allow the cells to recover. The cells are then transferred to 15 ml conical centrifuge tubes and washed once with culture medium. The cells are then cultured for 14 to 16 h at 37°C in medium containing 5 mM sodium butyrate to enhance the expression of the IL-2 receptor chimeras from the cytomegalovirus promoter.[15] Signaling experiments are performed from 15 to 36 h after transfection.

2. **Loading the magnetic beads with antibodies to the IL-2 receptor** — The magnetic sorting of positively transfected cells was performed by a modification of a previously described protocol.[17] The first step in this procedure is to load magnetic beads with antibodies to the IL-2 receptor. In most signaling experiments, we pool cells from 5 transfections. To magnetically sort this many cells requires 800 μl of beads which are first washed several times in 10 ml of PBS to remove the sodium azide (used as a preservative). After each wash, the beads are recovered using a magnet. The beads are then incubated with 10 μg of antibodies to the IL-2 receptor (7G7B6) in 1 ml of PBS at 37°C for 30 min. The beads are washed twice with PBS to remove the unbound antibody, once with sorting medium, and then resuspended in 5 ml of sorting medium.

* The period of time from the removal of the thymidine block to when the cells begin mitosis varies depending upon cell type, and should be determined empirically.
** We find that plasmid DNA purified on CsCl gradients gives the highest transfection efficiency.
*** The voltage and capacitance that give the highest transfection efficiency is cell-type specific, and therefore must also be determined empirically.

3. **Clustering the chimeric receptors on the cell surface** — Transiently transfected normal human fibroblasts are harvested with trypsin-EDTA. Once cells are dislodged from the tissue culture dishes, soybean trypsin inhibitor is added to inhibit further trypsin-mediated proteolysis. The transfected cells are washed twice with PBS, once with sorting medium, and then resuspended in 5 ml of sorting medium. To cluster the chimeric receptors on the cell surface, the cell suspension is mixed with the antibody-coated magnetic beads and then incubated at ambient temperature for 30 to 40 min with gentle mixing.
4. **Magnetic sorting of positively expressing cells and lysate preparation for the detection of phosphorylated proteins** — Using the magnet, the cells are washed twice with 10 ml of PBS. The cells are then resuspended in 1 ml of PBS and transferred to a 1.5 ml Eppendorf tube. Transfected cells coated with magnetic beads are recovered using a magnet specifically made for Eppendorf tubes and the PBS is removed. The tubes are then removed from the magnet, and the cells are incubated in 100 μl of lysis buffer on ice for 10 min. To remove lysate from the magnetic beads, the tubes are then placed on the magnet and the lysates are pipetted into clean Eppendorf tubes. Lysates are then centrifuged for 15 min at 4°C in a microcentrifuge and the supernatants are collected. A BCA protein assay is used to determine the protein concentration of each lysate.
5. **Western blotting for phosphotyrosine** — To analyze the tyrosine phosphorylation of FAK, 10 to 15 μg of cell lysate is separated by sodium dodecyl sulfate (SDS) polyacrylamide gel electrophoresis (PAGE) under reducing conditions.[18] In some cases, FAK is immunoprecipitated from the lysate prior to electrophoresis. The proteins are then transferred from the gel to nitrocellulose filters using a trans-blot cell apparatus. The filters are first incubated in blocking buffer for 2 h at 37°C to inhibit nonspecific protein interactions, and then for 1 h at ambient temperature with the mouse monoclonal antibody to phosphotyrosine, 4G10, diluted 1/2000 in blocking buffer. The filters are washed three times for 10 min in PBS containing 0.1% Tween-20 to remove unbound antibody. The filters are then incubated with horseradish peroxidase-conjugated antibodies to mouse IgG for 30 min and then washed again as described above. Antibody binding is visualized by enhanced chemiluminescence (ECL) following the protocol provided by the supplier. All incubations with antibodies and all washes are performed at ambient temperature on a rocking platform.
6. **Immunoprecipitation of FAK** — FAK is immunoprecipitated from 150 to 300 μg of cell lysate from transfected cells by incubation with 1 to 5 μg of polyclonal antibodies to FAK at 4°C on a rocker. Immune complexes are recovered with Protein A-Sepharose at 4°C on a rocker for 1 h. The beads are washed three times with immunoprecipitation buffer to remove unbound proteins and then the beads are resuspended in 30 μl of 2X sample buffer containing β-mercaptoethanol as a reducing agent. The samples are incubated at 100°C for 5 min and then analyzed by SDS PAGE. The phosphorylation of FAK is then assessed by Western blotting with antibodies to phosphotyrosine as described above.
7. **Stripping and reprobing Western blots** — After visualizing the phosphotyrosine signal, the Western blot is stripped of antibody and reprobed with antibodies to FAK to ensure that differences in signals for phosphotyrosine are not due to differences in protein loading onto the gels. To accomplish this, the filters are first rinsed briefly in PBS to remove excess ECL reagents, and then incubated in stripping buffer at 70°C for 60 min. The filters are then washed extensively with PBS containing 0.1% Tween-20 to remove residual SDS and β-mercaptoethanol. The filters are incubated in blocking

buffer and stained with the mouse monoclonal antibody to FAK (1/1000 dilution) as described for Western blotting with antibodies to phosphotyrosine.

C. Materials

1. **Normal human fibroblasts** — We obtained normal human foreskin fibroblasts from Vec Technologies (Rensselaer, NY). However, there are several protocols for the isolation of these cells in the literature.
2. **Antibodies to the human IL-2 receptor** — For clustering experiments, we currently use the mouse monoclonal antibody, clone 7G7B6. Purified 7G7B6 IgG is available commercially from Upstate Biotechnology Incorporated (Lake Placid, NY). In addition, one can purchase the 7G7B6 hybridoma cell line from American Type Culture Collection (Rockville, MD).
3. **Antibodies to phosphotyrosine** — We have generally used monoclonal antibody 4G10 from Upstate Biotechnology Incorporated for the detection of tyrosine phosphorylated proteins.
4. **Antibodies to FAK** — We find that the monoclonal antibody to FAK (clone 77) available from Transduction Laboratories (Lexington, KY) works well for Western blotting. To immunoprecipitate FAK, we have used rabbit polyclonal antibodies to FAK provided by Drs. Jun-Lin Guan (Cornell University) and Thomas Parsons (University of Virginia). Polyclonal antibodies to FAK are also available commercially from Upstate Biotechnology Incorporated.
5. **Bicinchoninic acid (BCA) protein assay** — We utilize the Micro BCA protein assay reagent kit from Pierce (Rockford, IL) to quantitate protein concentration in cell lysates.
6. **Magnetic beads and magnets** — Anti-mouse IgG conjugated magnetic beads (catalogue number 8-4340D) and the permanent magnets for magnetic separation (catalogue number 8-4101S, and catalogue number 8-MB4111S) can be purchased from PerSeptive Biosystems (Cambridge, MA).
7. **Trypsin/EDTA and Soybean Trypsin Inhibitor** — These reagents are from Life Technologies (Grand Island, NY) and Sigma (St. Louis, MO), respectively.
8. **Electroporator** — For transfecting our cells, we use a Bio-Rad gene pulser with a capacitance extender and electroporation cuvettes from Bio-Rad (Hercules, CA).
9. **Trans-blot cell apparatus** — For transferring proteins from SDS PAGE gels to nitrocellulose, we use a Bio-Rad trans-blot cell apparatus.
10. **ECL** — ECL reagents were purchased from Amersham (Arlington Heights, IL).

D. Buffers

1. **Culture medium** — Cells are cultured in Dulbeco's modified Eagle medium (DMEM) from Life Technologies (Grand Island, NY), supplemented with 10% fetal bovine serum (Hyclone Laboratories Inc., Logan, UT) in the case of fibroblasts and with 5% fetal bovine serum in the case of MG-63 cells.
2. **Electroporation buffer** — 20 mM HEPES (pH 7.05) containing 137 mM NaCl, 5 mM KCl, 0.7 mM Na_2PO_4, 6 mM dextrose, and 1 mg/ml bovine serum albumin (BSA).

3. **Sorting medium** — PBS containing 4 mM EDTA, 1 mM $MgCl_2$, 100 μg/ml chondroitin sulfate, 1 mg/ml nonfat dry milk, 10 μg/ml BSA, and 10 mM HEPES. The final pH is adjusted to 7.2, and the medium is filter sterilized.
4. **Modified RIPA buffer** — 50 mM Tris-HCl (pH 7.4) containing 1% Nonidet P-40, 0.25% sodium deoxycholate, 150 mM NaCl, 1 mM EDTA, and protease and phosphatase inhibitors (1 mM PMSF, 1 μg/ml leupeptin, 1 μg/ml aprotinin, 1 mM Na_3VO_4, and 1 mM NaF).
5. **Blocking buffer** — Blocking buffer contains 3% BSA and 0.1% Tween-20 in PBS.
6. **Transfer buffer** — 5 mM sodium borate.
7. **Immunoprecipitation buffer** — 20 mM Tris-HCl (pH 8.0) containing 1% Nonidet P-40, 10% glycerol, 1 mM Na_3VO_4, and protease and phosphatase inhibitors (1 mM PMSF, 20 μg/ml leupeptin, 20 μg/ml aprotinin, 1 mM Na_3VO_4, and 1 mM NaF).
8. **Stripping buffer** — 62.5 mM Tris-HCl (pH 6.8) containing 2% SDS and 100 mM β-mercaptoethanol.

III. Inhibition of Cell Attachment

A. Overview

Integrin-mediated cell attachment to the extracellular matrix (ECM) is a prerequisite for many cellular processes, including cell spreading, migration, proliferation, and differentiation. Cell attachment is associated with intracellular changes such as actin polymerization and strengthened interactions of integrins with the cytoskeleton.[19] Integrin β cytoplasmic domains have been implicated in the regulation of cell attachment, since recombinant heterodimeric integrins lacking the β subunit cytoplasmic domain or containing mutations in the β cytoplasmic domain are inhibited in their ability to mediate cell attachment.[4,20-22] Integrin β cytoplasmic domains are likely to regulate cell attachment by interacting with cytoplasmic proteins that, in turn, modulate the ability of integrins to interact extracellularly with the ECM and intracellularly with the actin cytoskeleton.[1,2,5,23]

We have recently demonstrated that expression of high levels of chimeric receptors containing specific integrin β cytoplasmic domains can inhibit cell attachment, and have identified regions within the β cytoplasmic domain that are involved in regulating β1 integrin-mediated cell attachment.* Since this dominant negative effect on cell attachment requires high levels of expression of the chimeric receptors, and high expressors represent only a small population of the transiently transfected cells, a standard cell attachment assay could not be used. For this reason, we developed a cell attachment assay which utilizes flow cytometry to correlate the level of expression of the chimeric receptor with the ability of transiently transfected cells to adhere to fibronectin, a β1 integrin-mediated process. The experimental strategy

* A. Mastrangelo, S. Homan, and S. LaFlamme, manuscript submitted.

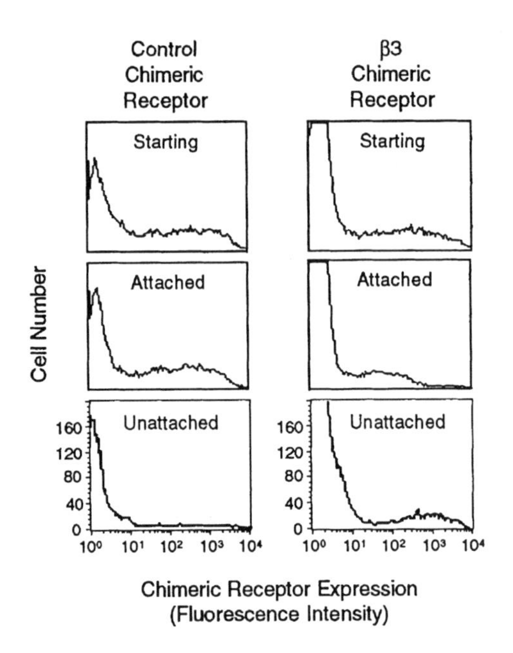

FIGURE 3

Expression of high levels of chimeric receptors containing β cytoplasmic domains inhibit cell attachment. Flow cytometry was used to examine the attachment properties of MG-63 cells that transiently expressed chimeric receptors containing either no cytoplasmic domain (control receptor) or the β3 cytoplasmic domain. Shown are representative histograms depicting the levels of chimeric receptor expression on the starting population, the attached cells, and the unattached cells. Notice that in the starting population the distribution ranges from cells that are negative for the chimeric receptor ($<10^1$ fluorescence units) to cells that express high levels of the chimeric receptor ($>10^3$ fluorescence units). Also notice that cells expressing high levels of the β3 chimeric receptor did not attach to fibronectin, as indicated by the absence of high expressing cells in the attached population.

is to examine the expression of the chimeric receptor in the attached and unattached cell populations using flow cytometry. Using this approach, we were able to quantitatively and qualitatively compare the expression of transfected chimeric receptors in attached and unattached populations of cells with that of the starting population. An example of the ability of chimeric receptors to inhibit attachment of MG-63* cells to fibronectin is provided in Figure 3. Notice that cells expressing high levels of the chimeric receptor containing the β1 cytoplasmic domain are present in the unattached population of cells, whereas cells expressing low to moderate levels of the β1 chimera maintain their ability to attach to fibronectin (Figure 3). This is also reflected by the mean level of expression of the chimeric receptor in the unattached population of cells, which is approximately twice that of the starting population, and by the mean expression level of the chimeric receptors in the attached population, which is approximately half that of the starting population. Overall, this cell attach-

* We have mainly used MG-63 cells in our cell attachment studies; however, similar results are obtained with normal human fibroblasts.

ment protocol is a rapid and highly reproducible method that allows the analysis of the attachment capabilities of a small population of transiently transfected cells.

B. Protocols

1. **Transient transfection** — The procedure for the transient transfection of the human osteosarcoma cell line, MG-63, is essentially the same as that for normal human fibroblasts described above. However, since we find that MG-63 cells are resistant to growth arrest in response to thymidine, we transfect subconfluent cultures which contain a moderate number of mitotic cells.
2. **Coating tissue culture dishes with fibronectin** — Human fibronectin is diluted in PBS to a final concentration of 10 μg/ml. Approximately 6 ml of the PBS/fibronectin solution is added per 100 mm tissue culture dish. The dishes are then incubated at 37°C for 1 h on a rocking platform. The solution of fibronectin is then removed and the dishes are washed several times with PBS before the attachment assay.*
3. **Cell attachment** — Transiently transfected cells are harvested 15 to 36 h after transfection using trypsin/EDTA. Once the cells are dislodged from the tissue culture surface, soybean trypsin inhibitor is added to inhibit further trypsin-mediated proteolysis. The cells are then washed several times with PBS. Approximately 2.0×10^6 cells per sample are resuspended in 1 ml of serum-free culture medium (pre-warmed to 37°C) and placed in 1.5 ml polypropylene** microfuge tubes. These cells are used in the attachment assay. In addition, approximately 1.0×10^6 cells from each transfection are resuspended in 1 ml of serum-free culture medium and placed in separate microfuge tubes. These cells represent the starting population of each transfection and are not used in an attachment assay, but are kept at 37°C until all samples are processed for flow cytometry. The microfuge tubes containing the cells for the attachment assay are incubated at 37°C for 15 min with the caps*** open to allow the cells to recover from the trypsinization. Subsequently, the cells are added to fibronectin-coated 100 mm tissue culture dishes that contain 9 ml of serum-free culture medium (pre-warmed to 37°C). Once the cells are added to the fibronectin-coated dishes, they are incubated at 37°C for 10 min. In order to recover the unattached cells, the dishes are rotated at 150 rpm on an orbital shaker for 30 sec at ambient temperature and the medium containing the unattached cells is then removed and placed in a 15 ml conical tube. To recover any remaining unattached cells, 5 ml of pre-warmed culture medium is gently added to each dish and the dishes are gently rocked three times by hand. This medium is added to the 15 ml conical tube that contains the previously collected unattached cells. Since the unattached cells are usually few in number, to quantitatively recover and analyze the unattached cells, 1.0×10^6 "carrier" non-transfected MG-63 cells are added to each sample containing unattached cells. Next, the attached cells are washed twice with PBS and removed from the dishes using trypsin/EDTA. The trypsin is inactivated with soybean trypsin inhibitor. The unattached cells and the attached cells are kept separate

* The dishes can be blocked with 1% heat-denatured BSA as described in the next section. However, we generally do not find this step necessary for our short attachment assays.

** Polypropylene tubes are preferred because cells tend to adhere less to polypropylene than to polystyrene.

*** The caps to the microfuge tubes are left open so that the sodium bicarbonate in the medium can be buffered by CO_2 in the incubator to a pH of 7.3 to 7.4.

and are washed twice with PBS. The starting population of cells is also washed twice with PBS. Each sample is then analyzed by flow cytometry for expression of the chimeric receptor.

4. **Flow cytometry** — Each sample, which contains approximately 1.0×10^6 cells, is centrifuged, resuspended in 100 µl of cold PBS containing 0.01% sodium azide, and placed in polypropylene microfuge tubes. Sodium azide is used to prevent antibody-mediated capping.[24] Phycoerythrin(PE)-conjugated mouse anti-human IL-2 receptor or an isotype-matched control antibody is added to the samples according to the manufacturer's instructions. The samples are then mixed by gentle vortexing and incubated at 4°C for 30 min, with gentle mixing every 10 min. After two washes with cold PBS/0.01% sodium azide, the cells are resuspended in 0.6 ml of PBS containing 1% formaldehyde and stored at 4°C in the dark for up to 48 h without loss of fluorescence signal or cell integrity. The cells can be analyzed by flow cytometry anytime within this 48 h period. Equal numbers of cells expressing the chimeric receptor are analyzed from the starting, attached, and unattached cell populations.

C. Materials

1. **MG-63 cells** — MG-63 cells are a human osteosarcoma cell line that can be purchased from American Type Culture Collection.
2. **Monoclonal antibodies to the human Interleukin-2 (IL-2) receptor** — For flow cytometric studies, we use a PE-conjugated mouse, anti-human monoclonal antibody (clone 2A3) from Becton Dickinson (San Jose, CA).
3. **Isotype-matched negative control antibody for flow cytometry** — Nonspecific antibody binding in our flow cytometric studies is determined using a nonspecific mouse antibody (Becton Dickinson) that is PE-conjugated and isotype-matched to the antibody specific for the human IL-2 receptor.
4. **Human plasma fibronectin** — was generously provided by Dr. Denise Hocking (Albany Medical College). Human plasma fibronectin is also commercially available from Life Technologies (Grand Island, NY).
5. **Flow cytometer** — We use the FACScan flow cytometer from Becton Dickinson for our experiments.

IV. Inhibition of Cell Spreading

A. Overview

Integrins are thought to function in cell spreading by forming transmembrane links between the extracellular matrix and the cytoskeleton, and by triggering signaling events that regulate cytoskeletal organization and cell shape. Recombinant heterodimeric integrins lacking β cytoplasmic domains or containing specific mutations within their β cytoplasmic domains are inhibited in their ability to mediate cell spreading, suggesting that integrin β cytoplasmic domains also play a role in regulating cell spreading.[22,25,26] Additionally, the observation that chimeric receptors

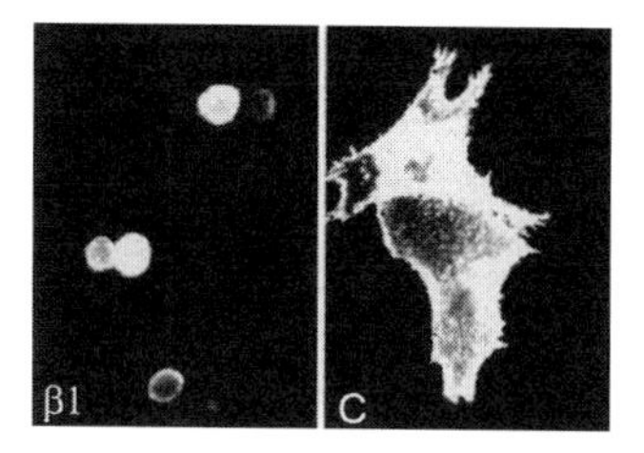

FIGURE 4
Chimeric receptors containing the β1 cytoplasmic domain inhibit cell spreading. Cells transiently expressing chimeric receptors containing the β1 cytoplasmic domain (β1) or the control receptor (C) were allowed to attach to collagen I coated coverglasses for 1.5 h, then fixed and stained with antibodies to the IL-2 receptor and analyzed by immunofluorescence microscopy. Notice that cells expressing the β1 chimeric receptors are round, whereas cells expressing the control receptor are well spread.

containing certain β cytoplasmic domains (β1 and β3) inhibit cell spreading, while others have no inhibitory effect (β4 and β5), supports the idea that β cytoplasmic domains regulate cell spreading by interacting with specific cytoplasmic proteins.[7] Understanding the molecular mechanisms by which these chimeric receptors inhibit cell spreading is likely to provide insights into the mechanism by which integrins regulate changes in cell shape.

To examine the ability of chimeric receptors to inhibit cell spreading on specific matrix proteins, normal human fibroblasts or MG-63 cells are transfected with the various chimeric receptors. 15 h post transfection, the transiently transfected cells are plated on coverslips coated with specific extracellular matrix proteins, such as fibronectin, laminin, vitronectin, and collagen I, and then incubated at 37°C for 1.5 h. After 1.5 h, the cells are fixed and stained with antibodies to the IL-2 receptor. The percentage of transfected cells inhibited in spreading is calculated following the examination of the cells by immunofluorescence microscopy. An example of the ability of the β1 chimeric receptor to inhibit cell spreading is provided in Figure 4.

B. Protocols

1. **Coating coverslips with matrix proteins** — Fibronectin, vitronectin, or collagen I are diluted to 20 μg/ml in PBS. Approximately 500 μl of each solution is pipetted onto individual glass coverslips and the coverslips are incubated in a moistened chamber overnight at 4°C. The coverslips are then washed with PBS and blocked for 30 min at ambient temperature with PBS containing 1% BSA, which has been heat-denatured at 80°C for 1 h. The coverslips are extensively washed with PBS to remove excess BSA.
2. **Cell spreading** — Normal fibroblasts or MG-63 cells are transfected by electroporation as described above. 15 h post transfection, the cells are harvested and resuspended in serum-free culture medium and allowed to recover at 37°C for 20 min. The cells are then plated on coverslips coated with the various matrix proteins in serum-free culture medium and allowed to spread at 37°C for 1.5 h. To assay cell spreading of transfected cells, the cells are stained with antibodies to the IL-2 receptor. The inhibition of cell spreading is analyzed in 10 groups of 10 randomly selected transfected and non-transfected cells by immunofluorescence and phase contrast microscopy. The inhibition

of cell spreading is then calculated from the ratio of the percentage of transfected to nontransfected cells that are inhibited in cell spreading.

3. **Staining cells for immunofluorescence microscopy** — To stain cells for immunofluorescence microscopy, the coverslips are washed several times with PBS, and then fixed with a 4% formaldehyde solution for 30 min at ambient temperature. The fixative is removed and the coverslips are again washed several times with PBS. We generally use a FITC-conjugated monoclonal antibody to the IL-2 receptor at the dilution suggested by the manufacturer. If nonspecific antibody binding is a problem, the cells can be blocked with a BSA-glycine solution prior to incubation with the antibody.

C. Materials

1. **Extracellular matrix proteins** — Fibronectin, vitronectin, and collagen I are available commercially from Life Technologies (Grand Island, NY).
2. **FITC-conjugated anti-IL2-receptor** — Although there are several FITC-conjugated antibodies available commercially, we have used the mouse anti-human IL-2 receptor antibody from Accurate Chemical & Scientific Corp. (Westbury, NY).

D. Buffers

1. **4% formaldehyde fix** — Cells are fixed for immunofluorescence microscopy in PBS containing 4% formaldehyde and 5% sucrose.
2. **BSA-glycine blocking solution** — Nonspecific antibody binding is inhibited by incubation with a solution containing 3% BSA and 0.1% glycine dissolved in H_2O (final pH 7.2).

V. Concluding Remarks

The value of the chimeric receptor approach is the ability to isolate the role of integrin β cytoplasmic domains from effects due to the rest of the integrin heterodimer, and the ability of β cytoplasmic domains expressed in chimeric receptors to interact with intracellular proteins in a manner functionally similar to β cytoplasmic domains present in heterodimeric integrin receptors. This approach is likely to continue to be useful in identifying the mechanisms by which integrin β cytoplasmic domains contribute to integrin function. For example, cell adhesion triggers a number of signals including the activation of protein kinase C, phosphoinositide-3 kinase, and small GTP binding proteins.[27-31] Chimeric receptors will be useful in defining the role of β cytoplasmic domains in the activation of these signals. Although the exact mechanism of the dominant negative effect is not yet known, the chimeric receptors are thought to function as dominant inhibitors, either (1) by sequestering cytoplasmic proteins required for endogenous integrin function; (2) by inhibiting signaling pathways required for integrin function; or (3) by activating signaling pathways that inhibit integrin function. This can be ascertained by co-expressing the chimeric receptors with proteins known to bind to β cytoplasmic domains and then assaying for a reversion of the inhibitory phenotype induced by the chimeric receptors. Similarly, the role of specific signaling pathways in the dominant negative phenotype can be tested by co-expressing dominant negative or active forms of signaling proteins suspected to be involved in regulating attachment and spreading, and then assaying for a reversion of the inhibitory phenotype. Moreover, these techniques can be easily adapted to investigate the signaling characteristics of non-integrin receptors and to study the effects of non-integrin proteins on these processes.

Acknowledgments

The authors thank Dr. Ken Yamada for his support during the development of the chimeric receptor system, and Dr. Jane Sottile for helpful comments during the preparation of this manuscript. The work in the authors' laboratory is supported by NIH Grants GM51540, T32HL07194, and T32HL07529, and AHA (NY-affiliate) Grants 960148 and 970151.

References

1. Yamada, K. M. and Miyamoto, S., Integrin transmembrane signaling and cytoskeletal control, *Curr. Opin. Cell Biol.,* 7, 681, 1995.
2. Dedhar, S. and Hannigan, G. E., Integrin cytoplasmic interactions and bidirectional transmembrane signaling, *Curr. Opin. Cell Biol.,* 8, 657, 1996.
3. LaFlamme, S. E. and Auer, K. L., Integrin signaling, *Sem. Can. Biol.,* 7, 111, 1996.
4. Sastry, S. K. and Horwitz, A. F., Integrin cytoplasmic domains: mediators of cytoskeletal linkages and extra- and intracellular initiated transmembrane signaling, *Curr. Opin. Cell Biol.,* 5, 819, 1993.
5. LaFlamme, S. E., Homan, S. M., Bodeau, A. L., and Mastrangelo, A. M., Integrin cytoplasmic domains as connectors to the cell's signal transduction apparatus, *Matrix Biol.,* 16, 153, 1997.
6. Akiyama, S. K., Yamada, S. S., Yamada, K. M., and LaFlamme, S. E., Transmembrane signal transduction by integrin cytoplasmic domains expressed in single-subunit chimeras, *J. Biol. Chem.,* 269, 15961, 1994.
7. LaFlamme, S. E., Thomas, L. A., Yamada, S. S., and Yamada, K. M., Single-subunit chimeric integrins as mimics and inhibitors of endogenous integrin function in receptor localization, cell spreading and migration, and matrix assembly, *J. Cell Biol.,* 126, 1287, 1994.
8. Lukashev, M. E., Sheppard, D., and Pytela, R., Disruption of integrin function and induction of tyrosine phosphorylation by the autonomously expressed β1 integrin cytoplasmic domain, *J. Biol. Chem.,* 269, 18311, 1994.
9. Smilenov, L., Briesewitz, R., and Marcantonio, E. E., Integrin β1 cytoplasmic domain dominant negative effects revealed by lysophosphatidic acid treatment, *Mol. Biol. Cell,* 5, 1215, 1994.

10. Blystone, S. D., Lindberg, F. P., LaFlamme, S. E., and Brown, E. J., Integrin β3 cytoplasmic tail is necessary and sufficient for regulation of α5β1 phagocytosis by αvβ3 and integrin associated protein, *J. Cell Biol.,* 130, 745, 1995.
11. Chen, Y.-P., O'Toole, T. E., Shipley, T., Forsyth, J., LaFlamme, S. E., Yamada, K. M., Shattil, S. J., and Ginsberg, M. H., "Inside-out" signal transduction inhibited by isolated integrin cytoplasmic domains, *J. Biol. Chem.,* 269, 18307, 1994.
12. Schaller, M. D. and Parsons, J. T., Focal adhesion kinase and associated proteins, *Curr. Opin. Cell Biol.,* 6, 705, 1994.
13. Guan, J.-L., Focal adhesion kinase in integrin signaling, *Matrix Biol.,* 16, 195, 1997.
14. Tahiliani, P. D., Singh, L., Auer, K. L., and LaFlamme, S. E., The role of conserved amino acid motifs within the integrin β3 cytoplasmic domain in triggering focal adhesion kinase phosphorylation, *J. Biol. Chem.,* 272, 7892, 1997.
15. Goldstein, S., Fordis, C. M., and Howard, B. H., Enhanced transfection efficiency and improved cell survival after electroporation of G2/M-synchronized cells and treatment with sodium butyrate, *Nucl. Acid Res.,* 17, 3959, 1989.
16. Giordano, T., Howard, T. H., Coleman, J., Sakamoto, K., and Howard, B. H., Isolation of a population of transiently transfected quiescent and senescent cells by magnetic sorting, *Exp. Cell Res.,* 192, 193, 1991.
17. Padmanabhan, R., Corsico, C., Holter, W., Howard, T., and Howard, B. H., Purification of transiently transfected cells by magnetic affinity cell sorting, *Anal. Biochem.,* 170, 341, 1988.
18. Laemmli, U. K., Cleavage of structural proteins during the assembly of the head of bacteriophage T4, *Nature,* 227, 680, 1970.
19. Lotz, M., Burdsal, C., Erickson, H., and McClay, D., Adhesion to fibronectin and tenascin: Quantitative measurements of initial binding and subsequent strengthened response, *J. Cell Biol.,* 109, 1795, 1989.
20. Hibbs, M. L., Jakes, S., Stacker, S. A., Wallace, R. W., and Springer, T. A., The cytoplasmic domain of the integrin lymphocyte function-associated antigen 1β subunit: sites required for binding to intercelluler adhesion molecule 1 and the phorbol ester-stimulated phosphorylation site, *J. Exp. Med.,* 174, 1227, 1991.
21. Hibbs, M. L., Xu, H., Stacker, S. A., and Springer, T. A., Regulation of adhesion to ICAM-1 by the cytoplasmic domain of LFA-1 integrin β subunit, *Science,* 251, 1611, 1991.
22. Filardo, E. J., Brooks, P. C., Deming, S. L., Damsky, C., and Cheresh, D. A., Requirement of the NPXY motif in the integrin β3 subunit cytoplasmic tail for melanoma cell migration *in vitro* and *in vivo*, *J. Cell Biol.,* 130, 441, 1995.
23. O'Toole, T. E., Integrin signaling: building connections beyond the focal contact? *Matrix Biol.,* 16, 165, 1997.
24. Bourguignon, L. Y., Jy, W., Majercik, M. H., and Bourguignon, G. J., Lymphocyte activation and capping of hormone receptors, *J. Cell. Biochem.,* 37, 131, 1988.
25. Ylanne, J., Chen, Y., O'Toole, T. E., Loftus, J. C., Takada, Y., and Ginsberg, M. H., Distinct functions of integrin α and β cytoplasmic domains in cell spreading and formation of focal adhesions, *J. Cell Biol.,* 122, 223, 1993.
26. Ylanne, J., Huuskonen, J., O'Toole, T. E., Ginsberg, M. H., Virtanen, I., and Gahmberg, C. G., Mutation of the cytoplasmic domain of the integrin β3 subunit, *J. Biol. Chem.,* 270, 9550, 1995.

27. Clark, E. A. and Hynes, R. O., Ras activation is necessary for integrin-mediated activation of extracellular signal-regulated kinase 2 and cytosolic phospholipase A2 but not for cytoskeletal organization, *J. Biol. Chem.,* 271, 14814, 1996.
28. King, W. G., Mattaliano, M. D., Chan, T. O., Tsichlis, P. N., and Brugge, J. S., Phosphatidylinositol 3-kinase is required for integrin-stimulated AKT and raf-1/mitogen-activated protein kinase activation, *Mol. Cell. Biol.,* 17, 4406, 1997.
29. Khwaja, A., Rodriguez-Viciana, P., Wennstrom, S., Warne, P. H., and Downward, J., Matrix adhesion and Ras transformation both activate a phosphoinositide 3-OH kinase and protein kinase B/Akt cellular survival pathway, *EMBO J.,* 16, 2783, 1997.
30. Chun, J.-S., Ha, M.-J., and Jacobson, B. S., Differential translocation of protein kinase C ε during HeLa cell adhesion to a gelatin substratum, *J. Biol. Chem.,* 271, 13008, 1996.
31. Vuori, K. and Ruoslahti, E., Activation of protein kinase C precedes $\alpha 5\beta 1$ integrin-mediated cell spreading on fibronectin, *J. Biol. Chem.,* 268, 21459, 1993.
32. Leonard, W. J., Depper, J. M., Crabtree, G. R., Rudikoff, S., Pumphrey, J., Robb, R. J., Kronke, M., Svetlik, P. B., Peffer, N. J., Waldmann, T. A., and Greene, W. C., Molecular cloning and expression of cDNAs for the human interleukin-2 receptor, *Nature,* 311, 626, 1984.

Chapter 2

Using Synthetic Peptides to Mimic Integrins: Probing Cytoskeletal Interactions and Signaling Pathways

Carol A. Otey

Contents

I. Introduction ... 20
II. Peptides in Affinity Chromatography ... 20
A. Overview ... 20
B. Protocol ... 21
III. Small-Scale Peptide Resin Pull-Down Experiments ... 23
A. Overview ... 23
B. Protocol ... 24
IV. Peptides in Microtiter-Well Binding Assays ... 25
A. Overview ... 25
B. Protocol ... 25
V. Peptides in Single-Cell Microinjection ... 26
A. Overview ... 26
B. Protocol ... 26
VI. Peptides on Pins and on Paper ... 29
VII. Materials and Instruments ... 30
Acknowledgments ... 31
References ... 31

0-8493-3385-7/99/$0.00+$.50

I. Introduction

Integrins are transmembrane proteins, with the largest portion of the molecule oriented to the extracellular side of the membrane. This creates a challenge for researchers whose interests are focused on the cytoplasmic domains, as this portion of the integrin molecule is relatively quite small. Numerous approaches have been used to probe the function of integrin cytoplasmic domains in different experimental systems, and several of these methods are discussed elsewhere in this volume. Fusion proteins of various types have been used to mimic integrin tails, but the fusion partners can create problems of non-specific binding in certain types of assays. To make a more "pure" integrin cytoplasmic domain, synthetic peptides can be utilized. Synthetic peptides have been used to identify novel cytoskeletal binding partners for integrin cytoplasmic domains, to map the binding sites for integrin-associated proteins, to estimate the affinity of these binding interactions, and even to probe the functions of integrin binding partners in living cells. The following sections describe methods for using synthetic peptides in a variety of solid-phase binding assays, and in single-cell microinjection experiments.

II. Peptides in Affinity Chromatography

A. Overview

Many important functions of integrins are mediated by proteins that bind to the short cytoplasmic domains of the two integrin subunits. These cytoplasmic domains serve to physically connect integrins to the actin cytoskeleton, forming an important mechanical linkage between the outside of the cell and the inside. In addition, integrin cytoplasmic domains interact with catalytic binding partners to initiate outside-in signaling pathways. It is likely that cytoplasmic binding partners are also involved in regulating the affinity of integrins through the mechanism known as inside-out signaling. One way of identifying novel cytoplasmic binding partners for a specific integrin subunit is to use synthetic peptides in an affinity chromatography column, to "fish out" proteins that bind specifically and with high affinity to a particular integrin tail. The integrin-derived peptides are immobilized on a column matrix, and lysates of cultured cells or tissues are passed over the column. After the column is washed extensively, any bound proteins are eluted and further characterized. This technique is a good choice for making a preliminary investigation into the cellular binding partners of a particular integrin subunit; however, there is a potential for non-specific binding in these assays (see Section II.B, Step 6, Interpreting the Results). As a first approximation, however, this method can be very effective.

We used this technique to identify cytoplasmic binding partners for β_1 integrin, and found that a small number of proteins from a fibroblast lysate consistently bound to the β_1 integrin peptide. Some of these proteins have not been characterized as yet, but one protein was identified by Western blot as alpha-actinin.[1] This was considered a surprising result at the time; however, recent studies have identified alpha-actinin as a binding partner for the cytoplasmic tails of a variety of transmembrane proteins, including the neutrophil β_2 integrin subunit,[2] the leukocyte adhesion molecule L-selectin,[3] the intercellular adhesion molecule ICAM,[4] and even a neurotransmitter receptor, the NMDA receptor,[5] suggesting that alpha-actinin may play a conserved role in linking the actin cytoskeleton to transmembrane receptor molecules.

B. Protocol

1. **Synthesizing the peptide** — Many different companies sell custom peptides, and they will provide these as crude peptides, which can be purified by reverse-phase HPLC, or as purified peptides (see Section VII, Materials and Instruments). The synthetic peptide can be designed to correspond to one part of an integrin cytoplasmic tail (for example, a highly conserved sequence) or to the entire cytoplasmic domain. The peptide should be synthesized with an extra cysteine residue, at the N-terminal end of the peptide, which will be used to immobilize it on the affinity resin. Using an extra cysteine for attaching the peptide also serves to orient the peptide with its C-terminus free. When purchasing a synthetic peptide, remember that it is important to store the peptide as directed by the manufacturer. If stored properly, synthetic peptides are stable for years.
2. **Conjugating the peptide to the affinity matrix** — Weigh out 0.5 g of Thiopropyl Sepharose 6B and transfer it to a 15 ml screw-cap tube. Fill the tube with de-gassed Tris-buffered saline and allow the resin to swell at room temperature for 1 h. Wash the resin three times, 15 ml each, by spinning at low speed and resuspending the resin pellet in fresh TBS. After the last wash, there should be about 2 mls of loosely-packed resin in the tube. Divide the resin equally into two 15 ml tubes: batch 1 will be conjugated to peptide, and batch 2 will be left unconjugated and used as a pre-column.

   If plenty of peptide is available, it can be coupled at a high concentration, such as 10 mg/ml. If peptide is limiting, a concentration of 1 mg/ml is adequate. Weigh out the desired amount of peptide and dissolve it in 1 ml of TBS. Spin down batch 1 of resin, aspirate off the wash buffer, and add the peptide solution. Place the tube on a rocker and allow the conjugation to proceed at 4°C for 1 h. Remove the tube from the rocker and allow the resin to settle. Carefully remove the supernatant, which contains some unbound peptide. This can be stored frozen, and the unbound peptide can be recovered by re-purification on HPLC.

   Wash both batches of resin three times by gently pelleting at low speed in a clinical centrifuge and resuspending the resin in TBS (fill the tube with buffer). Follow this with three washes in acetate buffer (0.1 M sodium acetate, pH to 4.5 with acetic acid). To reduce any unreacted sulphydryl groups, resuspend the resins in blocking buffer (0.1% β mercaptoethanol, made fresh in 10 ml of acetate buffer). Rock the resin in blocking buffer for 5 min, then wash by gently pelleting the resin and resuspending in acetate buffer. Wash for a total of two times in acetate buffer, followed by two times in TBS. The resin can be stored in TBS with azide at 4°C.

3. **Preparing the cell lysates** — For a 1 ml column, a minimum of 10 confluent 100-mm dishes of cells will be needed. The cells are extracted in 5 ml of lysis buffer such as 1% Triton X-100 in column wash buffer (50 mM Tris acetate, 50 mM NaCl, 10 mM EGTA, 2 mM $MgCl_2$), with 125 µg/ml aprotinin and leupeptin as protease inhibitors. Extract the cells either by adding the cold lysis buffer directly to the culture dish and allowing the dish to sit on a bed of ice for 10 min, or by scraping the cells into the buffer using a Teflon spatula. Spin the lysate for 30 min at 100,000 × *g* to remove any insoluble material. Save a 20 µl sample to run on the gel.
4. **Running the column** — Pour two small columns, one with the unconjugated resin (the pre-column) and one with the peptide-conjugated resin. Pass the cell lysate through the pre-column and collect the flowthrough (the pre-column serves to remove any proteins that bind non-specifically to the resin rather than specifically to the integrin peptide). Load the flowthrough from the pre-column onto the peptide column, and run it slowly through the peptide column (one drop per second).

   Wash both the pre-column and the peptide column extensively by flowing through column wash buffer (at least 50 ml of wash buffer per column). Elute bound proteins by passing 2 ml of 300 mM NaCl through the column. Collect the eluate from both the pre-column and the peptide column, for comparison.
5. **Identifying bound proteins** — Dialyze the column eluates into buffered saline to remove the excess salt, then concentrate the samples and analyze by loading 20 µl onto a polyacrylamide gel. Compare the pattern of proteins eluted from the pre-column and the peptide column, and look for proteins that bind specifically to the integrin peptide. These proteins should be further characterized. Compare the molecular weight of the eluted proteins to that of known cytoskeletal proteins, then use Western blot analysis of the eluted fraction to see if any of the bands can be identified as known proteins. This approach was used to identify alpha-actinin in the eluate from a β_1 peptide column (Figure 1). If there are integrin-binding proteins in the eluate that cannot be identified by Western blot, two approaches can be used to obtain information about these potential integrin-binding partners. Individual protein bands can be excised from the stained gel and sent to a microsequencing lab, to try to obtain some partial protein sequence or a tryptic map; or, if this is unsuccessful, then one can take the more time-consuming route of excising individual bands for use in immunizing animals to raise specific antibodies to the unknown proteins. These antibodies can then be used to screen an expression library or to affinity-purify the novel protein.
6. **Interpreting the results: cautionary tales** — Affinity chromatography utilizing synthetic peptides is an efficient method for "fishing out" integrin-binding proteins from a whole-cell lysate, but it is not without pitfalls. Once a putative integrin-binding protein is identified, it is important to follow up with additional experiments to confirm that the interaction is biologically relevant. Several criteria should be used to judge the potential relevance of the interaction; for example, the putative binding partner should co-localize with integrins. A molecule that is found in a separate compartment of the cell, distant from sites of cell adhesion, is unlikely to have an important role in integrin function. Also, the putative binding interaction should be detectable with an intact integrin heterodimer, not only with an integrin-derived peptide representing the cytoplasmic tail of a single integrin subunit. This can be tested through co-immunoprecipitation experiments, or by incorporating isolated integrins into synthetic lipid vesicles and using the vesicles in a binding assay.

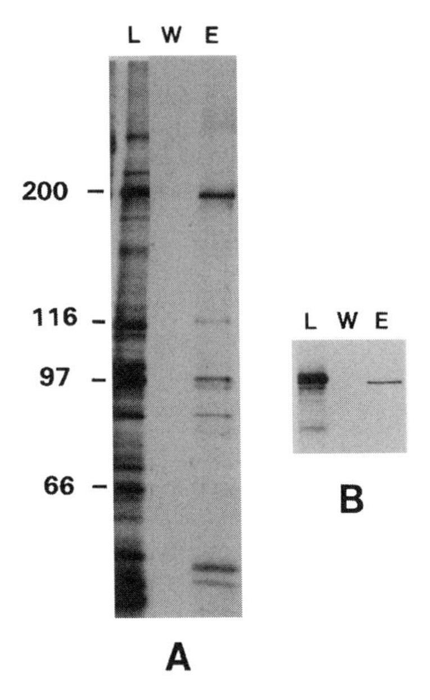

FIGURE 1
Peptide affinity column. Lysates of cultured chick fibroblasts were passed over an affinity column conjugated to a synthetic peptide representing the cytoplasmic domain of β_1 integrin. Panel A: Coomassie blue-stained gel of the column load (L), the last wash (W), and the salt-eluted fraction (E). Panel B: Western blot of the samples shown in A, stained with an antibody to alpha-actinin.

III. Small-Scale Peptide Resin Pull-Down Experiments

A. Overview

Synthetic peptides are expensive, and the cost of purchasing many milligrams of a long peptide can be prohibitive for many researchers. However, with a small amount of peptide, one can make a batch of peptide-coated resin using the same conjugation protocol that was described for the affinity chromatography experiments. Even 0.5 ml of peptide-coated resin (which requires only 5 mg of peptide) is adequate to perform a small number of precipitations from a cell lysate. The resin can then be boiled directly in gel sample buffer to release the bound proteins, and analyzed by Western blot. This technique will not provide enough eluted material to attempt the identification of a novel integrin binding protein, but it does make it possible to ask very focused questions about the binding of specific proteins to particular integrin tails. For example, this approach was used successfully to demonstrate that endogenous FAK co-precipitates with β_1 integrin. By using short peptides, we mapped the binding site to a region near the transmembrane domain.[6] Information derived from these small-scale experiments was subsequently utilized to design single-cell microinjection experiments to investigate the down-stream function of FAK.[7]

B. Protocol

1. **Prepare the affinity resin** as in Section II, Step 1. To make 0.5 ml of peptide-conjugated resin, weigh out 0.18 g of dry resin. Swell and wash the resin as described, and conjugate to 5 mg of peptide dissolved in 0.5 ml of TBS. After incubating the peptide solution with the resin at 4°C for 1 h, carefully pipet off the supernatant, and save it to use in monitoring the amount of peptide that bound to the resin (see Step 6). Then continue to wash and block the peptide-conjugated resin as described in Section II.
2. **Extract 1 to 2 dishes of cells** in 1 ml of lysis buffer as described in Section II, Step 3. Clear the lysate by centrifugation at 100,000 × *g* for 30 min. Save 50 µl of lysate as a positive control for the Western blot.
3. **Perform the pull-down** — Mix 50 µl of conjugated resin with 1 ml of lysate, and rock at 4°C for 2 h. Spin down the resin for 2 min at the lowest speed of a microfuge. Resuspend the pellet in 1 ml of TBS. Wash the resin for a total of six times by resuspending in fresh TBS and re-pelleting in the microfuge.
4. **Elute the bound proteins** — After the last wash has been aspirated, add 25 µl of gel sample buffer to the tube, and boil the resin in the sample buffer for 5 min to release the bound proteins. Spin the resin at high speed for 2 min, then collect the sample buffer and resolve by SDS-PAGE.
5. **Identify the precipitated proteins** — Transfer the gel to nitrocellulose, and stain the Western blot with an antibody specific to the protein of interest. Be sure to include both negative controls (antibodies to proteins that do not associate with focal adhesions) and positive controls (antibodies to proteins that are known to bind directly and with high affinity to the integrin subunit).
6. **Monitor the amount of peptide coupled** — Once a protein has been shown to bind to one integrin tail, it may be desirable to ask if this binding potential is shared with other integrin subunits. This is easily done by making different peptides to represent the different integrin tails. However, peptides do not always bind equally well to the Thiopropyl Sepharose resin, especially if the peptides have been stored for a long time, or stored improperly. If a number of different peptides are going to be used in a pull-down experiment, then the efficiency of coupling of the peptide to the resin should be monitored by measuring the amount of an off-product released by the Thiopropyl Sepharose. After coupling the peptides to the resin, collect the supernatant. Prepare a 1:10 and a 1:100 dilution of this sample, to provide a range of readings. If your spectrophotometer is equipped with a microcuvette, you will need only 100 µl of each dilution; otherwise, make 1 ml of each. Read the absorbance at a wavelength of 343, and calculate the concentration of the thiopyridine off-product using the formula A_{343} divided by 8.3×10^3 = molar concentration of peptide bound. If you find that one peptide binds to the resin with a greater efficiency than other peptides, you can adjust the amount of peptide resin used in a pull-down assay to compensate for this difference.

IV. Peptides in Microtiter-Well Binding Assays

A. Overview

Once a cytoplasmic protein has been identified as a potential binding partner for an integrin subunit, then it becomes necessary to characterize the interaction more carefully. To investigate if the protein has bound to the integrin peptide directly, or as part of a macromolecular complex, the putative binding partner can be used in a microtiter well binding assay. This approach can also be used to estimate the affinity of the interaction. For this type of solid-phase binding assay, the integrin peptide is immobilized through adsorption onto plastic microtiter wells, and the candidate binding partner is added to the well in solution, either as a purified endogenous protein or as a fusion protein. The greatest sensitivity is obtained by first radiolabeling the binding partner with I^{125}. We used this method to characterize the binding of alpha-actinin to β_1 integrin.[1] We added I^{125}-labeled alpha-actinin to removable microtiter wells that had been coated with integrin peptide, and the amount of labeled protein that bound to each well was measured by placing individual wells into gamma counter tubes and counting the I^{125}. Binding of the I^{125}-protein was displaced by the addition of "cold" unlabeled protein. The results from a representative binding assay were analyzed by Scatchard plot to determine the affinity of the interaction.

B. Protocol

1. **Coating the microtiter wells** — Prepare a solution of 1 mg/ml peptide in PBS, with azide. This can be stored at 4°C indefinitely. Add 50 µl of peptide solution to each well, and incubate overnight at 4°C. Collect the peptide solution, and store at 4°C (unbound peptide can be recycled). Add 120 µl of 2% BSA in PBS to each well to block any unbound sites.
2. **Performing the binding assay** — Dilute the I^{125}-labeled protein to a concentration of 1 ng/µl in PBS, and add 10 µl of I^{125}-protein solution to the peptide-coated wells. Also, prepare a stock solution of unlabeled protein at a concentration of 10 ng/µl. Add increasing amounts of unlabeled protein to the wells: 0 µl to well #1, 10 µl to well #2, 20 µl to well #3, etc., to a maximum of 100 µl. Add PBS to each well to bring the total volume to 110 µl: 100 µl to well #1, 90 µl to well #2, 80 µl to well #3, etc. Incubate the wells at 37°C for 2 h, then rinse four times by flooding the wells with rinse buffer (0.1% BSA in PBS) and then inverting the plate.
3. **Measuring the amount of bound protein** — Snap apart the microtiter wells, place each well in a numbered gamma counter tube, and count.
4. **Calculating the affinity of binding** — Many computers are equipped with software for performing a Scatchard plot analysis to determine the affinity of binding, or this calculation can be done manually as follows. Calculate the amount of labeled and unlabeled protein added to each well. Prepare a table, and enter the following data for each well: total amount of added protein, amount of bound protein, amount of free

protein, and the ratio of bound/free. On a piece of graph paper, plot the amount of bound protein on the x axis and the ratio of bound/free on the y axis. Draw a line through all of the points, and calculate the slope of the line. The Kd for the binding interaction is 1/slope.

V. Peptides in Single-Cell Microinjection

A. Overview

In some cases, it is possible to interfere with an interaction between an integrin tail and a cytoplasmic binding partner by microinjecting synthetic peptides directly into living cells. We used this approach to try to block the activation of FAK by integrin. In these experiments, a synthetic peptide was used to mimic only the membrane-proximal region of the β_1 cytoplasmic domain. This short, 13 amino acid peptide acted as a dominant negative: FAK bound to the peptide but was not activated by it. The result was that the FAK-inhibited cells rapidly underwent apoptotic cell death. Similar results were obtained in the Cance lab,[8] using antisense oligonucleotides to reduce the expression of FAK in cultured cells. A role for FAK in the inhibition of apoptosis was also found by Frisch and co-workers,[9] who showed that constitutive activation of FAK protected epithelial cells from apoptosis when the cells were held in suspension.

B. Protocol

For microinjection experiments, care should be used in selecting the type of cultured cell. Some cells are particularly delicate and do not survive microinjection well. The cells do not have to be completely spread at the time of injection, but they have to be at least lightly attached or else the needle will knock them off the coverslip. We have successfully injected a variety of fibroblasts, including the chick embryo fibroblasts described below.

1. **Preparing primary chick fibroblasts** — Chicks of embryonic day 8 to 11 are optimal. The head, wings, limbs, and internal viscera are removed, and the remaining embryonic tissue is incubated in trypsin-EDTA solution for 15 min at 37°C. Tissue is dispersed by trituration, and then diluted with complete medium (Dulbecco's modified essential medium supplemented with 10% fetal bovine serum, 100 units/ml penicillin, and 100 µg/ml streptomycin) to inhibit further digestion. The isolated cells are collected by centrifugation, resuspended in complete medium, transferred to tissue culture dishes, and maintained in complete medium at 37°C. Chicken embryo fibroblasts (CEFs) cannot be maintained indefinitely in culture, so cells from the second to twelfth passage are normally used in microinjection experiments.
2. **Preparing peptides for microinjection** — It is desirable to conjugate small peptides to a carrier protein before injecting the peptides. As was described in Section II, coupling of peptides is more easily accomplished if the peptides are synthesized with a terminal cysteine residue, which also serves to orient the peptide. Peptides can be

conjugated to bovine serum albumin (BSA) at a ratio of 70 to 100 moles of peptide per mole of BSA using a heterobifunctional coupling agent called sulfo-MBS, following the protocol supplied by the manufacturer. Some companies will provide synthetic peptides already conjugated to carrier protein (see Section VII, Materials and Instruments). For microinjection studies, it is critically important to have at least one control peptide, in which the sequence of the integrin peptide is scrambled. Since the interpretation of microinjection data may be rather subjective, it is a good idea to perform the experiments "blindly." One member of the lab can aliquot the integrin peptide conjugate and the control peptide conjugate into coded tubes, so that the individual performing the injections does not know what he/she is injecting, and the results can be decoded at the end of the experiment.

Dialyze the peptide-BSA conjugates into injection buffer (75 mM KCl, 10 mM potassium phosphate buffer, pH 7.5), and concentrate them to approximately 2 mg/ml using a Centricon-30 apparatus. At this stage, the peptides can be aliquotted and stored at –20°C. Immediately prior to injection, centrifuge the peptides at high speed in a microfuge to remove any aggregates, and then filter-sterilize the peptides.

3. **Preparing coverslips** — It is important to keep track of the cells that have been microinjected, and this can be accomplished in several ways. The cells can be injected with an inert fluorescent tracer, which can be mixed in with the peptide-BSA solution. However, if the ultimate goal of the experiment involves immunofluorescent staining of focal adhesions, stress fibers, or other subcellular structures, then it is desirable to use a method other than fluorescent labeling to identify the injected cells. One approach is to use a diamond pencil to inscribe a very small circle in the center of each coverslip, and then to inject all of the cells within that circle. One can then compare the morphology of cells within the circle to the uninjected control cells outside the circle. A more elegant way to keep track of injected cells is to plate the cells onto CELLocate coverslips, which are etched with a lettered grid. Gridded sheets of paper are supplied along with the coverslips, so that a permanent map can be kept to record the injected cells from each experiment.

   After the coverslips are marked, autoclave the coverslips and coat them with an appropriate matrix protein. We typically use fibronectin at a concentration of 50 μg/ml. Coat the coverslips overnight at 37°C.

4. **Conditions for microinjection** — As discussed elsewhere in this volume, growth factors and other molecules present in serum can act cooperatively with integrins. To investigate integrin-mediated signaling independently from growth-factor signaling, microinjection experiments should be performed in the absence of serum. Fibroblasts can live without serum for 24 h with no ill effects. To begin the microinjection experiment, trypsinize the cells and wash them free of serum by pelleting and resuspending twice in serum-free, HEPES-buffered, CO_2-independent medium. Plate the cells onto the matrix-coated coverslips. To minimize the trauma, keep the cells on a heated stage during the microinjection, and return them to the incubator as soon as injection is completed.

   While the cells are pelleting, adjust the settings on the microinjector. These will vary depending on the type of cell to be injected and the type of microinjector system. Typically, a maximum volume of 10% of the cell's total volume can be injected successfully. With the Eppendorf Transjector 5246, we use an injection pressure of 115 hectapascals, a constant pressure of 150 to 200 hectapascals, and an injection time of 0.4 sec to microinject fibroblasts. When the cells have been plated, load the needle with freshly centrifuged, sterile peptide solution. To prevent clogging of needles, we use the Eppendorf microloading tips to backfill the needles, and we lower the filled needles into the cell culture medium as quickly as possible.

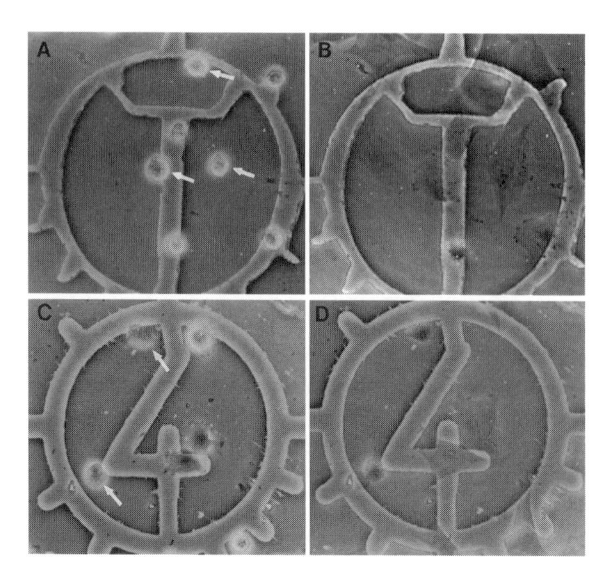

FIGURE 2

Effect of microinjected integrin peptides on cell spreading. Panels A, C: Phase contrast image of newly plated fibroblasts just prior to injection. Cells marked with arrows were injected. Panels B, D: Same cells, 4 h after injection of either control scrambled peptide (B) or short integrin peptide (D). Note that the cells injected with the control peptide spread normally, but the cells injected with the integrin peptide remained rounded.

5. **Interpreting the results** — The final step in the experiment will be to assay for changes within the peptide-injected cell. If the morphological changes in the injected cells are dramatic, such as a complete failure to spread, then these will be apparent just by observing the cells in a light microscope (Figure 2). If the integrin-derived peptide represents a crucial site for cytoskeletal anchoring, then one might assay for stress fiber formation, focal adhesion formation, and/or changes in cell shape. To analyze stress fiber formation, cells are fixed by immersion of the coverslips in 4% formaldehyde. The cells are then permeabilized in 0.2% Triton X-100, and labeled with rhodamine-conjugated phalloidin (Sigma, St. Louis, MO). If the microinjected peptide is interfering with integrin-mediated signaling processes, then the long-term effect for the cell may be the onset of apoptotic cell death. Apoptosis of injected cells can be detected in the light microscope by staining cells with reagents that fluorescently label cleaved DNA. One such product is the ApopTag *In Situ* Detection Kit. Step-wise protocols for using this kit are provided by the manufacturer (see Section VII, Materials and Instruments).

   Another type of phenotypic change that might be expected in a microinjected cell is an alteration in surface morphology. In a cell undergoing apoptosis, for example, a dramatic increase in surface blebbing can be observed. Other changes in surface morphology that might occur after peptide microinjection could include an increase or decrease in the number of filopodia or lamellipodia. Details of cell shape and surface morphology can be visualized with good resolution by scanning electron microscopy. For SEM experiments, CELLocate coverslips are highly recommended. The etched grid is still clearly visible after the coverslip is coated with gold-palladium, so that the injected cells are easily located (see Figure 3, panel A).

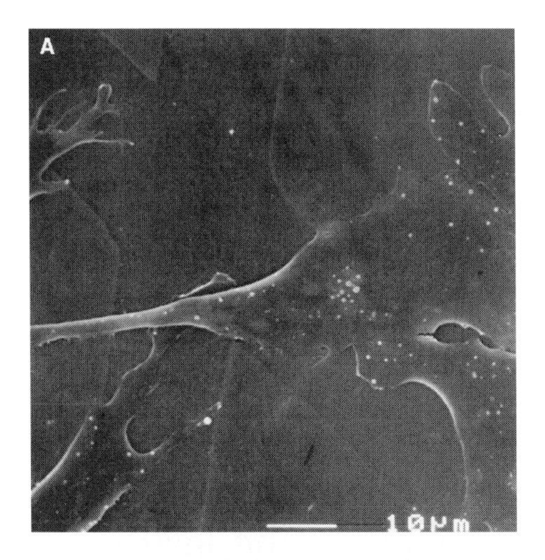

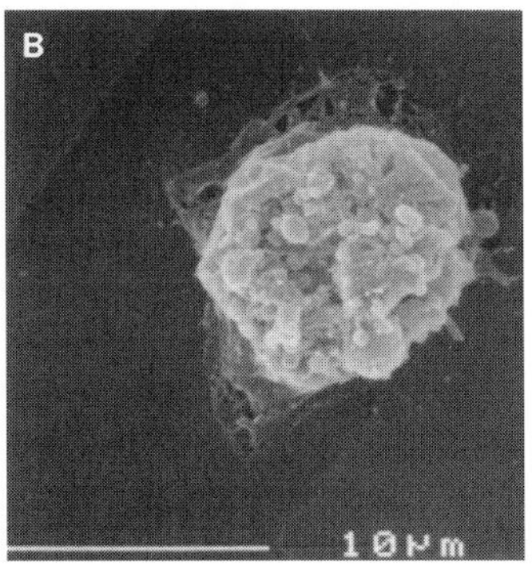

FIGURE 3
Scanning electron microscopy of microinjected fibroblasts. Cells were injected with control peptide (A), or integrin peptide (B). At 4 h post-injection, the cells were fixed and processed for SEM. Panel A shows a low-magnification view (900 ×) in which the etched "X" grid of the CELLocate coverslip is visible. The morphology of the control-injected cell is normal. Panel B shows a high-magnification view (3000 ×) of a cell injected with the integrin peptide, which displays the surface blebbing that is characteristic of apoptotic cell death.

VI. Peptides on Pins and on Paper

For mapping the binding site of a protein in great detail, one can make a series of peptides, 10 to 13 amino acids in length, offset from the previous peptide by a single amino acid. We mapped the binding site for alpha-actinin on the cytoplasmic tail of β_1 integrin using a set of these short overlapping peptides, which were synthesized directly on the heads of plastic pins, and the pins themselves served as a solid-phase binding surface.[10] The pins are held in an 8 × 12 array, so that they fit into a 96-well microtiter plate. Once the peptides have been assembled, the binding of I^{125}-labeled proteins can be measured by immersing the heads of the pins into a protein solution. The pin method for peptide synthesis was originally described by Geysen et al.,[11] and it requires many organic solvents that are not commonly available in a cell biology or biochemistry lab. Recently, a simpler technique has been developed, in which multiple peptides are synthesized on cellulose membranes. This method, called spot synthesis, has been described in detail,[12,13] so it is possible to purchase the individual amino acids and perform the synthesis manually, following the procedures described in the literature. Another option is to purchase the SPOTs kit from Genosys, which includes a pre-derivatized membrane and all of the necessary amino acids, as well as protocols on software. Genosys will also generate custom peptides on membranes, if desired. An advantage of the spot synthesis technique is that the peptides are easily phosphorylated on the paper filters.[14]

VII. Materials and Instruments

1. Peptides. We have purchased most of our synthetic peptides from Quality Controlled Biochemicals (Hopkinton, MA). This company is accommodating to the needs of the researcher, and its staff is happy to prepare peptides which have been optimized for a specific purpose. If the peptide is to be used for microinjection studies, QCB will either provide the purchaser with BSA that has been pre-activated for conjugation to peptides or send the peptides already conjugated to carrier. If one specifies that the peptides will be used for microinjection, QCB will clean the peptides using an additional round of purification with "cell friendly" buffers to remove any traces of organic solvents.
2. Thiopropyl Sepharose 6B can be purchased from Pharmacia (Piscataway, NJ).
3. Removable microtiter wells (trade name Immulon Removawell Strips) and well-holders can be purchased from Dynatech Laboratories (Chantilly, VA).
4. Prepulled microinjection needles and microloading tips are available from Eppendorf (Madison, WI). If the cells to be injected are particularly delicate or if needles of unusual dimensions are required, custom-designed prepulled needles can be purchased from World Precision Instruments (Sarasota, FL). WPI provides their customers with free samples of different types of needles to try.
5. Gridded CELLocate coverslips are available from Eppendorf.
6. Matrix proteins, such as human plasma fibronectin or vitronectin, are available from Sigma.
7. Tissue culture supplies are all from Gibco.
8. Microinjector. We have successfully used many different types of microinjectors, and our favorite is the Eppendorf Transjector 5246 and Micromanipulator 5171, which we use in combination with a Zeiss Axiovert 135 microscope.
9. ApopTag *In Situ* Detection Kit is available from Oncor (Gaithersburg, MD).
10. Bifunctional coupling agent sulfo-MBS is available from Pierce Chemical Company (Rockville, IL).
11. Centricon concentrators are available from Amicon Corp. (Danvers, MA).
12. Custom peptides synthesized on cellulose membranes or kits for assembling peptides on membranes (SPOTs kit) can be purchased from Genosys (The Woodlands, TX).

Acknowledgments

The author thanks Mana Parast for help with the figures and Marc Lotano for critical reading of the manuscript. The author's lab is supported by NIH Grant GM50974.

References

1. Otey, C. A., Pavalko, F. M., and Burridge, K., An interaction between α-actinin and the β_1 integrin subunit *in vitro*, *J. Cell Biol.*, 111, 721, 1990.
2. Pavalko, F. M. and LaRoche, S. M., Activation of human neutrophils induccs an interaction between the integrin β_2-subunit (CD18) and the actin binding protein alpha-actinin, *J. Immunol.*, 151, 3795, 1993.
3. Pavalko, F. M., Walker, D. M., Graham, L., Goheen, M., Doerschuk, C. M., and Kansas, G. S., The cytoplasmic domain of L-selectin interacts with cytoskeletal proteins via alpha-actinin: receptor positioning in microvilli does not require interaction with alpha-actinin, *J. Cell Biol.*, 129, 1155, 1995.
4. Carpen, O., Pallai, P., Staunton, D. E., and Springer, T. A., Association of intercellular adhesion molecule (ICAM-1) with actin-containing cytoskeleton and alpha-actinin, *J. Cell Biol.*, 118, 1223, 1992.
5. Wyszynski, M., Lin, J., Rao, A., Nigh, E., Beggs, A., Craig, A. M., and Sheng, M., Competitive binding of alpha-actinin and calmodulin to the NMDA receptor, *Nature*, 385, 439, 1997.
6. Schaller, M. D., Otey, C. A., Hildebrand, J., and Parsons, J. T., Focal adhesion kinase and paxillin bind to peptides mimicking β integrin cytoplasmic domains, *J. Cell Biol.*, 130, 1181, 1995.
7. Hungerford, J. E., Compton, M. T., Matter, M. L., Hoffstrom B. G., and Otey, C. A., Inhibition of $pp125^{FAK}$ in cultured fibroblasts results in apoptosis, *J. Cell Biol.*, 135, 1383, 1996.
8. Xu, L.-H., Owens, L. V., Sturge, G. C., Yang, X. Y., Liu, E. T., Craven R. J., and Cance, W. G., Attenuation of the expression of the focal adhesion kinase induces apoptosis in tumor cells, *Cell Growth Diff.*, 7, 413, 1996.
9. Frisch, S. M., Vuori, K., Ruoslahti, E., and Chan-Hui, P.-Y., Control of adhesion-dependent cell survival by focal adhesion kinase, *J. Cell Biol.*, 134, 793, 1996.
10. Otey, C. A., Vasquez, G. B., Burridge, K., and Erikson, B. W., Mapping of the α-actinin binding site within the β_1 integrin cytoplasmic domain, *J. Biol. Chem.*, 268, 21193, 1993.
11. Geysen, H. M., Meloen, R. H., and Barteling, S. J., Use of peptide synthesis to probe viral antigens for epitopes to a resolution of a single amino acid, *Proc. Natl. Acad. Sci. USA*, 81, 3998, 1984.
12. Bartl, R., Hoffman, S., Tegge, W., and Frank, R., Solid phase peptide synthesis on cellulose carriers, *Innovation and Perspectives in Solid Phase Synthesis*, Second International Symposium, Epton, R., Ed., Intercept Ltd., Andover, 1992, p. 281.
13. Frank, R., Spot-synthesis: An easy technique for the positionally addressable, parallel chemical synthesis on a membrane support, *Tetrahedron*, 48, 9217, 1992.

14. Edlund, M., Wikstrom, K., Toomik, R., Ek, P., and Obrink, B., Characterization of protein kinase C-mediated phosphorylation of the short cytoplasmic domain isoform of C-CAM, *FEBS Lett.*, 425, 166, 1998.

Chapter 3

Use of the Yeast Two-Hybrid System for the Characterization of Integrin Signal Transduction Pathways: Identification of the Integrin-Linked Kinase, $p59^{ILK}$

Gregory E. Hannigan and Shoukat Dedhar

Contents

I. Introduction 34
II. Preparation of Bait Plasmids for Interaction Screening 35
 A. Overview 35
 B. Protocols 35
 C. Technical Comments 37
III. Testing of Baits for Autoactivation and Fusion Protein Expression 38
 A. Overview 38
 B. Protocols 38
IV. Conducting cDNA Library Screens with Confirmed Baits 40
 A. Overview 40
 B. Protocols 40

0-8493-3385-7/99/$0.00+$.50

V. Rescue of Interaction Plasmids and Confirmation of the Interaction 43
A. Overview 43
B. Protocols 43
C. Technical Comments 44
VI. Materials and Reagents 45
A. Yeast Growth Media (Ref. 16) 45
B. Other Reagents 46
VII. Concluding Remarks and Future Directions 47
Acknowledgments 47
References 48

I. Introduction

Specific protein-protein interactions underlie many fundamental cellular processes, such as gene transcription, mRNA splicing, and signal transduction. In the latter case, intracellular protein complexes are commonly formed in response to ligand activation of cell surface receptors, such as the receptor tyrosine kinases, and discrete protein domains mediate interactions which stabilize the complex and/or propagate the signal.[1] Similarly, occupation and clustering of integrin receptors induces the assembly of focal adhesion complexes, formation of which is dependent on protein interactions of integrin cytoplasmic domains.[2-4]

Many biochemical, genetic, and molecular biological techniques have provided the opportunity for investigators to identify specific protein interactions. As a result, great strides have been made over the past decade toward understanding the signal transduction mechanisms regulating cellular responses to environmental stimuli. The focus of this chapter is the application of the yeast two-hybrid genetic screen in the isolation of a novel protein serine/threonine kinase, ILK, mediating integrin signal transduction.

We selected the LexA-based interaction trap, developed in the laboratory of Roger Brent,[5] as the two-hybrid system of choice for analysis of integrin signal transduction pathways. This system and its use have recently been described in detail by its developers,[5] and LexA-based two-hybrid systems are now commercially available, e.g., LexA MatchMaker® (Clontech, Inc., Palo Alto, CA), Hybrid Hunter® (Invitrogen Corp., Carlsbad, CA), and DupLEX-A® (OriGene Technologies, Inc., Rockville, MD). Excellent manuals are available for downloading as .pdf files from the company web sites (www.Clontech.com; www.invitrogen.com; www.origene.com). These manuals detail use of the two-hybrid system components, and provide general protocols for yeast growth, transformation, and selection. These companies are rapidly developing interaction libraries from a variety of human tissues from various organisms, greatly increasing the versatility of the system for use in tissue-specific and developmental stage-specific screens. We will limit ourselves to a practical discussion of the interaction trap as it is used in our laboratories, for the isolation of cellular proteins which bind to integrin cytoplasmic domains. This is exampled by the identification of the integrin-linked kinase,

ILK, as a cytoplasmic partner of the β_1 integrin subunit.[6] Also, the utility of the system in identifying additional, "downstream" effectors of integrin signaling will be appreciated.

II. Preparation of Bait Plasmids for Interaction Screening

A. Overview

The two-hybrid strategy exploits the modularity of protein structure, particularly the independent action displayed by DNA-binding and transactivating domains of transcription factors.[5,7] A cDNA fragment encoding a protein or domain of interest is fused to a DNA-binding (DNAB) domain, such that the hybrid "bait" fusion protein will bind to the DNAB's cognate operator sequence. The DNA library, to be screened in the interaction hunt, has been cloned in-frame with a transcriptional activating domain. Library cDNAs are sometimes referred to as "prey" in the two-hybrid vernacular. Specific pairing of a library-encoded protein sequence with the bait-encoded protein reconstitutes a hybrid transcription factor in the yeast nucleus, which is thus capable of activating a specific operator-driven reporter gene. Reporter genes are of two forms: 1) selectable auxotrophic markers, such as LEU2, which confer interaction-dependent growth in media lacking the critical amino acid leucine; 2) β-galactosidase, transcription of which results in production of blue colonies when grown under appropriate selection, on plates containing the chromogenic β-galactosidase substrate, 5-bromo-4-chloro-3-indolyl-β-D-galactopyranoside (X-gal).

Crucial to the success of an interaction trap experiment is the nature of the bait construct used to program the screen. Most importantly, bait fusion proteins must be transcriptionally inert. In the case of proteins which are not transcription factors, such as cell surface receptors, acidic sequences expressed as part of the bait may be capable of non-specific activation of the LexAop-driven reporter system (this is also true for non-LexA-based systems). The size of the non-LexA moiety of the bait-encoded protein can vary over a wide range. We have successfully conducted screens with protein baits, comprising 47 up to 452 amino acids, fused to the LexA DNA binding domain. The sensitivity of the reporter requires that each bait be tested for activation using the exact conditions under which the screen will be performed.

B. Protocols

1. Plasmid construction

The cytoplasmic domain of the β_1 integrin subunit was amplified by the polymerase chain reaction (PCR) for use as a bait in the screen. Primers were designed such that an EcoRI linker was placed 5′, and an XhoI linker 3′, of sequences encoding amino acid residues 738 to 798 (i.e., carboxy terminal 60 residues) of the β_1 subunit.

TABLE 1
Yeast expression plasmids used in the two-hybrid screen for integrin interactors

Plasmid	Selection	Use
pEG202β_1/α_5INT	*HIS3*, Amp^r	integrin bait expression
pJG4-5	*TRP1*, Amp^r	library fusion expression
pSH18-34	*URA3*, Amp^r	reporter lacZ gene with 8 LexA operators
pJK101	*URA3*, Amp^r	repression assay for nuclear localization of bait protein
pRFHM1	*HIS3*, Amp^r	non-activating fusion for negative activation and positive repression controls
pSH17-4	*HIS3*, Amp^r	positive activation control

A similar set of primers was designed for amplification of the α_5 integrin cytoplasmic domain (residues 1022 to 1049), which was used in the construction of a specificity control. The 5′ and 3′ termini of each amplification primer contained a GGCC quadruplet to extend the product for efficient digestion by the restriction enzymes. These primers were used to amplify the cytoplasmic sequences from full length β_1 and α_5 integrin cDNA templates. Thermostable polymerases such as Pwo (Boehringer-Mannheim, Indianapolis, IN) or Pfu (Stratagene, Inc., La Jolla, CA) provide 3′ to 5′ exonuclease (proofreading) activity and are good choices for this amplification, since they effect accurate amplification of the templates. Products were digested with EcoRI and XhoI prior to ligation into the double-digested bait plasmid, pEG202. This resulted in an in-frame fusion of the 47 amino acid integrin domain with the LexA DNA-binding domain. The DNA sequence of each integrin bait construct was confirmed prior to its use in the screen. These bait plasmids are designated pEG202β_1 INT and pEG202α_5INT (Table 1). Alternatively, we have cloned PCR products into T-vector (Novagen, Madison, WI) with subsequent digestion and purification of the fusion fragment from this plasmid. The former method eliminates a round of ligation and fragment isolation. In-frame fusion junctions are always confirmed by DNA sequencing.

2. Electroporation of Yeast strain EGY48

a. Grow an overnight 5 ml culture, from a single colony of yeast, in appropriate medium.

b. Dilute saturated overnight culture to 50 ml medium, grow 4 to 5 h until OD_{600} is 1.2 (ca. 2×10^7/ml).

c. Harvest cells 5 min at 4000 × *g* and resuspend in 50 ml sterile ice cold water.

d. Repeat Step c.

e. Resuspend cells in 25 ml ice-cold 1 M sorbitol, 5 min at 4000 × *g*, 4°C.

f. Resuspend pellet in 0.25 ml ice-cold 1 M sorbitol.

g. Transfer 40 μl cell suspension from Step 6 to a 1.5 ml microcentrifuge tube on ice.

h. Add ≤5 ng plasmid DNA/tube in maximum 5 μl volume to cells, mix with pipette.

i. Pipette 45 μl cell/DNA mix into gap of 0.2 cm cuvette (Strategene #400924).

j. Discharge current (1000V for Stratagene 1000).

k. Add 1 M ice-cold sorbitol to cuvette, transfer to sterile 1.5 ml microcentrifuge tube.
l. Plate 200 μl suspension immediately on appropriate CM dropout (complete synthetic medium, see Section VI, Materials and Reagents) selective plate, incubate at 30°C.
m. Colonies will be evident 36 to 72 h later.

C. Technical Comments

All of the bait testing and library screening is done in the host *Saccharomyces cerevisiae* strain, EGY48 (*MATα, his3, trp1, ura3*, LexAop$_6$::*LEU2*). The LEU2 gene is an integrated reporter that confers interaction-dependent growth in medium lacking leucine. We routinely employ two methods for transformation of the yeast strain, EGY48, with plasmid DNA. The first method is the lithium acetate (LiAc) transformation procedure, and the second is a simple electroporation protocol. Each method has advantages for specific applications, and either method works well with the EGY48 host. The LiAc method is somewhat more cumbersome than electroporation; moreover, it is prone to lower and more variable transformation efficiencies (5×10^4 to 5×10^6 transformants/μg) in our hands. The LiAc method is also less expensive, not requiring additional financial outlay for electroporation apparatus and cuvettes. Nonetheless, affordable basic electroporators are currently available, and the time savings inherent to electroporation protocols contribute to their economical application. Both procedures can be used to efficiently transform yeast cells with single plasmids, or for cotransformation with two plasmids. Note that either method results in significant (≥10-fold) decreases in co-transformation efficiency, and we generally use sequential transformations to develop stock strains carrying multiple plasmids. Routine transformations are accomplished by electroporation in our lab. It is, however, important to note that the much lower amounts of plasmid DNA used in electroporation will significantly compromise representative coverage of clones in a library screen. It is thus essential to use the LiAc protocol for library transformations, in order to ensure adequate cDNA representation in the interaction screen.

The above transformation protocol is adapted from Reference 8, and is applicable to EGY48 grown in YPD, or EGY48 carrying a selectable plasmid (Table 1) under appropriate growth conditions. We routinely use electroporation to establish stock strains carrying specific baits and reporters. Electroporation is not used to transform yeast with cDNA libraries for interaction screens. We find that 50 ml of culture yields sufficient yeast for 4 to 5 transformations. We use the inexpensive Stratagene 1000 electroporator (10 μF capacitor) which can be set to charge over a range of 200 to 2500 V. This is a compact, integrated unit that provides ease of use with multiple samples. The operator selects the voltage via a front panel LED display, inserts the cuvette holder, and discharges the voltage.

We find this setup minimizes the time between electroporation and plating of cells, which is a critical factor for the successful recovery of transformants. As with bacteria, a number of factors affect electroporation and the resulting transformation efficiency of the yeast cells. In addition to the electroporation parameters, success of the procedure is affected by the quality and concentration of plasmid DNA, the ionic strength of the resuspension medium of cells and DNA, and the yeast strain.[8] The advantages of electroporation over the LiAc method relate to higher, more consistent transformation efficiencies, ease of manipulation, and less susceptibility to lot variations in reagents such as PEG. Preparation of electro-competent yeast cells is readily accomplished in the user's laboratory. We do not freeze stocks of competent cells, as freezing is reported to decrease transformation efficiencies >10 fold.

TABLE 2
Transformation efficiencies for electroporation* of EGY48 host cells

	200V	500V	1000V	1500V	2000V
0.5 ng	860	850	890	655	660
20 ng	180	855	860	1015	635

* Standard electroporation protocol using the Stratagene Electroporator 1000. Numbers are colonies/indicated weights of plasmid used in transformation.

We find that 1000V, 0.5 to 1 ng plasmid/transformation routinely yields efficiencies of $>10^6$ colonies/μg. Although various protocols suggest using <100 ng DNA for a single transformation, it can be seen from our data that, for transformation of EGY48, even 20 ng reduces the transformation efficiency significantly (Table 2).

III. Testing of Baits for Autoactivation and Fusion Protein Expression

A. Overview

It is essential to test all bait plasmids for intrinsic transactivation in the yeast system and nuclear expression of the fusion protein, prior to attempting a screen. Even weak activity of the bait in this assay will negate its use in an interaction screen. This test assays for transactivation of the reporter plasmid, pSH18-34, when co-transformed with a bait plasmid, such as pEG202β_1INT. Note that transformation of a plasmid-bearing strain routinely yields efficiencies of ca. 10^6/μg plasmid, using the electroporation protocol outlined here.

B. Protocols

1. Autoactivation protocol

a. Transform EGY48 with the reporter pSH18-34, by electroporation (see above). Grow a 50 ml culture overnight at 30°C, in CM[U]-medium. Maintain a stock of the reporter strain, EGY48/pSH18-34 at –70°C, in YPD/20% glycerol.

b. Dilute 5 ml of EGY48/pSH18-34 strain to 50 ml YPD. Grow 3 to 4 h, until OD_{600} is ca. 1.2. Transform equal aliquots of culture with 5 ng of bait plasmid, or 5 ng pSH17-4 activation positive control (see electroporation protocol, above). Plasmids used for transformation must be clean. Prepare mini-preps using a column purification kit such as those made by Qiagen or Sigma.

c. Select EGY48/pSH18-34/pEG202$_x$INT on CM[HU]-agar plates. Use the same selection for the pSH17-4 control. Pick 3 to 6 colonies from each transformant plate and grow a 5 ml overnight culture in CM[HUT]-medium.

d. Using a bacterial loop, streak each clone out on the following selective agars:

   i) CM[HUT]-/galactose/X-gal

   ii) CM[HUTL]-/galactose

   Bait transformants should be white on (i) and not grow on (ii) plates. The pSH17-4 control should be blue on (i) and show good growth on (ii) plates.

2. Western immunoblotting for protein expression in yeast

a. Grow 5 ml yeast culture with appropriate selection (e.g., CM[H]-for pEG202-based bait vectors).

b. Spin 5 ml culture at 900 × *g*, 5 min (r.t.).

c. Discard supernatant and freeze pellet at –80°C, or in dry ice/ethanol bath, 10 min.

d. Thaw pellet in 100 µl cracking buffer (60 mM Tris pH 6.8, 10% (v/v) glycerol, 1% (w/v) sodium dodecyl sulfate, 1% (100% = 14.3 M)) β-mercaptoethanol.

e. Transfer suspension to 1.5 ml conical tube containing 0.15 g (ca. 100 µl) acid-washed glass beads.

f. Heat bead/suspension mixture to 70°C for 10 min.

g. Vortex 1 min to mechanically disrupt the yeast cell membrane. Time can be extended as required, to improve the extraction efficiency and protein yield.

h. Centrifuge at 200 × *g*, 5 min at r.t.

i. Supernatant is used for SDS-PAGE and Western detection. Fusions can be detected with an anti-LexA monoclonal antibody according to vendor instructions (Clontech #K1609-1) or antibodies specific for the bait protein.

This method for Western immunoblotting of yeast lysates is adapted from one described in the Invitrogen Hybrid Hunter® manual (www.invitrogen.com), and in our hands provides quick, efficient extraction of EGY48 cells for use in standard Western blot protocols. This is necessary to demonstrate expression of the bait fusion protein in EGY48.

3. Assay for nuclear expression of the bait fusion protein

The LexA system[5] is designed so that bait proteins can be tested for nuclear expression, by assaying bait-dependent repression of the lacZ reporter gene. This is an important test to perform subsequent to confirming that an expressed bait is non-activating, in the event that this results from a failure of the LexA bait fusion protein to reach the nucleus and/or interact with the LexA operator. The repression plasmid, pJK101, carries the lacZ reporter gene under control of the GAL1 upstream activating sequence (UAS). Two LexA operator sequences are placed between the GAL1 UAS and lacZ structural gene, such that binding of LexA fusion protein inhibits GAL1-driven reporter transcription. pRFHM1 produces a nonactivating fusion of LexA with the *Drosophila* bicoid homeodomain, and acts as a positive control for the repression assay. The repression assay involves cotransformation of EGY48 with pJK101 and pRFHM1, pJK101 with bait plasmid(s), and transformation with pJK101 alone, as a transcription control. These transformants are plated on CM[HU]-selective agar. After 2 to 3 days growth at 30°C, 6 individual colonies from each transformation plate are patched onto CM[HU]-/X-gal agar. Incubate at 30°C for 12 to 18 h for optimal visualization of differential β-gal expression. In our hands, repression by baits is generally in the range of 50% to 80% of pJK101 autonomous activity, so we do not incubate plates for longer than 24 h, to achieve optimal resolution of the differential β-galactosidase expression. This differential is readily seen on plate assays, so we do not routinely use the more quantitative solution β-gal assay. For example, both pEG202β_1INT and pEG202α_5INT caused significant (>50%) reduction in β-galactosidase activity in the pJK101 repression assay (our unpublished data), confirming that both bait proteins were synthesized and transported to the nucleus.

IV. Conducting cDNA Library Screens with Confirmed Baits

A. Overview

This method is adapted from a protocol by Geitz and Woods,[9] and represents a scaled-up version of their published protocol. This method is sufficient for up to 16 transformations of 2×10^5 each, such that 12 plates yield 2.4×10^6 library transformants. All procedures are undertaken using aseptic technique, with sterilized supplies and reagents. Identical screens were performed with pEG202β_1INT and pEG202α_5INT, of a HeLa cell-derived cDNA library in pJG4-5. It is important to note that pJG4-5 contains a GAL4 UAS which renders expression of the library cDNAs galactose dependent. This confers an additional level of selection on the interaction screen, as interactions should not be evident on selective plates containing glucose as the sole carbon source (Figure 1). Interactions are thus scored on plates containing galactose rather than glucose. Unless otherwise indicated here, growth and selective media are assumed to contain glucose as the carbon source.

B. Protocols

1. Library screening

a. Grow 20 ml culture of EGY48/pSH18-34/pEG202$_x$INT from a single colony, in CM[HU]-for 4 to 5 h at 30°C.

b. Dilute culture to 100 ml CM[HU]-, grow overnight.

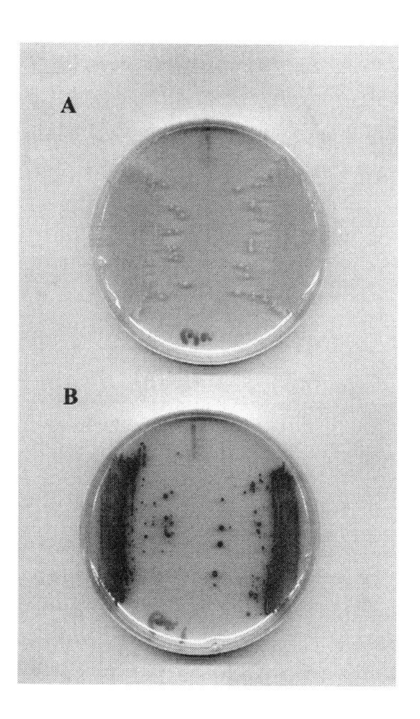

FIGURE 1
Confirmation of an interaction by plating on differential indicator media. A pEG202 bait construct was used to isolate cDNAs from a pJG4-5 interaction library. To confirm a putative interaction, the rescued pJG4-5 library clone was used to transform an EGY48 host strain carrying both the pEG202-based bait and pSH18-34 lacZ reporter. Two independent clones from this re-transformation were streaked on: A) CM[HUT]-/glucose/X-gal plates, and B) CM[HUT]-/galactose plates. A and B show the same two clones. Expression of the lacZ reporter is seen only when the pJG4-5 cDNA is induced on the galactose-containing plates.

c. Dilute to 500 ml with YPD, grow 3 to 4 h until OD_{600} is 1.0 to 1.2 (ca. 2×10^7/ml). Determine relationship between OD_{600} and cell number by growing a culture in YPD. At hourly intervals, determine OD_{600} and plate dilutions to enumerate colonies.

d. Harvest cells 10 min at $2000 \times g$, r.t.

e. Resuspend, wash cell pellet in 250 ml sterile dH_20.

f. Resuspend cells in 100 mM LiAc (dilute fresh from 1 M stock solution LiAc to a concentration of 2×10^9/ml. Incubate at 30°C for 15 min.

g. Dispense plasmid DNA (pJG4-5 based cDNA library) sufficient for 2×10^5 transformants into 15 ml conical centrifuge tubes. Volume of DNA should be ≤20 µl. Library DNA is prepared using, e.g., ProMega or Qiagen plasmid purification kits. [To determine transformation efficiency, scale this method down such that 50 µl cells (from a 50 to 60 ml culture) are added to 300 µl PEG/LiAc and transformed with 100 ng (<5 µl) library DNA. After spinning transformed cells out of PEG/LiAc, resuspend in 1 ml ddH_20 and plate serial 100-fold dilutions on CM[HUT]-for colony counting. Ideally, for library transformations, the efficiency should be around 5 to 8×10^5 colonies/µg. We have screened productively with efficiencies as low as 5×10^4/µg.]

h. Add 300 µg boiled carrier salmon sperm DNA to each tube, mix. Carrier DNA has been sonicated, extracted by phenol/chloroform, ethanol-precipitated and resuspended in distilled water, to a stock concentration of 10 mg/ml. We confirm that sonicated DNA runs on a 1% agarose gel as fragments <1 kb.

i. Dispense a 300 µl aliquot of yeast (6×10^8 cells in 100 mM LiAc) into each tube containing the DNA.

j. Add 1.8 ml PEG/LiAc to each tube, mix well with 5 ml pipette (or vortex briefly). Incubate at 30°C for 30 min (PEG/LiAc is 40% PEG 3350 in 100 mM LiAc). Note that PEG must be 3350, not 8000, as the latter greatly decreases transformation efficiencies.

k. Heat shock in 42°C water bath for 20 min.

l. Spin cells out of PEG/LiAc ($4000 \times g$, 10 min). Use pipette tip to remove PEG/LiAc.

m. Resuspend pellets in 200 µl ddH_2O/tube. Mix well (or briefly vortex) and spread on CM[HUT]-selective agar in 500 cm^2 plates (Nunc #166508).

n. Invert and incubate at 30°C for 48 to 72 h. Individual colonies should cover the surface of the plates quite densely after 3 days.

ILK was identified in a screen using pEG202β_1INT against a HeLa cDNA interaction trap library. This library has been described[10] and was constructed by EcoRI/XhoI directional cloning of cDNAs into pJG4-5. The pEG202β_1INT and pEG202α_5INT baits were screened identically against the pJG4-5 HeLa cDNA library. Only the β_1 screen identified ILK, supporting the specificity of the β_1-ILK interaction.

2. Harvesting library transformants

a. Using a sterile microscope slide, scrape each plate containing library transformants into CM[HUT]-liquid medium. Dilute transformant pool to 500 ml CM[HUT]-and grow at 30°C for 4 h. Although aseptic technique is adhered to, this step is the most likely to introduce bacterial contamination. We find that supplementing selective media and plates to 1X concentration of tissue culture Penicillin/Streptomycin (GIBCO/LifeTech) or 50 µg/ml ampicillin provides a simple measure to counter bacterial growth.

b. Harvest culture at $4000 \times g$, 10 min. Concentrate to 50 ml in YPD/15% glycerol. Determine the plating efficiency of the concentrated pool by plating appropriate serial dilutions on CM[HUT]-/galactose agar (100 µl of 10^{-4} to 10^{-9} dilutions). Store 5 ml aliquots of the concentrated pool at –70°C. This provides a large stock of library transformants for replating, if required.

c. Plate 5 to 10 colonies per original transformant colony on CM[HUTL]-/galactose agar. After 2 to 3 days, patch colonies (sterile toothpick) onto CM[HUT]-/galactose/X-gal. At 24 to 36 h, positive colonies will appear blue on these plates. These colonies when streaked on CM[HUT]-/glucose/X-gal should appear white.

We have found that if plating efficiency is determined on plates containing X-gal, a small number of positives may be isolated at this earlier step. Further economies can be introduced by plating the trapped transformants directly onto CM[HUTL]-/galactose/X-gal agar and selecting for blue colonies. Care must be taken to confirm that blue colour does not develop in these colonies on glucose containing plates (Figure 1). Color development on glucose indicates a false positive.

V. Rescue of Interaction Plasmids and Confirmation of the Interaction

A. Overview

It is imperative that interactions scored in the yeast two-hybrid screen are confirmed in an independent experimental setting, such as *in vitro* binding reactions with fusion proteins, or co-immunoprecipitation from mammalian cell lysates (in the case of mammalian protein screens). In order to characterize an interacting clone, the library plasmic must be rescued from the yeast clone carrying bait, reporter, and library plasmids. Protocols for rapid plasmid extraction are derived from Reference 11.

B. Protocols

1. **Isolate plasmids from each positive clone by rapid plasmid extraction**—
   a. Grow positives overnight in 2 ml cultures of CM[T]-medium, to enrich for trp+ library plasmids. Harvest in 1.5 ml microcentrifuge tubes, 4000 × *g*, 5 min.
   b. Resuspend pellet in 200 µl of 100 mM NaCl/10 mM Tris-HCl pH 8.0/1 mM EDTA/1% sodium dodecyl sulfate/2% Triton X-100, add 0.33 g acid-washed glass beads (455 to 600 µm, Sigma).
   c. Add 200 µl saturated phenol:chloroform:isoamyl alcohol. Vortex 2 min.
   d. Spin 1500 × *g*, 10 min.
   e. Collect aqueous phase and ethanol precipitate. Resuspend in 20 µl water.
2. **Transformation of *E. coli*** — Electroporate *E. coli* DH5α with 1 to 2 µl plasmid prep, pick 3 to 5 colonies per plate. Note that chemical transformation of *E. coli* with yeast rapid prep plasmid DNA yields very low (to nil) numbers of colonies; electroporation is thus strongly recommended for this step, based on our experiences.
3. **DNA preparation and analysis** — Mini-prep (Sigma PlasmidPure) and restriction digest (EcoRI and XhoI) of rescued plasmids with pEG202, pSH18-34 and pJG4-5 plasmids as controls. Select clone showing pJG4-5 pattern with insert (usually 600 to 1000 bp from the HeLa library).
4. **Sequencing of the cDNA insert** — As these are often not representative of full length transcripts, trapped inserts are used to screen appropriate cDNA libraries for isolation and characterization of full length cDNAs.
5. **Confirmation of interaction** — Retransform EGY48/pEG202$_x$INT/pSH18-34 with the rescued interacting library plasmid, and select on CM[HUT]-agar plates. Confirm interaction by patching transformant colonies onto CM[HUT]-/galactose/X-gal and scoring blue colonies after 24 to 36 h at 30°C. These should also grow on CM[HUTL]-/galactose plates.

C. Technical Comments

Ideally, a screen for protein interactors should be performed at 3- to 4-fold library redundancy, such that interacting clones will usually be represented more than once in the population of confirmed interactors. The isolation of two non-identical cDNAs coding for the same protein strengthens the assignment of this protein as a binding partner of the bait protein. However, the fact that a given screen has yielded multiple isolates of the same cDNA should not be interpreted as evidence against the validity of the interaction.

Importantly, the specificity of identified protein interactions can be directly tested in the yeast system. This is simply done by testing an interacting clone against bait constructs unrelated to the original bait, in a directed two-hybrid interaction assay. The stringency of the specificity test can be increased according to the nature of the interaction being investigated. For example, ILK contains in its N-terminal half a domain comprised of four ankyrin-like repeats.[6] Such repeats are present in a large number of cell cycle regulatory and signaling molecules, and are involved in specifying protein-protein interactions. Notably, $p16^{INK4}$ is an inhibitor of cyclin-dependent kinases, which is comprised solely of four ankyrin-like repeats.[12] Cytoplasmic ankyrin-like repeats in the *Drosophila* Notch transmembrane protein are known to mediate interactions of Notch with two downstream effectors, Delta and Suppressor-of-Hairless.[13] Interacting clones identified in a two-hybrid screen baited with the ankyrin repeat region of $p59^{ILK}$ could be stringently tested against baits expressing $p16^{INK4}$ and Notch ankyrin repeats. If these interactions were also positive the indication would be that the $p59^{ILK}$ interaction is non-specific. Clearly, the more specificity controls that are included in the test, the stronger the conclusions can be regarding the relevance of the identified interaction. Three is a reasonable number.[5]

Further confirmation of an identified protein interaction requires analysis of the association under conditions that are independent of the yeast system. The two-hybrid screen is performed in a physiologic environment that is, nonetheless, highly engineered and often foreign (from either an organismal or a cellular localization standpoint) to the proteins being studied. In particular, we showed that β_1 immunoprecipitates of rat intestinal epithelial cell (IEC18) lysates contained significant amounts of $p59^{ILK}$, as tested by affinity-purified anti-ILK antibodies.[6] Additionally, tests of cellular co-localization using immunocytochemistry, confocal microscopy, or electron micrographic approaches all provide valid criteria for assessing the physiological relevance of an identified interaction. Finally, experiments should be designed to analyze the new protein/interaction in the context of a known biological function of the protein used to bait the two-hybrid screen. Thus, overexpression of $p59^{ILK}$ in the IEC18 cells was shown to have significant effects on their adhesion to extracellular matrix proteins, and that this effect might contribute to cellular transformation.[6] Further work indicated that ILK overexpression had induced the IEC18 cells to synthesize and deposit a fibronectin matrix,[14] which has previously been shown to require functional $\alpha_5\beta_1$ integrin.[15] This result strongly suggests a role for ILK in mediating "inside-out" integrin signaling, and provides an important functional correlate to the observed association of integrins and $p59^{ILK}$ in the two-hybrid and integrin co-immunoprecipitation experiments.

VI. Materials and Reagents

A. Yeast Growth Media (Ref. 16)

1. YPD

Per liter:
10 g yeast extract
20 g peptone
20 g D-glucose (filter sterilize a 20% stock, add to 2% after autoclaving other components)
pH to 6.0 with NaOH
add 20 g agar for plates

2. Complete Minimal Drop-Out Medium (CM)

Per liter:
1.3 g drop-out powder (see 4 below)
1.7 g Yeast Nitrogen Base (minus amino acids)
5 g $(NH_4)SO_4$
20 g D-glucose
pH to 6.0 with NaOH
add 20 g agar for plates

3. CM/X-gal Indicator Plates

as for CM plates (minus glucose), make to 800 ml and autoclave
add 100 ml 20% galactose
add 100 ml of 0.7 M potassium phosphate (pH 7.0)
2 ml of 20 mg/ml X-gal

4. Drop-Out Powder

Mix the following weights (g) of nutrients, and grind together into a homogeneous powder using a clean, dry mortar and pestle. Store at room temperature.

L-arginine	1.2
L-aspartic acid	6.0
L-glutamic acid	6.0
L-lysine	1.8
L-methionine	1.2
L-phenylalanine	3.0
L-serine	22.5
L-threonine	12.0
L-tyrosine	1.8
L-valine	9.0
Adenine	2.5

CM media are supplemented with 8.3 ml/l of stock solutions of uracil, L-tryptophan, L-leucine, and/or L-histidine according to selection requirements, e.g., CM[HU]-refers to CM drop-out lacking L-histidine and uracil.

Uracil	2.4 mg/ml
L-tryptophan	4.8 mg/ml
L-histidine	2.4 mg/ml
K-leucine	7.2 mg/ml

B. Other Reagents

1. Sigma (St. Louis, MO)

Adenine	#A-9126
Uracil	#U-0750
L-valine	#V-0500
L-tryptophan	#T-0254
L-threonine	#T-8625
L-tyrosine	#T-3754
L-serine	#S-4500
L-phenylalanine	#P-2126
L-methionine	#M-9625
L-lysine	#L-5626
L-leucine	#L-8000
L-histidine	#H-8125
L-arginine	#A-5006
L-aspartic acid	#A-9256
Yeast Nitrogen Base (-amino acids)	#Y-0626
Ammonium Sulfate	#A-2939
D(+)-glucose	#G-5400
D-galactose	#D-5388
D-sorbitol	#S-1876
Polyethylene Glycol 3350	#P-3640
Salmon sperm DNA	#D-1626
Peptone#P-5905	
Bacto Agar (Difco)	
Glass beads (acid-washed)	#G-8772

2. Unipath Ltd. (Basingstoke, Hampshire, U.K.)

Oxoid Yeast Extract	#L21

3. Boehringer Mannheim (Indianapolis, IN)

X-gal (5-bromo-4-chloro-3-indolyl-β-D-galactopyranoside)	#703-729

VII. Concluding Remarks and Future Directions

The major complication in employing a yeast two-hybrid screen for protein-protein interactions is the isolation of false positives. Rules for reducing the proportion of false positive isolates from a given screen are not established, thus each bait plasmid must be empirically evaluated for its autoactivating function prior to using it in a screen. The LexA system, as described here, provides a thorough set of controls which we have found useful in reducing false positive rates. Time spent conducting these control experiments will improve the specificity of an interactor hunt, and result in significantly less labor for the investigator at the point of sorting through and confirming a collection of putative interacting clones (possibly a hundred or so). After transformation of EGY48 with the bait plasmid, selection of clones which express higher levels of the bait fusion protein could prove to be a critical step in determining the successful outcome of a screen. The pJK101 repression assay is a particularly elegant means of confirming that a bait fusion protein is being efficiently targeted to the yeast nucleus, thus saving investigators time in avoiding screens using expressed baits that are not able to interact with LexAop sequences. We cannot overemphasize another key issue in reducing wasted time and effort in conducting an interaction trap experiment. Always maintain EGY48 strains under appropriate auxotrophic selection. When growing up transformed stocks of EGY48, we routinely include control incubations in order to ensure that subsequent selective procedures are in fact selective for the desired marker. For example, if a bait/reporter strain is to be grown and selected in CM[HU]-, we set up a parallel incubation of this strain in CM[HUT]-, and only proceed with subsequent transformation and selection when the control culture is negative for growth.

In conclusion, the two-hybrid screen is an extremely powerful protocol for the elucidation of protein-protein interactions underlying signal transduction, indeed for the dissection of multi-protein complexes active in a number of biological functions, such as mRNA splicing and gene transcription. We anticipate that judicious application of this tool will continue to provide biologists with key insights into the complex question of signal transduction by cell adhesion molecules, as well as probing the mechanisms of molecular integration of these signals with those emanating from other growth regulatory and development receptor systems.

Acknowledgments

Work in the authors' laboratories is supported by grants from the Medical Research Council of Canada (S. D.), National Cancer Institute of Canada (G. H., S. D.), and the U.S. Army Breast Cancer Research Program (G. H.). G. H. is a Scholar of the MRC Canada, and S. D. is a Terry Fox Scientist of the NCI Canada.

References

1. Pawson, T. and Scott, J. D., Signaling through scaffolding, anchoring and adaptor proteins, *Science*, 278, 2075, 1997.
2. Dedhar, S. and Hannigan, G. E., Integrin cytoplasmic interactions and bidirectional transmembrane signal transduction, *Curr. Opin. Cell Biol.*, 8, 657, 1996.
3. Clark, E. A. and Brugge, J. S., Integrins and signal transduction pathways: the road taken, *Science*, 268, 233, 1995.
4. Schwartz, M. A., Schaller, M. D., and Ginsberg, M. H., Integrins: emerging paradigms of signal transduction, *Annu. Rev. Cell Biol. Dev. Biol.*, 11, 549, 1995.
5. Golemis, E. A., Gyuris, J., and Brent, R., Interaction trap/two-hybrid system to identify interacting proteins, in *Current Protocols in Molecular Biology*, Vol. 2, Ausubel, F. M., Brent, R., Kingston, R. E., Moore, D. D., Seidman, J. G., Smith, J. A., and Struhl, K., Eds., John Wiley & Sons, 1997.
6. Hannigan, G. E., Leung-Hagesteijn, C., Fitz-Gibbon, L., Coppolino, M. G., Radeva, G., Filmus, J., Bell, J., and Dedhar, S., Regulation of cell adhesion and anchorage-dependent growth by a new β1 integrin-linked protein kinase, *Nature*, 379, 91, 1996.
7. Bai, C. and Elledge, S. J., Gene identification using the yeast two-hybrid system, *Methods Enzymol.*, 283, 141, 1997.
8. Becker, D. M. and Lundblad, V., Introduction of DNA into yeast cells, in *Current Protocols in Molecular Biology*, Vol. 2, Ausubel, F. M., Brent, R., Kingston, R. E., Moore, D. D., Seidman, J. G., Smith, J. A., and Struhl, K., Eds., John Wiley & Sons, 1997.
9. Gietz, R. D. and Woods, R. A., High efficiency transformation with lithium acetate, in *Molecular Genetics of Yeast*, Johnston, J. R., Ed., IRL Press, New York, 1994.
10. Zervos, A. S., Gyuris, J., and Brent, R., Mxi1, a protein that specifically interacts with Max to bind Myc-Max recognition sites, *Cell*, 72, 223, 1993.
11. Hoffman, C. S., Preparation of yeast DNA, in *Current Protocols in Molecular Biology*, Vol. 2, Ausubel, F. M., Brent, R., Kingston, R. E., Moore, D. D., Seidman, J. G., Smith, J. A., Struhl, K., Eds., John Wiley & Sons, 1997.
12. Carnero, A. and Hannon, G. J., The INK4 family of CDK inhibitors, *Curr. Top. Microbiol. Immunol.*, 227, 43, 1998.
13. Matsuno, K., Diederich, R. J., Masahiro, J. G., Blaumueller, C. M., and Artavanis-Tsakonas, S., Deltex acts as a positive regulator of Notch signaling through interactions with the Notch ankyrin repeats, *Development*, 121, 2633, 1995.
14. Wu, C., Keightley, S. Y., Leung-Hagesteijn, C., Radeva, G., Coppolino, M., Goicoechea, S., McDonald, J. A., and Dedhar, S., Integrin-linked kinase regulates fibronectin matrix assembly, E-cadherin expression, and tumorigenicity, *J. Biol. Chem.*, 273, 528, 1998.
15. Wu, C., Bauer, J. S., Juliano, R. L., and McDonald, J. A., The $\alpha_5\beta_1$ integrin receptor, but not the α_5 cytoplasmic domain, functions in an early and essential step in fibronectin matrix assembly, *J. Biol. Chem.*, 268, 21883, 1993
16. Treco, D. A. and Lundblad, V., Basic techniques of yeast genetics, in *Current Protocols in Molecular Biology*, Vol. 2, Ausubel, F. M., Brent, R., Kingston, R. E., Moore, D. D., Seidman, J. G., Smith, J. A., Struhl, K., Eds., John Wiley & Sons, 1997.

Chapter 4

Focal Adhesion Kinase and its Associated Proteins

Jill K. Slack, Judith Lacoste, and J. Thomas Parsons

Contents

I. Introduction 49
II. Protein-Protein Interactions 51
A. Preparation of Cell Lysates 51
B. Co-Immunoprecipitation of FAK and Associated Proteins 52
C. *In vitro* Protein Association Using GST Fusion Proteins 55
D. Far-Western Analysis 56
III. Subcellular Localization 57
A. Immunofluorescence in Fixed Cells 57
B. Use of GFP Fusion Protein in Living Cells 58
IV. Epilogue 59
Acknowledgments 59
References 59

I. Introduction

Cellular adhesion to the extracellular matrix (ECM) is a critical regulator of fundamental processes including growth, differentiation, death, and migration. Loss of adhesive influences stimulates apoptosis in many non-transformed cells.[1] Indeed, a hallmark of transformation is the loss of adhesion-dependent growth control, which is often accompanied by increased cellular migration. Cells recognize and respond to the ECM through specialized adhesive structures called focal adhesions. In addition to linking structural components of the cytoskeleton to the underlying matrix, focal adhesions likely organize

0-8493-3385-7/99/$0.00+$.50

a competent signal transduction cascade initiated by receptor-mediated recognition of ECM components (e.g., fibronectin, laminin, collagen).

The ability of focal adhesions to organize a competent signal transducing complex is suggested by the observed interaction of heterodimeric transmembrane integrin receptors with focal adhesion associated tyrosine phosphorylated proteins.[2] Clustering integrin molecules without necessarily occupying the ligand binding domain results in increased tyrosine phosphorylation.[3,4] Among the proteins phosphorylated as a result of integrin clustering is a protein tyrosine kinase found within the focal adhesion (focal adhesion kinase or FAK).[5-10] Because integrins themselves lack catalytic activity, the stimulation of FAK kinase activity upon integrin occupancy suggests a role for FAK in the signaling that underlies integrin-mediated cell adhesion and migration. Indeed, the results from several experimental systems provide evidence for such a role for FAK in cell adhesion and migration. Mouse embryos deficient for FAK expression die early in development and cells derived from such early embryos exhibit reduced mobility *in vitro*.[11] The stable overexpression of FAK in Chinese hamster ovary cells stimulates cell migration on fibronectin.[12] Recent studies have shown that overexpression of FRNK (FAK-related non-kinase), the autonomously expressed C-terminal domain of FAK,[13] in chick embryo fibroblasts reduces FAK tyrosine phosphorylation concomitant with a delay in cell spreading on fibronectin.[14] Similar findings were obtained by microinjecting BALB/c 3T3 cells and human umbilical vein endothelial cells with a glutathione-*S*-transferase (GST) fusion protein containing the FAK C-terminal domain.[15] In human cancer cells, invasive tumor cells overexpress FAK relative to normal control cells.[16,17] Human cells expressing high levels of FAK lose attachment and undergo apoptosis when treated with FAK antisense oligonucleotides.[18] Taken together, these results implicate FAK in events necessary for cell spreading and migration in both normal and malignant cells.

Progress toward understanding how FAK mediates integrin signaling has come from analysis of FAK activity, subcellular localization, and binding partners. Integrin crosslinking stimulates FAK autophosphorylation activity[5-9] possibly through the interaction of FAK with cytoplasmic domains of β_1, β_2, and β_3 integrins.[2] Tyrosine 397 is a major site of autophosphorylation,[19] and promotes SH2-mediated binding of Src family kinases, Src and Fyn,[19,20] potentially generating a multi-enzyme signaling complex. Recently, the regulatory subunit of phosphatidylinositol 3-kinase (PI3K, p85 subunit) was reported to bind Tyr-397 through an SH2-mediated interaction.[21] Additional tyrosine phosphorylation sites have been described (Tyr-407, Tyr-576, Tyr-577 and Tyr-861), two of which (Tyr-576 and Tyr-577) have been suggested to play a role in regulating FAK kinase activity.[22,23] Phosphorylation of Tyr-926 creates a binding site for GRB2,[24] an SH2/SH3 adaptor protein that links growth factor receptor tyrosine kinase signal transduction to the Ras/MAPK pathway through the Ras GDP/GTP exchange protein SOS. Indeed, FAK stimulation as a consequence of ECM-integrin interaction does coincide with MAPK phosphorylation.[24-26] The C-terminal portion of FAK also directs the interaction of FAK with additional signaling molecules. Proline rich motifs located in the C-terminal portion of FAK direct the SH3-mediated interaction of additional signaling molecules. The sequence $P^{712}PKP^{715}SR$ binds the SH3 domains of the Crk-associated substrate

($p130^{CAS}$),[27,28] and may interact with the SH3 domain of the p85 regulatory subunit of PI3K as well.[29] FAK also binds the SH3 domain of Graf, a GTPase-activating protein for Rho and Cdc42, through the proline rich motif $P^{875}KKP^{878}PR$.[30] This interaction suggests a mechanism for crosstalk between integrin initiated signaling events and growth factor-stimulated actin reorganization.[30] A structural domain located within the C-terminus (amino acids 904 to 1040) allows targeting of FAK to focal adhesions,[31,32] possibly through its interaction with the focal adhesion proteins paxillin (residues 904 to 1053) and talin.[32-33] The interaction of FAK with paxillin and $p130^{CAS}$ recruits other adapter molecules to focal adhesions such as Crk,[34-36] which can also link to the Ras-MAP kinase pathway because of the ability of Crk to bind C3G and SOS, two guanine nucleotide exchange proteins for Ras.[37,38] Considering the nature of the proteins recruited/activated by FAK (directly or indirectly), FAK likely plays an important role in integrin-mediated signaling.

The purpose of this article is to review methodology used to identify FAK binding proteins. We discuss the identification of *in vivo* and *in vitro* protein interactions, as well as methods for detecting co-localization of FAK and its binding partners in intact cells using immunofluorescence.

II. Protein-Protein Interactions

Transmission of signals from extracellular cues to intracellular compartments involves protein phosphorylation necessitating protein-protein interactions. Such interactions have been detected using several methodologies, each with its unique advantages and disadvantages. It is possible to determine whether known proteins interact with each other or to identify novel proteins that interact with a protein of interest. For the purposes of this review, we will focus on methods employed to identify interactions between known proteins.

A. Preparation of Cell Lysates

FAK has been detected in a variety of cell types[8,39,40] and endogenous levels of FAK are easily detected using the assays described below. Because of its linkage to integrins and cell adhesion, FAK is usually examined in adherent cells. Subconfluent cultures of adherent cells are placed on ice and washed twice in CMF-PBS (calcium, magnesium free phosphate buffered saline, 137 mM sodium chloride [NaCl], 2.7 mM potassium chloride [KCl], 4.3 mM disodium phosphate [$Na_2HPO_4 \cdot 7H_2O$], and 1.4 mM potassium phosphate [KH_2PO_4, pH 7.2]). These are then lysed by scraping in supplemented radioimmunoprecipitation (S-RIPA) lysis buffer (50 mM *N*-2-hydroxyethylpiperazine-*N′*-2-ethanesulfonic acid [HEPES, pH 7.2], 150 mM NaCl, 2 mM ethylenediamine tetraacetic acid [EDTA], 1% (v/v) Nonidet P-40 [NP-40], 0.5% (w/v) sodium deoxycholate [DOC, 5β-cholan-24-oic acid-3α, 12α-diol, $C_{24}H_{39}O_4Na$], 100 μM leupeptin, 10 μM pepstatin, 0.05 TIU/ml aprotinin, 1 mM phenylmethylsulfonyl fluoride [$C_7H_7FO_2S$, PMSF], 1 mM benzamidine [$C_7H_8N_2 \cdot HCl$], 1 mM sodium orthovanadate

[Na_3VO_4], 10 mM sodium pyrophosphate [$Na_4P_2O_7 \cdot 10H_2O$], 40 mM sodium *p*-nitrophenyl phosphate [$C_6H_4NO_6PNa_2 \cdot 6H_2O$], and 40 mM sodium fluoride [NaF]).[41,42] Leupeptin, pepstatin, aprotinin, PMSF, and benzamidine are used as protease inhibitors. Sodium orthovanadate, sodium pyrophosphate, sodium *p*-nitrophenyl phosphate, and sodium fluoride are used to inhibit tyrosine and serine/threonine phosphatases. Usually, 1 ml of RIPA lysis buffer is used per 100 mm plate (approximately 10^7 cells). Lysates are cleared by centrifugation in a microfuge (e.g., Eppendorf 5415C) at 4°C for 10 min at 16,000 × *g* to remove cellular debris and nuclei, and the supernatant is collected into a 1.5 ml microfuge tube. Protein concentrations are determined by the bicinchoninic assay (BCA assay, Pierce). Generally, 2 to 4 mg of lysates are obtained per 10^7 cells.

Cells in suspension (either adherent cells placed in suspension by trypsinization or treatment with 1 to 10 mM EDTA, or non-adherent cells such as lymphocytes), are collected by centrifugation, washed twice with CMF-PBS, and resuspended in S-RIPA lysis buffer (1 ml per 10^7 cells). Lysates are cleared by centrifugation as above. For cell adhesion/spreading experiments, cells are first placed in suspension (by trypsinization or treatment with EDTA) in order to reduce background tyrosine phosphorylation of FAK. The cells are then plated for specified times (usually 15 to 120 min) on bacterial culture plates (untreated or ECM-coated), and lysed in S-RIPA lysis buffer as described above. These lysates are suitable for immunoprecipitations or direct immunoblots.

For the preparation of whole-cell lysates, cells may be placed directly in 1X SDS sample buffer: 62.5 mM tris(hydroxymethyl)aminomethane hydrochloride [Tris-HCl, pH 6.8], 10% (v/v) glycerol, 1% (w/v) sodium dodecyl sulfate [SDS], 1% (v/v) β-mercapto-ethanol, 5 μg/ml bromo-phenol blue. For adherent cells, proceed as described above, except that cells are scraped in 1 ml of 1X SDS sample buffer and boiled for 10 min. To prepare lysates from suspension cells, proceed as above; resuspend the cell pellets in 1 ml 1X SDS sample buffer and boil for 10 min. These lysates can be used for direct immunoblot analyses. Because of the high concentration of SDS, these lysates are not suitable for immunoprecipitations experiments.

B. Co-Immunoprecipitation of FAK and Associated Proteins

Immunoprecipitation of FAK and associated proteins is a useful method to identify *in vivo* protein associations. In this approach, cells are lysed and incubated with antibody to FAK. Antibody bound FAK, together with any proteins that bind FAK, are separated from the remaining cell lysate and subjected to Western analysis, using antibodies for suspected FAK binding proteins. One drawback to this approach is the inability to rule out intermediary proteins that may mediate suspected interactions.

1. Antibodies

Table 1 provides a summary of FAK-specific antibodies routinely used in our laboratory in the assays described below. Commercially available antibodies to FAK are listed in Table 2. Monoclonal antibody 2A7 (Upstate Biotechnology, Cat. #05-182)

TABLE 1
Antibodies used for study of FAK protein expression and binding partners by immunoprecipitation, immunoblotting, and immunofluorescence

Type	Antibody	Epitope	Cross Reactivity	Source
Monoclonal	2A7	residues 904-1040 of chicken FAK	rat, mouse, hamster, human, dog	J. T. Parsons or Upstate Biotechnology, Cat. #05-182[32,43]
Polyclonal	BC3	residues 651-1028 of chicken FAK	rat, mouse, human, bovine	J. T. Parsons or Upstate Biotechnology, Cat. #06-543[10]
	HUB3	residues 542-880 of human FAK	chicken, bovine	J. T. Parsons[48]

TABLE 2
Commercially available FAK antibodies

Type	Antibody	Epitope	Cross Reactivity	Source
Monoclonal	FAK	residues 354-533 of chicken FAK	human, dog, rat, mouse, chicken	Transduction Laboratories, Cat. #F15020[50-52]
	FAK (H-1)	residues 903-1052 of human FAK	mouse, rat	Santa Cruz Biotechnology, Cat. #sc-1688
	FAK	N.D.	human, mouse, hamster, chicken	Chemicon, Cat. #MAB2516
Polyclonal	FAK (A-17)	residues 2-18 of human FAK	mouse, rat, human, chicken	Santa Cruz Biotechnology, Cat. #sc-557
	FAK (C-20)	residues 1033-1052 of human FAK	mouse, rat, human, chicken	Santa Cruz Biotechnology, Cat. #sc-558
	FAK (C-903)	residues 903-1052 of mouse FAK	mouse, rat, human	Santa Cruz Biotechnology, Cat. #sc-932
	FAK	N.D.	human, mouse	Chemicon, Cat. #AB1605
	FAK	residues 748-1053 of human FAK	human, mouse, rat, hamster	Upstate Biotechnology, Cat. #06-543

was isolated from a screen of monoclonal antibodies derived from mice immunized with partly purified tyrosine phosphorylated proteins from Src-transformed cells.[43] BC3 is a rabbit polyclonal antibody raised against a bacterial TrpE fusion protein containing residues 651 to 1020 of chicken FAK, which correspond to the C-terminal domain.[10] HUB3 is a polyclonal antibody generated against a GST-fusion protein containing residues 542 to 880 of human FAK. This segment encompasses part of the kinase and C-terminal domains.[48] In addition to the antibodies we have generated to chicken and human FAK, antibodies specific for chicken, human, and mouse FAK are commercially available from Transduction Laboratories, Santa Cruz Biotechnology, Chemicon, and Upstate Biotechnology (Table 2). These companies provide a wide range of monoclonal and polyclonal antibodies specific for amino terminal, kinase, and carboxy terminal domains of FAK, which can be used effectively in the assays described below.

2. FAK immunoprecipitation/immunoblotting

For immunoprecipitation of FAK, 0.5 to 1.5 mg of lysate is incubated with the FAK-specific antibody for 1 h at 4°C using constant rotation. For monoclonal 2A7, 10 μg of protein A-purified antibody per mg of lysate are used. For polyclonal sera, 10 μl of BC3 or 10 μl of HUB3 per mg of cell lysate are used. Commercially available antibodies can be used according to manufacturer's recommendation. Immune complexes (IC) are recovered by the addition of 100 μl of protein A-sepharose beads (1:1 slurry, Pharmacia) followed by incubation for another hour at 4°C, while rotating. When immunoprecipitations with antibody 2A7 are performed, the beads are precoated with 20 μg of donkey anti-mouse immunoglobulin G (IgG, Jackson ImmunoResearch Laboratories, Cat. #715-005-150). IC are collected by centrifugation in a microfuge at 16,000 *g* for one min, then washed twice with 1 ml of ice-cold S-RIPA buffer and twice with 1 ml of ice-cold TBS (Tris-buffered saline, 10 mM Tris-HCl [pH 8.0], 150 mM NaCl). For each wash, the IC are centrifuged as above and resuspended by flicking the microtube. After the washes, IC are recovered by centrifugation in a microfuge and solubilized from the sepharose beads by boiling for 10 min in an equal volume (usually 50 μl) of 2X SDS sample buffer.

The solubilized proteins are resolved by electrophoresis on an 8% SDS polyacrylamide gel (0.8% cross-link *N*, *N'*-bismethylene acrylamide, 0.375 M Tris-HCl [pH 8.8], 0.1% (w/v) SDS) run at 105 mA in 1X SDS running buffer (25 mM Tris-HCl [pH 8.3], 0.2 M glycine, 0.1% (w/v) SDS). For direct immunoblot, 5 to 50 μg of lysates are loaded on the same type of gel and electrophoresis carried out as above. The amount of lysate used depends on the levels of expression (i.e., endogenous vs. overexpressed). Resolved proteins are transferred to a nitrocellulose membrane in transfer buffer (25 mM Tris-HCl [pH 8.3], 0.2 M glycine, 0.1% (w/v) SDS, 20% (v/v) methanol) for 1.2 h at 22 volts at 4°C. The membrane is then blocked for 1 h at room temperature in 5% (w/v) instant nonfat dry milk (we prefer Super G™, the Giant™ Supermarket brand)/TBST (0.5 M Tris-HCl [pH 7.4], 1.5 M NaCl, 1% (v/v) polyoxyethylenesorbitan monolaurate [Tween-20]) and immunoblotted for another hour in fresh blotting solution containing the primary antibody. When blotting for phosphorylated protein isoforms, it is necessary to block and blot in 1 to

5% BSA/TBST since milk contains intrinsic phosphatase activity. FAK can be detected on immunoblots using antibodies 2A7, BC3 or HUB3 at a dilution of 1:1000 in 25 ml of blotting solution. Following incubation with the primary antibody, the membrane is then quickly washed twice with TBST and incubated for another hour at room temperature with the secondary antibody (donkey anti-rabbit IgG, Amersham, Cat. #NA 934, or sheep anti-mouse IgG, Amersham, Cat. #NA 931) linked to horseradish peroxidase (HRP) in 5% (w/v) instant nonfat dry milk/TBST or 1 to 5% BSA/TBST at a 1:1000 dilution (25 ml total volume). HRP-antibody binding is visualized with enhanced chemiluminescence (ECL) following the manufacturer's instructions (Amersham).

C. *In vitro* Protein Association Using GST Fusion Proteins

In vitro protein association can be detected using GST fusion proteins. Cell lysates are incubated with fusion proteins containing glutathione S-transferase fused to regions of FAK. Proteins that interact with the FAK fusions are separated from the lysate using glutathione coupled beads, which bind the GST moiety, and subjected to Western analysis using antibodies to proteins suspected to interact with FAK. As with co-immunoprecipitation, this approach cannot rule out the ability of third-party proteins to mediate the interaction of interest. We have generated a GST-FAK fusion, termed GST-Cterm, that contains FAK amino acids 686-1053 fused to GST.[30] It has been used to identify interactions between FAK and Graf,[30] p130^{Cas}[27] and paxillin.[32]

1. Purification of GST fusion proteins

Plasmids encoding GST fusion proteins are grown in bacteria in LB broth with the appropriate antibiotic selection for 12 to 18 h, diluted 1:10 and grown an additional hour. Isopropyl β-D-Thiogalactopyranoside (IPTG) is added to a final concentration of 100 μM and the culture is grown an additional 3 to 4 h. The cells are collected by centrifugation for 15 min at 4°C at 4500 rpm using a GSA rotor in a Sorvall RC-5B refrigerated superspeed centrifuge. The pellet is resuspended in 1/10 the original volume NETN (20 mM Tris, 100 mM NaCl, 1 mM EDTA, pH 8.0 then 0.5% NP-40) supplemented with protease inhibitors (100 μM leupeptin, 10 μM pepstatin, 0.05 TIU/ml aprotinin, 1 mM phenylmethylsulfonyl fluoride [$C_7H_7FO_2S$, PMSF], 1 mM benzamidine [$C_7H_8N_2 \cdot HCl$]) and 1 mg/ml lysozyme. The bacteria are lysed by sonication at 80% duty cycle, 4 times, 1 min each. The lysate is cleared by centrifugation for 5 min at 4°C at 2000 rpm. Glutathione Sepharose beads (Pharmacia) are prepared by washing twice with TBS. The beads are reconstituted to a 70% slurry (v/v) in TBS. The lysate from each 500 ml bacterial culture is incubated with 1 ml glutathione beads for 1 h at 4°C with agitation. Beads are washed 4 times by centrifugation with NETN plus protease inhibitors. They can be stored as a 50% slurry in NETN plus protease inhibitors. We routinely recover >1 mg GST fusion protein per 500 ml bacterial culture. Recovery is dependent on the protein size. Typically, there is more degradation of proteins larger than 60 to 70 kDa resulting in lower recovery. The quantity and integrity of fusion proteins can be analyzed by SDS-PAGE. An aliquot (2-5 μl/lane) of protein coupled to beads is boiled in SDS sample buffer and loaded on a 12.5% SDS-PAGE gel. BSA standards (2, 4, and 8 μg) are electrophoresed through the same gel. The gels are stained with Coomassie Blue to visualize the proteins. An estimate of protein concentration can be determined by comparing the sample to the known BSA standard.

2. Isolation and immunoblotting of associated proteins

Equivalent amounts of GST fusion proteins (approximately 5 μg) or GST alone, immobilized on glutathione-Sepharose, are incubated with cell lysate for 1 h at 4°C with rotation. The amount of lysate added varies depending on the abundance of the protein expected to interact with the fusion. For abundant endogenous proteins such as FAK, 500 μg is sufficient. Less lysate is used for overexpressed proteins, typically 200 to 300 μg. If the abundance is unknown, 1 mg of protein is a reasonable amount with which to start. However, higher amounts of lysate increase non-specific interactions. The beads are washed twice with RIPA buffer with or without protease inhibitors and twice with TBS. Associated proteins are eluted by boiling in SDS sample buffer, resolved by SDS-PAGE and transferred to nitrocellulose for Western or Far-Western analysis. Since GST pull-down experiments determine whether two known proteins interact rather than identify novel proteins, Western blotting is performed using antibodies raised against the target protein of interest.

D. Far-Western Analysis

Direct protein-protein interactions can be detected *in vitro* using Far-Western analysis. Protein complexes, either from immunoprecipitations or whole cell lysates, are subjected to SDS-PAGE and transferred to nitrocellulose. Immobilized proteins are subjected to denaturation and renaturation before being probed with a radiolabeled GST fusion protein. Because the target proteins are immobilized on the nitrocellulose and probed with a single protein, this approach will not identify protein interactions that require an intermediary protein.

1. Generation of ^{32}P-labeled GST fusion proteins

Proteins to be used as probes in the Far-Western analysis are cloned in frame into a GST expression plasmid encoding the recognition sequence for the catalytic subunit of cAMP-dependent protein kinase (PKA).[44] The GST fusion proteins, expressed in *Escherichia coli,* are affinity purified using glutathione Sepharose as described above (see Section II.C.1, *Purification of GST fusion proteins)* and phosphorylated using heart muscle PKA.[44] All subsequent steps are performed at 4°C unless otherwise indicated. Approximately 3 mg of GST fusion proteins coupled to Sepharose beads are washed in heart muscle kinase (HMK) buffer (20 mM Tris [pH 7.5], 100 mM NaCl, 12 mM $MgCl_2$), resuspended in 40 μl of HMK buffer containing 1 mM dithiothreitol, 50 U PKA, and 100 μCi ^{32}P-γ-ATP (6000 Ci/mmol, DuPont NEN) and incubated for 45 min. The reaction is terminated by adding 1 ml HMK stop buffer (10 mM sodium phosphate [pH 8.0], 10 mM sodium pyrophosphate,

10 mM EDTA). The supernatant is removed and the beads are washed five times with NETN. The labeled GST fusion proteins are eluted from the Sepharose beads by washing the Sepharose for 10 to 15 min in 10 bed volumes of 20 mM glutathione, 100 mM Tris (pH 8.0), 120 mM NaCl at room temperature.

2. Protein denaturation/renaturation

Proteins from cell lysates or immunoprecipitations are resolved by SDS-PAGE and transferred to nitrocellulose as described above (see Section II.B.2, *FAK immunoprecipitation/immunoblotting)*. Proteins immobilized on the filters are denatured with guanidine hydrocloride (GuHCl) and renatured by washing the filters in sequentially diluted GuHCl. Briefly, the filters are incubated twice for 5 min each at 4°C in 100 ml hyb75 (20 mM *N*-2-hydrosyethylpiperazine-*N* = -2-ethanesulfonic acid [pH 7.7], 75 mM KCl, 0.1 mM EDTA, 2.5 mM $MgCl_2$, 1 mM dithiothreitol, 0.05% NP-40) plus 6 M GuHCl. The filters are washed four times for 10 min each at 4°C in 100 ml hyb75 containing GuHCl diluted in half from the previous wash step (i.e., 3 M, 1.5 M, 0.75 M, and 0.375 M GuHCl in each of the subsequent washes). The filters are transferred to new dishes, washed twice for 30 min in 100 ml hyb75, and blocked in 10 ml hyb75 containing 5% milk for 30 min, followed by a second 30 min in 10 ml hyb75 containing 1% milk. The blocked filters are blotted by adding 3×10^5 cpm/ml ^{32}P-labeled fusion protein (specific activity approximately 10^8 cpm/ug) and incubating at 4°C for 8 h. The filters are washed three times with 100 ml hyb75 containing 1% milk for 15 min at room temperature and exposed to film at –70°C with an intensifying screen.

III. Subcellular Localization

A. Immunofluorescence in Fixed Cells

Subcellular localization of proteins can be identified at a given time in intact, fixed cells using fluorescently tagged antibodies. Typically, proteins are recognized by a specific antibody, which, in turn, is recognized by a secondary antibody conjugated to a fluorophore (e.g., fluorescein, rhodamine, or Texas Red). Fluorescent images can be collected using a microscope equipped with filters to detect green or red fluorescent emissions. Co-localization of two different proteins can be detected using a monoclonal antibody for one protein and a polyclonal antibody for the second protein. Detection is accomplished using secondary antibodies, conjugated to different fluorophores, that differentially recognize the monoclonal or polyclonal primary antibody. Use of a filter that detects both green and red fluorescence will identify co-localization of the fluorophores; overlap of both emissions will appear yellow.

Cells are seeded on glass coverslips (either uncoated or coated with fibronectin [1 to 5 μg/cm^2] or with poly-L-Lysine [0.2–0.5 mg/ml] as a control), and grown overnight or for various periods of time. The coverslips containing adherent cells are washed in CMF-PBS twice and cells are fixed in 4% (w/v) paraformaldehyde/PBS (see below for preparation) for 10 min at room temperature. The coverslips are subsequently rinsed with CMF-PBS and the cells are permeabilized in CMF-PBS containing 0.5% (v/v) Triton X-100 for 5 min at room temperature. Coverslips are further rinsed three times with CMF-PBS and incubated with monoclonal antibody 2A7 (10 μg/ml) for 60 min at room temperature. Coverslips are rinsed and incubated with 5 μg/ml goat anti-mouse IgG (Jackson ImmunoResearch Laboratories, Cat. #115-005-003) for 60 min at room temperature. Finally, coverslips are incubated with 5 μg/ml fluorescein isothiocyanate (FITC)-conjugated donkey anti-goat IgG (Jackson ImmunoResearch Laboratories, Cat. #705-095-003) for 60 min at room temperature. Samples are visualized with a fluorescence microscope. Alternatively, the polyclonal antibody BC3 can be used at a 1:500 dilution, followed by an incubation with 5 μg/ml FITC-conjugated goat anti-rabbit IgG (Jackson ImmunoResearch Laboratories, Cat. #111-095-003). Note that both 2A7 and BC3 recognize phosphorylated and unphosphorylated forms of FAK and therefore these antibodies do not distinguish between active and inactive FAK.

For the preparation of the paraformaldehyde solution, 70 ml of deionized/distilled water is heated to 60°C. In a fume hood, 4 g of paraformaldehyde is added to the warm water and mixed on a stirring plate. To solubilize the paraformaldehyde, 1 to 2 drops of 1 N NaOH are added and the solution stirred for 20 to 30 min. The mixture is removed from the heat and 10 ml of 10X PBS are added. The pH of the solution is adjusted to pH 7.4 with HCl and deionized/distilled water is added to a final volume of 100 ml. The paraformaldehyde solution can be aliquoted and stored at –20°C for about 1 month.

B. Use of GFP Fusion Protein in Living Cells

With the identification of the gene encoding green fluorescent protein (GFP) from *Aequorea victoria*, it has become possible to determine protein subcellular localization in living cells. Technically, this approach is not considered immunofluorescence since antibodies are not used to stain the cells. Rather, cells express GFP fusion proteins. Expression constructs engineered to generate proteins containing the target protein of interest fused to GFP can be transfected or microinjected into cells. Alternatively, the fusion proteins can be introduced directly into cells by microinjection. Localization of expressed fluorescently tagged proteins can be monitored over time following various treatments. This approach has been used to visualize PLCγ membrane targeting following growth factor stimulation,[45] β-arrestin translocation to activated G protein-coupled receptors,[46] and glucocorticoid receptor translocation from the cytoplasm to the nucleus following hormone addition.[47] We have successfully tagged FRNK and paxillin with GFP and observed its localization in focal adhesions.[49]

IV. Epilogue

Progress toward understanding the role of FAK in proliferation, differentiation, apoptosis, and migration has been made by identifying FAK substrates and binding partners using the *in vitro* and immunofluorescence methods outlined here. Future experiments will need to address the temporal and spacial aspects of focal adhesion assembly and how FAK contributes to signaling in such compartments. The use of GFP-tagged FAK proteins coupled with more sophisticated image analysis, such as FRET, will be immensely important. Furthermore, characterization of additional FAK interacting proteins and establishment of parameters for protein complex assembly will be critical to our understanding of integrin-mediated signaling events.

Acknowledgments

J.T.P. is supported by grants from the DHHS-National Cancer Institute, CA29243, CA40042, and the Council for Tobacco Research U.S.A., Inc., #4491. We thank R. B. Adams, C. A. Borgman, M. T. Harte, A. Ma, R. K. Malik, J. Taylor, and W.-C. Xiong for sharing their versions of the protocols and K. Martin for sharing data.

References

1. Frisch, S. M. and Ruoslahti, E., Integrins and anoikis, *Curr. Opin. Cell Biol.*, 9, 701, 1997.
2. Schaller, M. D., Otey, C. A., Hildebrand, J. D., and Parsons, J. T., Focal adhesion kinase and paxillin bind to peptides mimicking β integrin cytoplasmic domains, *J. Cell Biol.*, 130, 1181, 1995.
3. Guan, J. L., Trevithick, J. E., and Hynes, R. O., Fibronectin/integrin interaction induces tyrosine phosphorylation of a 120 kDa protein, *Cell Regul.*, 2, 951, 1991.
4. Kornberg, L. J., Earp, H. S., Turner, C. E., Prockop, C., and Juliano, R. L., Signal transduction by integrins: increased protein tyrosine phosphorylation caused by clustering of β_1 integrins, *Proc. Natl. Acad. Sci. USA*, 88, 8392, 1991.
5. Kornberg, L., Earp, H. S., Parsons, J. T., Schaller, M., and Juliano, R. L., Cell adhesion or integrin clustering increases phosphorylation of a focal adhesion associated tyrosine kinase, *J. Biol. Chem.*, 267, 23439, 1992.
6. Burridge, K., Turner, C. E., and Romer, L. H., Tyrosine phosphorylation of paxillin and pp125FAK accompanies cell adhesion to extracellular matrix: a role in cytoskeletal assembly, *J. Cell Biol.*, 119, 893, 1992.
7. Lipfert, L., Haimovich, B., Schaller, M. D., Cobb, B. S., Parsons, J. T., and Brugge, J. S., Integrin dependent phosphorylation and activation of the protein tyrosine kinase $pp125^{FAK}$ in platelets, *J. Cell Biol.*, 119, 905 1992.
8. Hanks, S. K., Calalb, M. B., Harper, M. C., and Patel, S. K., Focal adhesion protein-tyrosine kinase phosphorylated in response to cell attachment to fibronectin, *Proc. Natl. Acad. Sci. USA,* 89, 8487, 1992.

9. Guan, J.-L. and Shalloway, D., Regulation of focal adhesion-associated protein tyrosine kinase by both cellular adhesion and oncogenic transformation, *Nature*, 358, 690, 1992.
10. Schaller, M. D., Borgman, C. A., Cobb, B. S., Vines, R. R., Reynolds, A. B., and Parsons, J. T., pp125^{FAK}, a structurally distinctive protein-tyrosine kinase associated with focal adhesions, *Proc. Natl. Acad. Sci. USA*, 89, 5192, 1992.
11. Ilic, D., Furuta, Y., Kanazawa, S., Takeda, N., Sobue, K., Nakatsuji, N., Nomura, S., Fujimoto, J., Okada, M., Yamamoto, T., and Alzawa, S., Reduced cell motility and enhanced focal adhesion contact formation in cells from FAK-deficient mice, *Nature*, 377, 539, 1995.
12. Cary, L. A., Chang, J. F., and Guan, J.-L., Stimulation of cell migration by overexpression of focal adhesion kinase and its association with Src and Fyn, *J. Cell Sci.*, 109, 1787, 1996.
13. Schaller, M. D., Borgman, C. A., and Parsons, J. T., Autonomous expression of a noncatalytic domain of the focal adhesion-associated protein tyrosine kinase pp125^{FAK}, *Mol. Cell. Biol.*, 13, 785, 1993.
14. Richardson, A. and Parsons, J. T., A mechanism for regulation of the adhesion-associated protein tyrosine kinase pp125^{FAK}, *Nature*, 380, 538, 1996.
15. Gilmore, A. P. and Romer, L. H., Inhibition of focal adhesion kinase (FAK) signaling in focal adhesions decreases cell motility and proliferation, *Mol. Biol. Cell*, 7, 1209, 1996.
16. Owens, L. V., Xu, L., Craven, R. J., Dent, G. A., Weiner, T. M., Kornberg, L., Liu, E. T., and Cance, W. G., Overexpression of the focal adhesion kinase (p125FAK) in invasive human tumors, *Cancer Res.*, 55, 2752, 1995.
17. Weiner, T. M., Liu, E. T., Craven, R. J., and Cance, W. G., Expression of focal adhesion kinase gene and invasive cancer, *Lancet*, 342, 1024, 1993.
18. Xu, L. H., Owens, L. V., Sturge, G. C., Yang, X. H., Liu, E. T., Craven, R. J., and Cance, W. G., Attenuation of the expression of the focal adhesion kinase induces apoptosis in tumor cells, *Cell Growth Differ.*, 7, 413, 1996.
19. Schaller, M. D., Hildebrand, J. D., Shannon, J. D., Fox, J. W., Vines, R. R., and Parsons, J. T., Autophosphorylation of the focal adhesion kinase, pp125^{FAK}, directs SH2-dependent binding of pp60^{src}, *Mol. Cell. Biol.*, 14, 1680, 1994.
20. Cobb, B. S., Schaller, M. D., Leu, T.-H., and Parsons, J. T., Stable association of pp60^{src} and pp50^{fyn} with the focal adhesion-associated protein tyrosine kinase pp125^{FAK}, *Mol. Cell. Biol.*, 14, 147, 1994.
21. Chen, H.-C., Appeddu, P. A., Isoda, H., and Guan, J.-L., Phosphorylation of tyrosine 397 in focal adhesion kinase is required for binding phosphatidylinositol 3-kinase, *J. Biol. Chem.*, 271, 26329, 1996.
22. Calalb, M. B., Polte, T. R., and Hanks, S. K., Tyrosine phosphorylation of focal adhesion kinase at sites in the catalytic domain regulated kinase activity: a role for Src family kinases, *Mol. Cell. Biol.*, 15, 954, 1995.
23. Calalb, M. B., Zhang, X., Polte, T. R., and Hanks, S. K., Focal adhesion kinase tyrosine 861 is a major site of phosphorylation by Src, *Biochem. Biophys. Res. Commun.*, 228, 662, 1996.
24. Schlaepfer, D. D., Hanks, S. K., Hunter, T., and Van der Geer, P., Integrin-mediated signal transduction linked to RAS pathway by GRB2 binding to focal adhesion kinase, *Nature*, 372, 786, 1994.

25. Chen, Q., Kinvh, M. S., Lin, T. H., Burridge, K., and Juliano, R. L., Integrin-mediated cell adhesion activates mitogen-activated protein kinases, *J. Biol. Chem.*, 269, 26602, 1994.
26. Morino, N., Mimura, T., Hamasaki, K., Tobe, K., Ueki, K., Kikuchi, K., Takehara, K., Kadawaki, T., Yazaki, Y., and Nijima, Y., Matrix/integrin interaction activates the mitogen-activated protein kinase, $p44^{erk-1}$ and $p42^{erk-2}$, *J. Biol. Chem.*, 270, 269, 1995.
27. Harte, M. T., Hildebrand, J. D., Burnham, M. R., Bouton, A. H., and Parsons, J. T., $p130^{Cas}$, a substrate associated with v-Src and v-Crk, localizes to focal adhesions and binds to focal adhesion kinase, *J. Biol. Chem.*, 271, 13649, 1996.
28. Polte, T. R. and Hanks, S. K., Interaction between focal adhesion kinase and Crk-associated tyrosine kinase substrate $p130^{Cas}$, *Proc. Natl. Acad. Sci. USA*, 92, 10678, 1995.
29. Guinebault, C., Payrastre, B., Racaud-Sultan, C., Mazarguil, H., Breton, M., Mauco, G., Plantavid, M., and Chap, H., Integrin-dependent translocation of phosphoinositide 3-kinase to the cytoskeleton of ADP-aggregated human platelets occurs independently of RhoA and without synthesis of phosphatidylinositol (3,4)-bisphosphate, *J. Cell Biol.*, 129, 831, 1995.
30. Hildebrand, J. D., Taylor, J. M., and Parsons, J. T., An SH3 domain-containing GTPase-activating protein for Rho and Cdc42 associates with focal adhesion kinase, *Mol. Cell. Biol.*, 16, 3169, 1996.
31. Hildebrand, J. D., Schaller, M. D., and Parsons, J. T., Identification of sequences required for the efficient localization of the focal adhesion kinase, $pp125^{FAK}$, to cellular focal adhesions, *J. Cell Biol.*, 123, 993, 1993.
32. Hildebrand, J. D., Schaller, M. D., and Parsons, J. T., Paxillin, a tyrosine phosphorylated focal adhesion-associated protein binds to the carboxyl terminal domain of focal adhesion kinase, *Mol. Biol. Cell*, 6, 637, 1995.
33. Chen, H.-C., Appeddu, P. A., Parsons, J. T., Hildebrand, J. D., Schaller, M. D., and Guan, J.-L., Interaction of focal adhesion kinase with cytoskeletal protein talin, *J. Biol. Chem.*, 270, 16995, 1995.
34. Birge, R. B., Fajardo, J. E., Reichman, C., Shoelson, S. E., Songyang, Z., Cantley, L. C., and Hanafusa, H., Identification and characterization of a high-affinity interaction between v-Crk and tyrosine-phosphorylated paxillin in CT10-transformed fibroblasts, *Mol. Cell. Biol.*, 13, 4648, 1993.
35. Sakai, R., Iwamatsu, A., Hirano, N., Ogawa, S., Tanaka, T., Mano, H., Yazaki, Y., and Hirai, H., Characterization, partial purification, and peptide sequence of p130, the main phosphoprotein associated with v-Crk oncoprotein, *EMBO J.*, 13, 3748, 1994.
36. Matsuda, M., Mayer, B. J., and Hanafusa, H., Identification of domains of the v-crk oncogene product sufficient for association with phosphotyrosine-containing proteins, *Mol. Cell. Biol.*, 11, 1607, 1991.
37. Tanaka, S., Morishita, T., Hashimoto, Y., Hattori, S., Nakamura, S., Shibuya, M., Matuoka, K., Takenawa, T., Kurata, T., Nagashima, K., and Matsuda, M., C3G, a guanine nucleotide-releasing protein expressed ubiquitously, binds to the Src homology 3 domains of CRK and GRB2/ASH proteins, *Proc. Natl. Acad. Sci. USA*, 91, 3443, 1994.
38. Matsuda, M., Hashimoto, Y., Muroya, K., Hasegawa, H., Kurata, T., Tanaka, S., Nakamura, S., and Hattori, S., CRK protein binds to two guanine nucleotide-releasing proteins for the Ras family and modulates nerve growth factor-induced activation, *Mol. Cell. Biol.*, 14, 5495, 1994.

39. Andre, E. and Becker-Andre, M., Expression of an N-terminally truncated form of human focal adhesion kinase in brain, *Biochem. Biophys. Res. Commun.*, 190, 140, 1993.
40. Turner, C. E., Schaller, M. D., and Parsons, J. T., Tyrosine phosphorylation of the focal adhesion kinase, pp125FAK during development: relation to paxillin, *J. Cell Sci.*, 105, 637, 1993.
41. Kanner, S. B., Reynolds, A. B., and Parsons, J. T., Tyrosine phosphorylation of a 120 kiladalton pp60src substrate upon epidermal growth factor and platelet-derived growth factor stimulation and in polyomavirus middle-T-antigen-transformed cells, *J. Immunol. Methods*, 120, 115, 1989.
42. Richardson, A., Malik, R. K., Hildebrand, J. D., and Parsons, J. T., Inhibition of cell spreading by expression of the C-terminal domain of focal adhesion kinase (FAK) is rescued by coexpression of Src or catalytically inactive FAK: a role for paxillin tyrosine phosphorylation, *Mol. Cell. Biol.*, 17, 6906, 1997.
43. Kanner, S. B., Reynolds, A. B., Vines, R. R., and Parsons, J. T., Monoclonal antibodies to individual tyrosine-phosphorylated protein substrates of oncogene-encoded tyrosine kinases, *Proc. Natl. Acad. Sci. USA,* 87, 3328, 1990.
44. Kaelin, Jr., W. G., Krek, W., Sellers, W. R., DeCaprio, J. A., Ajchenbaum, F., Fuchs, C. S., Chittenden, T., Li, Y., Farnham, P. J., Blanar, M. A., Livingston, D. M., and Flemington, E. K., Expression cloning of a cDNA encoding a retinoblastoma binding protein with E2F like properties, *Cell*, 70, 351, 1992.
45. Falasca, M., Logan, S. K., Lehto, V. P., Baccante, G., Lemmon, M. A., and Schlessinger, J., Activation of phospholipase Cγ by PI 3-kinase-induced PH domain mediated membrane targeting, *EMBO J.*, 17, 414, 1998.
46. Barak, L. S., Ferguson, S. S. G., Zhang, J., and Caron, M. G., A β-arrestin/green fluorescent protein biosensor for detecting G protein coupled receptor activation, *J. Biol. Chem.*, 272, 27497, 1997.
47. Rizzuto, R., Brini, M., DeGiorgi, F., Rossi, R., Heim, R., Tsien, R. Y., and Pozzan, T., Double labelling of subcellular structures with organelle-targeted GFP mutants *in vivo, Curr. Biol.*, 6, 183, 1996.
48. Schaller, M. D. and Parsons, J. T., unpublished observations.
49. Martin, K., Ma, A., and Parsons, J. T., unpublished observations.
50. Seufferlein, T. and Rozengurt, E., Sphingosine induces $p125^{FAK}$ and paxillin tyrosine phosphorylation, actin stress fiber formation, and focal contact assembly in Swiss 3T3 cells, *J. Biol. Chem.*, 269, 27610, 1994.
51. Greenberg S., Chang P., and Silverstein S. C., Tyrosine phosphorylation of the gamma subunit of Fc gamma receptors, p72syk, and paxillin during Fc receptor-mediated phagocytosis in macrophages, *J. Biol. Chem.*, 269, 3897, 1994.
52. Zhang C., Lambert M. P., Bunch C., Barber K., Wade W. S., Krafft G. A., and Klein W. L., Focal adhesion kinase expressed by nerve cell lines shows increased tyrosine phosphorylation in response to Alzheimer's A beta peptide, *J. Biol. Chem.*, 269, 25247, 1994.

Chapter

Analysis of Paxillin as a Multi-Domain Scaffolding Protein

Mary C. Riedy, Michael C. Brown, and Christopher E. Turner

Contents

I. Introduction 64
II. Transfection 64
  A. Overview 64
  B. Protocols 65
  C. Materials 66
III. Rate of Paxillin Localization to Focal Adhesions and Role in Cell Adhesion 66
  A. Overview 66
  B. Protocols 67
  C. Materials 71
  D. Buffers 71
IV. GST-Paxillin Precipitation Kinase Assays 72
  A. Overview 72
  B. Protocols 72
  C. Materials 73
  D. Buffers 74
V. Cell Migration 74
  A. Overview 74
  B. Protocols 74
  C. Materials 75

0-8493-3385-7/99/$0.00+$.50

VI. Immunoprecipitation and Western Blotting 76
A. Overview 76
B. Protocols 76
C. Materials 77
D. Buffers 78
References 78

I. Introduction

Cells communicate with their extracellular environment via a variety of cellular receptors. In one example, integrins associate with the extracellular matrix (ECM), initiating a cascade of intracellular signaling events including association of these receptors with the actin cytoskeleton, cytoskeletal reorganization, and protein phosphorylation.[1] Cytoskeletal proteins that become phosphorylated include the focal adhesion kinase (FAK), p130^{cas}, and paxillin. These proteins are also phosphorylated during stimulation by growth factors that include angiotensin II, platelet derived growth factor (PDGF), activin A, bombesin, nerve growth factor, and TGF-β.[2-6]

Paxillin co-localizes with FAK and vinculin, along with several other structural and regulatory proteins, to points of contact of the cell membrane with the ECM, termed focal adhesions.[7-9] The formation of these structures is stimulated by both cell adhesion and growth factors, and they likely act as important nucleation centers for efficient signal transduction. Paxillin is composed of multiple protein binding sites such as SH3- and SH2-binding domains, five LD motifs, and four LIM domains.[8-12] Additionally, sites which are phosphorylated by FAK and/or Src function as SH2-binding sites for the adaptor protein Crk and the tyrosine kinases Csk and Lyn.[12,13]

A series of paxillin-homologous proteins including Hic-5,[14] PaxB,[15] and leupaxin[16] is emerging. The methods described in this chapter have been used to define the role paxillin and related family members play as molecular adaptors, as well as to evaluate their participation in growth factor and adhesion-dependent signaling events.

II. Transfection

A. Overview

In an attempt to understand the role(s) of the focal adhesion-related protein paxillin in cellular function *in vivo*, it is necessary to be able to identify important domains of this protein that specify the role(s). This approach has been instrumental in the determination that the LIM domains of paxillin direct the protein to focal adhesions.[10] To accomplish this, mutations (such as deletions and substitutions) can be made in the cDNA encoding the protein prior to inserting this modified cDNA of paxillin into a cell. The transfected protein may be of a species diverse from the host cell

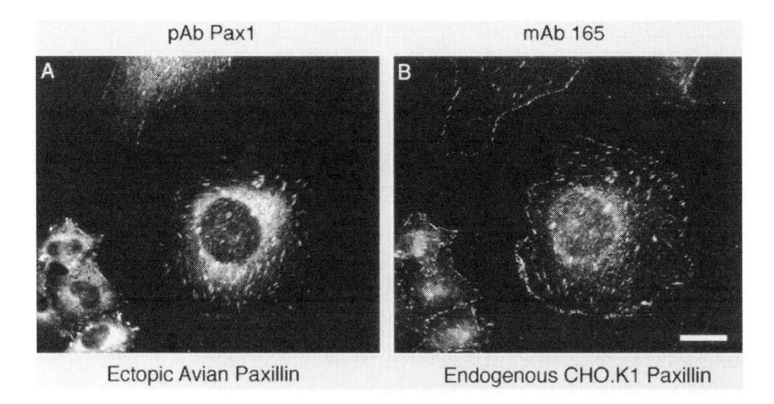

FIGURE 1

Focal adhesion localization of ectopic avian paxillin and endogenous paxillin in CHO.K1 cells. Double-labeled, indirect immunofluorescent analysis of CHO.K1 cells transfected with avian paxillin. A) Expression and localization of avian paxillin visualized with polyclonal rabbit anti-chicken paxillin antibodies (pAb Pax1) and B) Endogenous paxillin visualized with the monoclonal anti-paxillin antibody 165 (mAb 165). Both ectopic and endogenous localize to focal adhesions.

or contain a "tag," such as hemagglutinin (HA), Myc, xPRESS™, or T7. The species difference or the tag can be used to distinguish this protein from its endogenous counterpart by employing immunospecific antisera for identification.[10] Two successful protocols are outlined for transfection of chicken paxillin into such cells as CHO.K1 (LipoFECTAMINE™). Figure 1 illustrates CHO.K1 cells transfected with chicken paxillin, in comparison to endogenous rodent paxillin.

B. Protocols

(LipoFECTAMINE™: Works well for paxillin transfection of CHO.K1 cells.)

1. Cells are plated at a density of 3.5×10^5 cells/35 mm dish (one well of a 6-well dish) in serum-containing media, 24 h prior to transfection.
2. One microgram of the respective paxillin pcDNA3 construct and 10 µg of LipoFECTAMINE™ are diluted separately into 100 µl of serum- and antibiotic-free (unsupplemented) media in a 1.5 ml Eppendorf tube.
3. The DNA and lipid solution is mixed and incubated at room temperature for 15 min.
4. During this incubation, adherent cells were washed twice with prewarmed (37°C), unsupplemented medium.
5. Unsupplemented medium is then added to the DNA-lipid mixture to bring the volume to 1 ml and the mixture is overlaid onto the cells for a 6-h incubation at 37°C in 5% CO_2.
6. After 6 h, 1 ml of media supplemented with 20% FBS and 1% Penicillin-Streptomycin (complete medium) is added, and the cells are incubated overnight at 37°C in 5% CO_2.
7. The cells are washed twice in complete media, then changed to complete media with 2.0 mg/ml G418 (from 100 mg/ml stock) for approximately 2 weeks.

8. Following this initial phase of selection, the G418 concentration is reduced to 750 μg/ml for maintenance of enriched populations of transfected cells.
9. The cells are harvested using 2 ml of trypsin/EDTA, washed in complete media and plated onto sterile 13 mm glass coverslips in complete media (G418, 750 μg/ml), to be processed for immunofluorescence analysis after 24 h.
10. The cells are grown to a high density, whereupon a portion of the heterogeneous population is frozen, and the remainder plated at a 1:20000 dilution in complete media for cellulose disk subcloning.[17]
11. A hole punch is used to prepare several hundred "disks" of Whatman 3M paper. These disks are placed in a foil-lined glass petri dish and autoclaved.
12. To subclone, dishes containing approximately 50 individual colonies are washed twice with PBS. Under an inverted microscope, the colonies are marked with a marker pen on the bottom of the dish.
13. A Whatman 3M disk is soaked in trypsin-EDTA, the media is aspirated from the cells, the disks are placed on the demarcated colonies and incubated at 37°C for 5 min.
14. The disks are then removed to a 96-well dish containing 200 μl supplemented media with 1.0 mg/ml G418.
15. The colonies are expanded to 60 mm dishes containing a coverslip, whereupon clones are examined by immunofluorescence for transfected paxillin expression levels.
16. Clones are transferred to separate 100 mm dishes in 5 ml media with 750 μg/ml G418 to expand.

C. Materials

LipoFECTAMINE™ (GIBCO-BRL, 18324-012, Grand Island, NY), Penicillin/Streptomycin (Sigma Biologicals, P-3539), PBS (GIBCO-BRL, 21600), G418 (Mediatech, 61-234-RG, Herndon, VA), FBS (Summit Biotechnology, Ft. Collins, CO), and Whatman 3M paper (Fisher Scientific, Pittsburgh, PA).

III. Rate of Paxillin Localization to Focal Adhesions and Role in Cell Adhesion

A. Overview

As a cytoskeletal adaptor protein, paxillin is a biochemical-structural composite of protein recognition domains. In our efforts to understand the precise role of paxillin, we have engaged in a molecular dissection of these diverse domains. Following mutagenesis of discrete regions of the paxillin molecule and reintroduction of these mutant molecules into cells, effects on cellular functions can be examined. Such a scheme of truncation and site-directed mutagenesis has been used to determine the means by which paxillin targets to focal adhesions (see previous section).[10]

Using this approach, we have found that wild-type avian paxillin localizes more rapidly and efficiently to focal adhesions than does an avian paxillin molecule containing a non-phosphorylatable mutation within the principal focal adhesion targeting motif, LIM3.[9] This methodology can be extended to study the effect of overexpression of paxillin mutants on the ability of cells to efficiently form focal adhesions, actin stress fibers, and their effect on cell adhesion.

B. Protocols

Localization of Paxillin to Focal Adhesions

Glass Coverslip Preparation

1. Place a 100-pack of 12 mm circular glass coverslips into a covered glass vessel containing 95% EtOH for storage.
2. Immediately prior to plating cells on coverslips, remove from EtOH and flame-dry.
3. Working in a tissue culture hood, carefully place the coverslips in a tissue culture dish. For each timepoint, at least four coverslips are prepared. For example, duplicate coverslips are included for polyclonal (pAb) ectopic-avian paxillin/actin staining, and pAb avian paxillin/monoclonal (mAb) endogenous-paxillin or pAb avian ectopic-paxillin/mAb PY20.
4. For matrix binding experiments, following placement in tissue culture dishes, carefully add 100 μl of the diluted matrix component (see below) onto the coverslip so that the entire surface is covered. Incubate at 37°C for 3 h to overnight or allow to dry in the hood, followed by washing with PBS to remove the soluble matrix component. The coated slips can be stored under PBS at 4°C overnight.

Transfected Cell Preparation

1. Place cells, transiently or stably transfected with avian paxillin pcDNA3 vectors, in suspension using PBS containing 1 mM EDTA. Count the cells using a hemocytometer and resuspend at 2×10^6 cells/ml in serum-free DMEM containing 1% BSA. Using a platform rocker, maintain cells in suspension for 60 min at 37°C in a humidified chamber containing 5% CO_2.
2. Within a 35 mm^2 tissue culture dish, 1×10^5 cells offer an optimal density (CHO.K1, A431 or SVT2); a lower density will be required with larger Swiss 3T3, RASM or REF52 cells, for instance.
3. Add 2 ml serum-free medium with the cells.
4. Treat cells as required. For time course of adhesion/spreading and analysis of focal adhesion formation and protein localization, duplicate coverslips are removed at the indicated times for processing.
5. Transfer coverslips to Coors™ porcelain "boats." Remember orientation of cell side!

Fixation/Permeablization

1. Place boats into a glass slide-staining dish containing 250 ml fixation buffer.
2. Incubate 8 min; the time of fixation is dependent upon the antigen and antibody.
3. Remove boats to a glass dish containing 250 ml TBS, incubate 5 to 10 min.
4. Permeablize cells by transferring boat to a glass slide-staining dish containing 250 ml permeablization buffer.
5. Incubate 2 min, followed by transfer to a glass dish containing TBS.
6. Incubate 5 to 10 min in TBS (Note: The Tris-HCl will quench free formaldehyde groups, thus reducing background fluorescence.)

Slide Staining

1. Prepare circular Parafilm™ inserts for 100 or 150 mm^2 petri dishes, then cover and place the Parafilm™ in the dish and line the inside walls with wetted Kimwipes™ to humidify the "chamber."
2. Prepare the primary antibody dilution in TBS. Microfuge 5 min at 14,000 × g (to precipitate aggregated antibodies) and transfer supernatant to a fresh tube.
3. Place 30 µl of the antibody dilution on the Parafilm™.
4. Place the slip, cell side down, onto the Ab solution being careful not to introduce air bubbles.
5. Cover the "chamber" and place it into a 37°C incubator for 90 min or overnight at 4°C depending upon the antibody.
6. Transfer the slips back into the boat and wash the slips in 250 ml TBS for 5 to 10 min. REMEMBER to note orientation of cell side in boat.
7. Repeat this procedure with the secondary antibody dilution, incubating for 45 min (if using phalloidin, add with the secondary antibody). Dilutions of phalloidin (from Molecular Probes) range from 1:500 to 1:2000 and must be determined empirically for each cell type and stimulation condition.
8. Remove Airvol™ mounting reagent from 4°C and bring to room temperature.
9. Following the secondary Ab incubation, place the slips into a boat and submerge in 250 ml TBS. Wash 5 to 10 min.
10. Prepare slides by placing a dot (~10 µl) of Airvol™ onto the glass microscope slide.
11. Briefly rinse the slips in water to prevent salt crystal formation and place the slips cell side down onto mount. Air-dry, in the dark, overnight at room temperature.
12. For analysis, a Zeiss Axiophot fluorescent microscope is used to monitor the capacity of the transfected paxillin molecules to localize to focal adhesions at various timepoints, on various matrices, as well as the effect of expression of these molecules on cell morphology, endogenous focal adhesion localization, and actin stress fiber formation.

***Troubleshooting*:** The wide availability of coumarin-conjugated phalloidin and secondary antibodies allows for a more comprehensive analysis of protein subcellular localization within a single experiment. This assay is suitable for studying the range of paxillin superfamily members as well as other cytoskeletal-associated proteins. Fixation protocols, the need for pre-blocking cells, antifade additives in the mounting solution, and antibody/phalloidin concentrations must be optimized empirically.

Effect of Paxillin on Cell Adhesion

96-Well Plate Preparation

1. Dilution of extracellular matrix components (ECM): Dilute Fn, Ln, and Vn in serum-free medium or PBS. Dilute Collagens in 0.25% acetic acid. Prepare 2% Gelatin solution in dH_2O, autoclave. For Fn, Ln, Col I, Col II, Col IV, and Gelatin (denatured Type I Collagen), perform a dilution series between 1 and 25 μg/ml (e.g., Control, 1, 5, 25 μg/ml). For Vn, use 0.1-5 μg/ml. Control can include poly L-lysine (0.1 mg/ml in water) or 3% BSA.
2. Add 50 μl of the diluted ECM component of interest to each well of a 96-well plate.
3. Incubate the plate covered for 3 h at 37°C in a humidified chamber.
4. Wash each well three times with 100 μl PBS or EBSS. This is especially important for the collagens that are diluted in acetic acid. In this case, the phenol red in the EBSS is a helpful indicator of the effective removal of residual acetic acid.
5. Block exposed, non-specific binding sites present in each well, including ECM-negative control wells, with 100 μl 3% BSA in serum-free medium for 3 h at 37°C in a humidified chamber.
6. Wash three times with 100 μl PBS or EBSS. Aspirate remaining PBS or EBSS using the Vaccu-pette/96 Multiwell Pipetter available from Sigma (Z37,080-0).

Preparation of Transfected Cells

1. Resuspend cells, stably transfected with avian paxillin pcDNA3 vectors, using PBS containing 1 mM EDTA (no trypsin). May require incubation at 37°C for 10 to 15 min for all the cells to lift off the dish/flask.
2. Wash three times with 3 ml serum-free medium, centrifuging at 2500 rpm and discarding supernatant for each wash.
3. Count the cells with a hemocytometer.
4. Resuspend cells in serum-free medium containing 1% BSA at 2×10^6 cells/ml, such that 50 μl contains 1×10^5 cells in a 15 ml conical tube. For assays in which XTT will be used for quantitation, serum-free, phenol-red-free medium is used.
5. Rotate the cells, end-over-end for 1 h at 37°C in a humidified chamber with 5% CO_2.

Adhesion Assay

1. Add 50 μl of cells from Section B to each well. Optimally, 8 replicates (one column) are prepared per condition.
2. Place in a 37°C humidified chamber with 5% CO_2 and incubate for 30 min.

3. Remove plates, and subject to three 10 sec 300 RPM bursts on vortex shaker.
4. Shake out media. Add 200 µl PBS or EBSS and repeat Step 3 two more times.
5. Aspirate remaining PBS or EBSS, using the 96-well vaccu-pette.

Plate Processing and Data Analysis

1. Add 100 µl methanol per well. Fix for 10 min at room temperature.
2. Remove methanol and add 100 µl diluted Giemsa (1:10 in dH_2O) per well. Stain for 1 h at room temperature.
3. Aspirate stain using the vaccu-pette and wash three times with 100 µl dH_2O per well.
4. Read the plate at OD_{540nm}, preferably using an absorbance plate reader with well-scanning capability.
5. Alternatively, the stain can be eluted to a fresh 96-well plate using 100 µl 0.1M Na-Citrate pH 4.2 per well followed by plate scanning at OD_{540}.
6. The absorbance data can be normalized to the BSA-only or poly L-lysine control wells for each clone tested, followed by averaging the eight replicates per condition and clone. The effects of expressing different regions of paxillin on adhesion, relative to pcDNA3.1HisLacZ control (for instance) or to wild-type avian paxillin-expressing cells, can then be determined. Each experiment is repeated at least three times with two-independent clones.

MTT/XTT Assay

1. MTT or XTT assays can also be employed to quantify adhesion efficiency. Following Adhesion Assay steps above, add 100 µl phenol-red-free, serum-containing medium with 0.5 mg/ml MTT or XTT/0.025 mM phenazine methosulfate (PMS) to each well and incubate for 3 to 4 h at 37°C in a humidified chamber with 5% CO_2.
2. For MTT assays, aspirate the medium and add 100 µl of either 0.1N HCl in isopropanol, DMSO, or 0.01N HCl with 10% SDS (followed by incubation at 37°C for 8 to 12 h)[18] to solubilize the colored mitochondrial dehydrogenase-metabolized MTT formazan precipitate. Read the OD_{570} minus $OD_{630\text{-}690nm}$.
3. For XTT assays, the derivatized, colored formazan product is water soluble and the plate can be directly read at OD_{450nm}.
4. Data reduction is as in Data Analysis, above.

***Troubleshooting*:** One of the principal means of introducing variability is through the removal of adherent cells during aspiration, using a glass pasteur pipette. In order to avoid direct contact with the wells we use the vaccu-pette which, when seated on the 96-well plate, does not come into contact with the well bottom. This eliminates any opportunity for removing cells and scratching the well. Additionally, using a plate reader that has well-scanning capability will minimize well-to-well variability.

A modification of this standard adhesion assay can be performed that gauges adhesion efficiency/strength. Following the adhesion assay, completely fill all 96 wells with ice cold PBS or EBSS, then seal the plate with Mylar film (Linbro Scientific, Inc., Hamden, CT, 76-49-05). Invert the plate and centrifuge replicate plates at 10, 50, 100, 500, and 1000 *g* for 10 min at 4°C followed by processing as above.[19]

C. Materials

1. Reagents for paxillin localization

Assistant-Brand Circular 12 mm (1/2") Cover Glasses (Carolina Biological Supply, Burlington, NC, Cat. #63-3029); Glass Slides (Fisher Scientific, 12-550A); Glass Slide-Staining Dish (Fisher Scientific, 900200); Airvol (Air Products, Allentown, PA, 205); BSA (Sigma, St. Louis, MO, A6793); FBS (Summit Laboratories, Ft. Collins, CO, S100-65); Donkey Anti-rabbit or -mouse IgG-Fluoroscein (or Rhodamine) (Chemicon, Temecula CA, AP182F, AP182R, AP192F, AP192R); Rhodamine- or Fluorescein-Phalloidin (Molecular Probes, Eugene, OR, R415, F432); Primary Antibodies are generally obtained from Santa Cruz Biotechnology (Santa Cruz, CA), Sigma (St. Louis), or Transduction Laboratories (Lexington, KY). Polyclonal rabbit anti-avian-specific paxillin antibodies generated in our laboratory. Mammalian cells transfected with avian paxillin constructs. All other reagents are obtained from Sigma.

2. Reagents for cell adhesion

Bovine serum albumin (BSA, Sigma, Cat. #A6793); Collagen Type I (Col I), Rat Tail (Collaborative Biomedical Products, Bedford, MA, 40236); Collagen Type II (Col II), Chicken Sternal Cartilage (Sigma, C9301); Collagen Type IV (Col IV), Mouse (Gibco BRL, Gaithersburg, MD, 33018-011); Earle's Balanced Salt Solution (EBSS, Gibco BRL, 81100); Fibronectin (Fn), Laminin (Ln, mouse GIBCO-BRL, 23017-015), Human Plasma (Sigma, F2006); Gelatin, Bovine Skin (Sigma, G1393); Giemsa Stain (Sigma, GS500); MTT, 3-[4,5-Dimethylthiazol-2-yl]-2,5-diphenyltetrazolium bromide (Sigma, M5655); Poly L-lysine Hydrobromide, 70,000-150,000 MW (Sigma, P6282); PBS (Gibco BRL, 21600); PMS, Phenazine Methosulfate (Sigma, P5812); Tissue Culture-treated 96-well Plates (Costar, Cambridge, MA); Vitronectin (Vn), Rat Plasma (Sigma, V0132); XTT, 2,3-bis[2-Methoxy-4-nitro-5-sulfophyenyl]-2H-tetrazolium-5-carboxanilide (Sigma, X4626).

D. Buffers

1. Tris-Buffered Saline (TBS):	3 g/l	Tris-HCl
	10 g/l	NaCl
	Adjust the pH to 7.4 with HCl, QS to 1 liter	

2.	**Fixation Buffer:**	3.7%	Formaldehyde in PBS
3.	**Permeablization Buffer:**	0.2%	Triton X-100 in TBS
4.	**Earle's Balanced Salt Solution (EBSS):**	0.122 g/l	$NaH_2PO_4 \cdot H_2O$
		6.8 g/l	NaCl
		2.2 g/l	$NaHCO_3$
		0.4 g/l	KCl
		0.097 g/l	$MgSO_4$ (anhyd.)
		0.2 g/l	CaCl (anhyd.)
		1.0 g/l	D-glucose
		0.011 g/l	Phenol Red·Na
		Adjust pH to 7.4 with HCl, QS to 1 liter	

5. **MTT or XTT:** 5 mg/ml stock in phenol red- and serum-free medium filtered through a 0.2 μm filter, aliquoted and stored at –20°C.

IV. GST-Paxillin Precipitation Kinase Assays

A. Overview

The primary amino acid sequence of paxillin reveals a diversity of protein-protein interaction motifs, as well as the presence of multiple sites that may serve as targets of protein phosphorylation.[8] In order to clearly define areas of protein association, it is useful to express isolated regions of a protein. Paxillin was initially identified as a substrate of tyrosine kinase activity and was found to bind to the cytoskeletal structural protein vinculin.[7] Production of portions of paxillin as GST-fusion proteins has led to the identification of paxillin LD motif subdomains that support binding to the focal adhesion tyrosine kinase FAK, vinculin, and the E6 oncoprotein of Bovine Papillomavirus Type 1 (BE6) as well as the E6 protein of the cancer associated HPV-16 (16E6).[8,10,20] Additional efforts using GST-paxillin fusion proteins in GST-paxillin precipitation kinase assays has been used to identify regions of kinase association and sites of phosphorylation.[10,11,21] This technique is likely to be of general importance for the study of other multi-domain proteins.

B. Protocols

1. Generate GST and GST-fusion proteins following standard protocols provided by the manufacturer, or as previously described.[22] Prepare a lysate of tissue or cells by homogenizing the tissue/cells in 10 volumes (w/v) of lysis buffer. Replace the leupeptin and PMSF with 1X EDTA-free Complete™ protease inhibitors when using paxillin full-length or LIM domain GST-fusions.
2. Clarify the lysate by centrifuging at 14,000 × *g* for 15 min at 4°C. Remove supernatant to a fresh tube. Perform a MicroBradford protein assay for each of the lysates and/or cell treatments to be studied to determine protein concentration.[23]

3. In a total volume of 1 ml lysis buffer, 1 mg aliquotes of lysate are incubated with 5 μg of the various GST-paxillin fusion proteins coupled to approximately 10 to 20 μl of the glutathione-Sepharose 4B beads or with the control, GST-GSH-Sepharose 4B, for 90 min at 4°C, end-over-end.
4. Wash extensively (4X) using 1 ml lysis buffer for each wash, followed by washing with 1 ml kinase buffer. Centrifuge in a microfuge for 10 sec at 14,000 × *g*, followed by aspiration of lysis buffer with a disposable transfer pipette. Take care not to dislodge the small pellet.
5. Aspirate the kinase buffer using a disposable transfer pipette. Centrifuge as above; using a Hamilton syringe with the beveled edge down, aspirate the residual 50 to 100 μl of buffer. Resuspend the pellet in 15 μl kinase buffer containing 5 to 10 μCi of [^{32}P]-γ-ATP.
6. Allow the phosphorylation reaction to proceed at room temperature for 20 min, mixing the reaction contents every 5 min.
7. Terminate the reaction by boiling directly in 2X SSB. The samples are then processed by SDS-PAGE, stained with Coomassie blue to confirm equal fusion protein loading (if not performing a phospho-amino acid analysis), dried down on Whatman 3M filter paper and subjected to autoradiography using Kodak X-Omat film at –80°C for 10 min to 1 h.

***Troubleshooting*:** It is important that at least 5 μg of fusion protein is utilized in the assay. It is equally important that the volume of "beads" used is no more than 20 μl, and that each of the fusion proteins used in the assay are within a 10 to 20 μl bead range. Using less than 5 μg of fusion protein generally results in a very low signal; conversely, as the amount of GSH-sepharose 4B bead volume increases, the nonspecific binding of proteins to the sepharose results in high background.

The composition of the lysis buffer is flexible. Immunoprecipitation lysis buffers can be used when supplemented with a reducing agent. Additionally, the composition of the kinase buffer can be altered to target particular kinases, such as PKC; or, to broaden the spectrum of kinase activities supported by the buffer, for instance, by including both 5 mM $MgCl_2$ and 5 mM $MnCl_2$. It is thought that Mn^{2+} is more selective for tyrosine kinases; however, using the HEPES/Mn^{2+} buffer above, we have identified tyrosine, threonine, and serine kinases binding to and phosphorylating paxillin.[10,11,21]

C. Materials

GST (pGEX-2T) Gene Fusion Vector (Pharmacia, Piscataway, NJ, Cat. #27-4801-01); Glutathione Sepharose 4B resin (Pharmacia, 17-0756-01); [γ-^{32}P]-ATP ›4000 Ci/mmol (ICN, Costa Mesa, CA, 35001X); Complete™ protease inhibitors (Boehringer Mannheim, Indianapolis, IN, 1-873-580 or 1-697-498); Coomassie Brilliant Blue R250 (Sigma, B0149); GelCode Blue (Pierce, Rockford, IL, 24590). All other reagents were obtained from Sigma.

D. Buffers

1. Lysis Buffer:

50 mM	Tris-HCl, pH 7.6
50 mM	NaCl
1 mM	EGTA
2 mM	$MgCl_2$
0.1%	mercaptoethanol (or 1-10 mM DTT)
1%	Triton X-100
14 μM	leupeptin
0.5 mM	phenylmethylsulfonyl fluoride or PMSF

2. Kinase Buffer:

10 mM	HEPES, pH 7.5
3 mM	$MnCl_2$

3. 2X SDS-PAGE Sample Solubilizing Buffer (2X SSB):

0.125M	Tris-HCl
20%	Glycerol
4.6%	SDS
0.001%	bromphenol blue pH to 6.8
8 μl	2-β-mercaptoethanol/ml sample buffer

V. Cell Migration

A. Overview

Cell migration is a complex, integrated process of assembly and disassembly of the structures found at the sites of contact of the membrane with its extracellular matrix (ECM), termed focal adhesions. Paxillin, a focal adhesion-related adaptor protein is found localized to focal adhesions in adherent cells.[7] Due to its cellular location, paxillin may play a role in cell migration and this aspect can be evaluated using a modified Boyden chamber. Cells can be stimulated to migrate by growth factors, extracellular matrix components, and during wound repair. Cells such as vascular smooth muscle cells (VSMC) will migrate to a gradient of angiotensin II (ATII), PDGF and insulin.[24,25] ECM components such as fibronectin or vitronectin will induce the migration of fibroblasts.

B. Protocols

1. Detach cells from petri dishes or flasks using PBS with 1 mM EDTA and place in 15 ml conical tube, wash once with 15 ml pre-warmed serum-free medium and resuspend cells at a concentration 1×10^4 cells/50 μl in serum-free media containing 1% BSA.
2. Maintain cells, by rocking, in suspension for 60 min in a 37°C water bath.

3. Add chemotactic agent in serum-free media to lower chamber that has been blocked with 3% BSA for 2 h at 37°C. One row of the chamber is filled with serum-free media, without cells, as a control for background Giemsa staining of filter (see Step 16). The volume of the lower chamber is approximately 30 to 50 µl, but a *slight* positive meniscus (over the well) is required. If the meniscus is too large, upon unit assembly, a tight seal between the filter and the lower chamber will be prevented and the cells may migrate into the space between the filter and the lower chamber. If there is no positive meniscus, an air bubble will be trapped under the filter, which precludes migration.
4. Carefully place filter (adhesive side down) over the wells of the lower chamber, being careful to avoid air bubbles.
5. Align gasket onto underside of upper chamber, using the five metal pins; close the manifold, twisting the screws just until tight.
6. Add 50 µl cells to top chamber wells. Place the tip in the middle of the well and expel the cells in a smooth motion to avoid introducing air bubbles. Touching the pipet tip to the filter can easily cause a puncture.
7. After adding cells, loosely cover upper chamber in parafilm.
8. Incubate at 37°C for 4 to 6 h (Note: in approximately 10 min, check for bubble formation in upper wells. Carefully remove any bubbles with a gel loading tip.)
9. After the designated incubation period (4 to 6 h), remove cells from upper chamber by vigorously shaking apparatus into sink to force cells out.
10. Wash with 100 µl 1X PBS per well by pipetting up and down in each well to dislodge remaining cells from the top of the filter.
11. Open chamber. DO NOT REMOVE FILTER from plate.
12. Take lower chamber out with filter attached and centrifuge unit at 2000 rpm for 3 min.
13. Remove filter, carefully scrape cells off the TOP of the filter with rubber scraper (supplied with apparatus), being careful not to scrape cells from the underside.
14. Submerge filter in methanol for 10 min at room temperature to fix cells.
15. Place filter in 50 mls of 1:10 Giemsa:H_2O for 1 h at room temperature.
16. Destain in H_20 for 1 h, replacing with fresh H_2O every 15 min.

***Troubleshooting*:** Avoid introducing air bubbles to either chamber by carefully and smoothly pipetting into the center of the wells. Air bubbles will interfere with normal migration; it will be necessary to discard the data from those wells. Make sure the chamber is well sealed to eliminate a space around the well. A lack of a tight seal will allow the cells to migrate laterally from the boundary of the well.

C. Materials

Modified Boyden chamber, 96-well, 8 µm pore-size filters (MBC96, Neuroprobe, Cabin John, MD), bovine serum albumin (BSA, Sigma Biologicals), PBS (GIBCO BRL, 21600) and trypsin/EDTA (Sigma), methanol (J. T. Baker, Philipsburg, PA) and Giemsa Stain (Sigma GS500).

VI. Immunoprecipitation and Western Blotting

A. Overview

The ability to isolate endogenous or transfected (avian) paxillin from cells allows for a means to study this protein under conditions such as growth factor stimulation and adhesion. Paxillin has been shown to be differentially phosphorylated on tyrosine, serine, and threonine residues during adhesion or growth factor stimulation.[2,3,6,21,26] Once isolated, paxillin can then be analyzed for phosphorylated residues via phosphoamino acid analysis or anti-phosphotyrosine Western blotting. Denatured paxillin migrates as a diffuse band when subjected to SDS-polyacrylamide gel electrophoresis (SDS-PAGE). The extensive phosphorylation of paxillin likely contributes to this mobility shift. This protocol can also be used to study proteins that associate with paxillin, assuming that the immunoprecipitating antibody does not displace this interaction.

B. Protocols

Immunoprecipitation

1. Remove media from cells, wash 100 mm dish in 5 ml PBS (with 1 mM Na_3VO_4). Add 250 µl of lysis buffer and scrape cells from the dish using a rubber policeman. Remove 25 µl (or 10% of lysate volume) for whole cell lysate.
2. Transfer remaining lysate to a 1.5 ml Eppendorf tube.
3. Centrifuge at 14,000 × g for 15 min to clarify the lysate.
4. Transfer the supernatant to a clean Eppendorf and add an appropriate amount of antibody (approximately 1 to 5 mg).
5. Incubate at 4°C for 2 h (rotating).
6. Add 20 µl of Protein A/G agarose to the immunoprecipitations (Santa Cruz Biotechnology).
7. Incubate 1 h at 4°C while rotating.
8. Centrifuge immunoprecipitations for 10 sec at 14,000 × *g*, remove supernatant and wash protein A/G agarose by adding 1 ml lysis buffer and centrifuging for 10 sec as above. Remove supernatant and repeat the wash three times.
9. Boil agarose, to denature and release bound proteins, in 50 µl 2X SDS-sample buffer for 5 min, then place on ice for 2 min.
10. Load samples onto a 7.5 to 10% PAGE.

Western Blotting

1. Transfer proteins from gel onto a nitrocellulose membrane (0.2 µm supported nitrocellulose such as Immobilon-NC [Millipore, Bedford, MA]).
2. Block membrane in 5 ml blocking buffer for 2 h at room temperature (this incubation can proceed overnight at 4°C in covered container).
3. Dilute primary antibody in blocking buffer (enough to cover nitrocellulose membrane), incubate at room temperature for 2 h.
4. Wash membrane in 5 ml wash buffer at room temperature for 10 min, discard wash and repeat four times.
5. Make a dilution of the secondary antibody, conjugated with horseradish peroxidase, in blocking buffer; incubate with nitrocellulose membrane at room temperature for 1 h.
6. Wash membrane as in Step 4.
7. Dab edge of nitrocellulose membrane to remove excess wash buffer before placing into 2 ml ECL™ (1:1 of the two reagents, Amersham).
8. Wrap the nitrocellulose membrane in Saran Wrap™, smooth out air bubbles, and expose to film (Xomat AR, Kodak).

***Troubleshooting*:** Reboil samples that were stored at –20°C prior to running on gel. Paxillin isolated by immunoprecipitation or from whole cell lysates is to be run denatured on a 7.5% SDS-PAGE to resolve paxillin isoforms (approximately 68 kDa) from immunoglobulin heavy chain. For immunoprecipitations, adjust lysis buffer conditions to be compatible with the requirements for antibody binding. When detecting phosphotyrosine with anti-phosphotyrosine antibodies (4G10, Upstate Biotechnology Institute, PY20 or RC20, Transduction Laboratories) it is important to keep the NaCl concentration of the wash buffer and blocking buffer below 150 mM (we recommend using 100 mM NaCl). For a more detailed analysis of paxillin protein phosphorylation, it is necessary to metabolically label the protein and perform two-dimensional phospho-peptide mapping of the immunoprecipitated protein. Such analysis is described elsewhere.[11,20]

C. Materials

Triton-X 100, NaCl, Tris-HCl, DOC, EDTA, Leupeptin, PMSF, β-glycerophosphate, NaF, Na_2VO_4, p-nitrophenylphosphate, Na-Pyrophosphate, Microcystin LR, Tween-20, bovine serum albumin (BSA), SDS, 2-β-mercaptoethanol (all purchased from Sigma Biologicals) and Complete™ protease inhibitors (Boehringer Mannheim, Indianapolis, IN, 1-873-580 or 1-697-498). Most antibodies are purchased from Santa Cruz (Santa Cruz, CA), Transduction Laboratories (Lexington, KY), and Sigma Biologicals.

D. Buffers

1. Lysis Buffer:	1%	Triton X-100
	150 mM	NaCl
	10 mM	Tris-HCl pH 7.6
	0.1%	DOC
	1 mM	EDTA
	2 mM	Na_3VO_4
	5 μg/ml	leupeptin (protease inhibitor)
	1 mM	PMSF (phenylmethylsulfonyl fluoride)
2. Phosphatase Inhibitors:	25 mM	β-glycerophosphate
	25 mM	NaF
	1 mM	Na_3VO_4
	1 mM	p-nitrophenylphosphate
	2 mM	Na-pyrophosphate
	0.2 μM	Microcystin LR (add separately)
3. Blocking Buffer:	20 mM	Tris-HCl pH 7.6
	100 mM	NaCl
	0.2%	Tween-20
	3%	BSA
4. Wash Buffer	20 mM	Tris-HCl pH 7.5
	100 mM	NaCl
	0.1%	Tween-20
5. Antibody Dilutions:	prepare in blocking buffer	
6. 2X SDS-PAGE Sample Solubilizing Buffer:		
	0.125M	Tris-HCl
	20%	Glycerol
	4.6%	SDS
	0.001%	bromphenol blue pH to 6.8
	8 μl	2-β-mercaptoethanol/ml sample buffer

References

1. Schwartz, M. A., Schaller, M. D., and Ginsberg, M.H., Integrins: emerging paradigms of signal transduction, *Annu. Rev. Cell Dev. Biol.*, 11, 549, 1995.
2. Zachary, I., Sinnett-Smith, J., Turner, C. E., and Rozengurt, E., Bombesin, vasopressin, and endothelin rapidly stimulate tyrosine phosphorylation of the focal adhesion-associated protein paxillin in Swiss 3T3 cells, *J. Biol. Chem.*, 15, 22060, 1993.

3. Rankin, S. and Rozengurt, E., Platelet-derived growth factor modulation of focal adhesion kinase (p125FAK) and paxillin tyrosine phosphorylation in Swiss 3T3 cells. Bell-shaped dose response and cross-talk with bombesin, *J. Biol. Chem.*, 269, 704, 1994.
4. Turner, C. E., Pietras, K. M., Taylor, D. S., and Molloy, C. J., Angiotensing II stimulation of rapid paxillin tyrosine phosphorylation correlates with the formation of focal adhesions in rat aortic smooth muscle cells, *J. Cell Sci.*, 180, 333, 1995.
5. Melamed, I., Turner, C. E., Aktories, K., Kaplan, D. R., and Gelfand, E. W., Nerve grow factor triggers microfilament assembly and paxillin phosphorylation in human B lymphocytes, *J. Exp. Med.*, 181, 1071, 1995.
6. Riedy, M. C., Brown, M. C., Molloy, C. J., and Turner, C. E., Activin A and TGF-β stimulate phosphorylation of focal adhesion proteins and cytoskeletal reorganization in rat aortic smooth muscle cells, *J. Biol. Chem.*, 1998.
7. Turner, C. E., Glenney, J. R., Jr., and Burridge, K., Paxillin: a new vinculin-binding protein present in focal adhesions, *J. Cell Biol.*, 111, 1059, 1990.
8. Turner, C. E. and Miller, J. T., Primary sequence of paxillin contains putative SH2 and SH3 domain binding motifs and multiple LIM domains: identification of a vinculin and pp125Fak-binding region, *J. Cell Sci.*, 107, 1583, 1994.
9. Brown, M. C., Perrotta, J. A., and Turner, C. E., Serine and threonine phosphorylation of the paxillin LIM domains regulates paxillin focal adhesion localization and cell adhesion to fibronection, *Mol. Biol. Cell*, 9, 1998.
10. Brown, M. C., Perrotta, J. A., and Turner, C. E., Identification of LIM3 as the principal determinant of paxillin focal adhesion localization and characterization of a novel motif on paxillin directing vinculin and focal adhesion kinase binding, *J. Cell Biol.*, 135, 1109, 1996.
11. Bellis, S. L., Miller, J. T., and Turner, C. E., Characterization of tyrosine phosphorylation of paxillin *in vitro* by focal adhesion kinase, *J. Biol. Chem.*, 270, 17437, 1995.
12. Schaller, M. D. and Parsons, J. T., pp125FAK-dependent tyrosine phosphorylation of paxillin creates a high-affinity binding site for Crk, *Mol. Cell Biol.*, 15, 2635, 1995.
13. Clark, E. A. and Brugge, J. S., Integrins and signal transduction pathways: the road taken, *Science*, 268, 233, 1995.
14. Schibanuma, M., Mashimo, J., Kuroki, T., and Nose, K., Characterization of the TGF beta 1-inducible hic-5 gene that encodes a putative novel zinc finger protein and its possible involvement in cellular senescence, *J. Biol. Chem.*, 269, 26767, 1994.
15. Mazaki, Y., Hashimoto, S., and Sabe, H., Monocyte cells and cancer cells express novel paxillin isoforms with different binding properties to focal adhesion proteins, *J. Biol. Chem.*, 272, 7437, 1997.
16. Lipsky, B. P., Beals, C. R., and Staunton, D. E., Leupaxin is a novel LIM domain protein that forms a complex with PYK2, *J. Biol. Chem.*, 273, 11709, 1998.
17. Domann, R. and Martinez, J., Alternative to cloning cylinders for isolation of adherent cell clones, *Biotechniques*, 18, 594, 1995.
18. Pasqualini, R. and Hemler, M. E., Contrasting roles for integrin beta 1 and beta 5 cytoplasmic domains in subcellular localization, cell proliferation, and cell migration, *J. Cell Biol.*, 125, 447, 1994.

19. Bauer, J. S., Varner, J., Schreiner, C., Kornberg, L., Nicholas, R., and Juliano, R. L., Functional role of the cytoplasmic domain of the integrin alpha 5 subunit, *J. Cell Biol.*, 122, 209, 1993.
20. Vande Pol, S. B., Brown, M. C., and Turner, C. E., Association of Bovine Papillomavirus Type 1 E6 oncoprotein with the focal adhesion protein paxillin through a conserved protein interaction motif, *Oncogene*, 8, 43, 1998.
21. Bellis, S. L., Perrotta, J. A., Curtis, M. S., and Turner, C. E., Adhesion of fibroblasts to fibronectin stimulates both serine and tyrosine phosphorylation of paxillin, *Biochem. J.*, 325, 375, 1997.
22. Turner, C. E. and Brown, M. C., Purification and assays for paxillin, *Meth. Enz.*, 298, 77, 1998.
23. Bradford, M. M., A rapid and sensitive method for the quantitation of microgram quantities of protein utilizing the principle of protein-dye binding, *Anal. Biochem.*, 72, 248, 1976.
24. Bornfeldt, K., Raines, E., Nakano, T., Graves, L., Krebs, E., and Ross, R., Insulin-like growth factor and PDGF-BB induce directed migration of human arterial smooth muscle cells via signaling pathways that are distinct from those of proliferation, *J. Clin. Invest.*, 93, 1266, 1994.
25. Griendling, K. K., Tsuda, T., Berk, B. C., and Alexander, R. W., Angiotensin II stimulation of vascular smooth muscle, *J. Card. Pharm.*, 14, S27, 1989.
26. Burridge, K., Turner, C. E., and Romer, L. H., Tyrosine phosphorylation of paxillin and pp125FAK accompanies cell adhesion to extracellular matrix: a role in cytoskeletal assembly, *J. Cell Biol.*, 119, 893, 1992.

Chapter

p130^{Cas} in Integrin Signaling

Fabrizio Dolfi, Miguel Garcia-Guzman, and Kristiina Vuori

Contents

I. Introduction 82
II. Tyrosine Phosphorylation of Cas 83
  A. Overview 83
  B. Protocols 83
  C. Materials 84
  D. Buffers 85
III. Protein-Protein Interactions Mediated by Tyrosine-Phosphorylated Cas 86
  A. Overview 86
  B. Protocols 86
  C. Materials 89
  D. Buffers 89
IV. Functional Studies on the Cas-Crk Signaling Complex 89
  A. Overview 89
  B. Protocols 91
  C. Materials 95
  D. Buffers 96
Acknowledgments 97
References 97

0-8493-3385-7/99/$0.00+$.50

I. Introduction

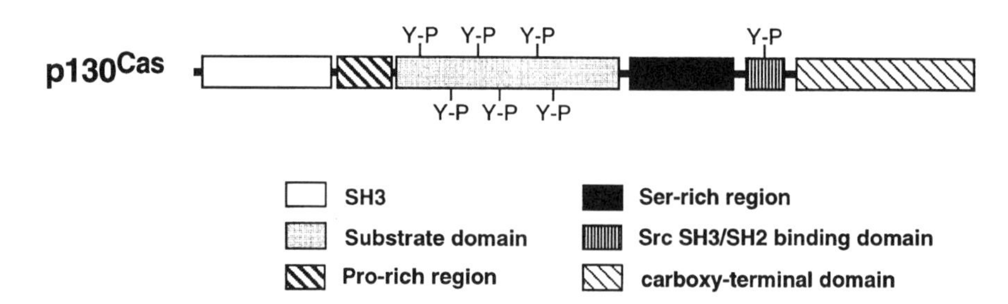

FIGURE 1
Schematic representation of the structure of Cas. The distinct domains of Cas are indicated; for details, see the text. Y-P, phosphorylated tyrosine-residue.

p130^{Cas} (Cas—Crk-Associated Substrate) was originally identified as a major tyrosine-phosphorylated protein in cells transformed by the p47$^{v\text{-}crk}$ (*v-crk*)[1,2] and p60$^{v\text{-}src}$ (*v-src*) oncoproteins.[3,4] The tyrosine phosphorylation levels of Cas correlate well with the transforming phenotypes of cells and are therefore thought to play a role in the process of cellular transformation. Molecular cloning of Cas revealed that Cas lacks any enzymatic activity but contains multiple domains suitable for protein-protein interactions (Figure 1).[5] Cas has a SH3-domain in its amino-terminus; recent studies have demonstrated that the SH3-domain interacts directly with a proline-rich region in the focal adhesion kinase FAK, and also binds to two protein tyrosine phosphatases, PTP1B and PTP-PEST.[6-8] The Src family tyrosine kinases, in turn, bind to a region near the carboxy-terminus of Cas via their SH2- and SH3-domains.[9] Interactions mediated by these two domains of Cas may therefore be responsible for the regulation of the tyrosine phosphorylation status of Cas and also target Cas to focal adhesions, which are sites of close cell-extracellular matrix (ECM) interactions.[10-12] Additionally, Cas has a cluster of multiple putative SH2-binding motifs (substrate domain; Tyr^{377}-Tyr^{414});[5] nine of these are YDV/TP sequences that conform to the binding motif for the adaptor protein Crk SH2-domain. These structural characteristics indicate that Cas is a docking molecule, which can assemble and transmit cellular signals via interactions through the SH2- and SH3-domains of a wide variety of signaling proteins. Two recently cloned molecules, Efs (Embryonal Fyn-associated substrate)/Sin (Src-interacting or signal-integrating protein)[13,14] and HEF1 (Human Enhancer of Filamentation 1)/Cas-L,[15,16] have similar primary structure to Cas and are assumed to comprise a new family of docking proteins.

Consistent with a role as a signaling molecule, Cas has been reported to become tyrosine-phosphorylated in response to a number of different cellular stimuli, many of which affect the assembly of focal adhesions and stress fibers. These stimuli include integrin-mediated cell adhesion[17-19] and ligand engagement of a variety of different receptors, such as G-protein-coupled receptors and receptor tyrosine kinases.[20-24] Following tyrosine phosphorylation, Cas interacts with SH2-containing signaling molecules, such as the adaptor proteins Crk and Nck, possibly recruiting these molecules to focal adhesions and activating downstream signaling pathways.[10,25] Although the molecular interactions mediated by Cas are relatively well characterized by now, the functional significance of these interactions has remained largely unknown. In this chapter, we describe experimental protocols for studies on tyrosine phosphorylation of Cas in response to integrin-mediated cell adhesion and for detection of tyrosine phosphorylation-dependent interactions mediated by Cas. More recently, we have focused our attention on defining the functional significance of the Cas-Crk complex formation in intracellular signaling; these studies are described in the latter part of this chapter.

II. Tyrosine Phosphorylation of Cas

A. Overview

Cas has been found to become tyrosine-phosphorylated following cell adhesion to ECM substrates, but not to polylysine, which is consistent with the phosphorylation being mediated by integrins. This is further supported by the observation that cell adhesion to immobilized anti-integrin antibodies also results in elevated tyrosine phosphorylation of Cas.[18] The time course of Cas tyrosine phosphorylation in response to integrin ligand binding is relatively slow and persistent; maximal phosphorylation in fibroblasts occurs at around 15 min after plating, which coincides with cell spreading and actin filament reorganization, and the phosphorylation levels are maintained with a slow decline for the duration of cell adhesion.[18] This is in contrast to the growth factor-induced tyrosine phosphorylation of Cas, which has been shown to be rapid and transient.[20-24] It has been shown that integrin-induced Cas phosphorylation is greatly reduced in fibroblasts lacking Src, but not FAK, suggesting that Src is responsible for integrin-mediated tyrosine phosphorylation of Cas.[10] It is possible, however, that a close homologue of FAK, Pyk2,[26-28] may take over the role FAK in FAK —/— cells. A dual-kinase model for Cas phosphorylation by FAK and Src was recently suggested; Tachibana and co-workers found that FAK initiates the tyrosine phosphorylation of Cas by directly phosphorylating Cas at the Src-binding site. Upon binding to this site, Src family kinases would then carry out the bulk of tyrosine phosphorylation of Cas.[11] A protocol for studies on integrin-induced tyrosine phosphorylation of Cas is described next, and examples of these analyses can be found in several original publications.[17-19]

B. Protocols

1. **Cell culture and adhesion on ECM proteins**— For the experiments, monolayer cultures of rodent or human fibroblasts are grown in Dulbecco's modified Eagle's medium (DMEM)/10% fetal calf serum supplemented with penicillin and streptomycin. Prior to adhesion experiments, cells can be serum-starved for 24 h in DMEM/0.5% fetal calf serum to avoid inadvertent induction of tyrosine phosphorylation of Cas by lysophosphatidic acid present in the serum. We have carried out experiments with and without serum starvation and no significant differences have been found between these

two conditions. The cells are then detached by brief trypsinization followed by washing with soybean trypsin inhibitor. Alternatively, 5 mM EDTA can be used to detach cells. The cells are washed twice with DMEM containing 0.5% BSA, and cell suspensions are incubated in DMEM, 0.5% BSA at 37°C/5% CO_2 on a rotator. For most untransformed cell types we have tested, a rapid dephosphorylation of Cas is detected immediately following cell detachment as a result of integrin disengagement. For REF-52 cells, a 15 min incubation period in suspension is recommended for a complete dephosphorylation of Cas. Cells are then plated onto the dishes coated with various ECM substrates to induce integrin ligation. Cells are incubated at 37°C/5% CO_2 for 45 min (or any other time period desired); control cells referred to as suspended cells are held in suspension for an additional 20 min. We typically use 6-cm dishes that have been coated with 10 to 20 µg/ml of ECM proteins or polylysine overnight and blocked with 1% BSA in PBS for 1 h prior to plating the cells.

2. **Cell lysis, immunoprecipitation and immunoblotting** — Cells are washed with ice-cold phosphate-buffered saline (PBS) and lysed in 300 µl of modified RIPA buffer/6-cm dish. Lysates are clarified by centrifugation at 15,000 *g* for 15 min. A bicinchoninic acid (BCA) protein assay is used to determine the protein concentration of each lysate. Cas is immunoprecipitated from 100 to 200 µg of cell lysate by incubation with 1 to 5 µg of antibodies to Cas for 1 h on a rocker at 4°C. To precipitate the antibody-antigen complexes, Gammabind Sepharose for mouse antibodies or protein A-Sepharose for rabbit antibodies is added to the lysates, and rotation is continued for 2 h. The immunoprecipitates, pelleted by microcentrifuging, are washed three times in wash buffer. The pellets are boiled in Laemmli sample buffer containing β-mercaptoethanol for 3 min and electrophoresed on 4 to 12% precast SDS-PAGE gels. After electrophoresis, proteins are transferred to nitrocellulose or Immobilon-membrane using a trans-blot cell apparatus. The filters are first incubated in a blocking buffer for 2 h, followed by incubation with horseradish peroxidase-conjugated anti-phosphotyrosine py20 antibody, diluted 1/2000 in blocking buffer, for 1 h. The filters are washed three times for 10 min in PBS/0.1% Tween-20 to remove unbound antibody. Antibody binding is visualized by enhanced chemiluminescence (ECL) detection following the protocol provided by the supplier. All incubations and washes are performed at ambient temperature on a rocking platform.
3. **Stripping and reprobing immunoblots** — After immunoblot analysis with the anti-phosphotyrosine antibodies, the immunoblot is stripped of the antibody and reprobed with antibodies to Cas to ensure equal amount of loading. The filters are incubated in the stripping buffer at 50°C for 60 min. If necessary, more stringent stripping conditions at 70°C for 60 min can be used. The filters are then washed four times for 10 min in PBS/0.1% Tween-20, followed by incubation in a blocking buffer and antibody staining with anti-Cas antibodies (1/200—1/1000 dilution) as described for immunoblotting above. The filters are washed three times for 10 min in PBS/0.1% Tween-20 to remove unbound antibody, and blotted with a horseradish peroxidase-conjugated protein A (for rabbit antibodies, 1/1000 dilution) or anti-mouse Ig (for mouse antibodies, 1/1000 dilution) for 2 h, followed by washes and ECL detection as above.

C. Materials

1. **Cells and ECM proteins** — We have obtained rodent and human fibroblasts from American Type Culture Collection (Rockville, MD) and from Coriell (Camden, NJ). ECM

proteins fibronectin, vitronectin, laminin, and type I collagen can be obtained from several commercial sources, including GIBCO-BRL (Grand Island, NY) and Sigma (St. Louis, MO). Established protocols for purification of these proteins also exist in the literature; we have purified fibronectin and vitronectin from human plasma as described.[29,30]

2. **Antibodies** — We have used two commercially available antibodies, a polyclonal rabbit antibody C-20 (Santa Cruz Biotechnology, Santa Cruz, CA) and a monoclonal mouse antibody (Transduction Laboratories, Lexington, KY) against Cas in both immunoprecipitation and immunoblotting. We typically use 1 to 5 μg/IP of either one of the antibodies, and 1/200 dilution for the polyclonal antibody and 1/1000 dilution for the monoclonal antibody in immunoblotting. We have not been successful in using the polyclonal N-17 anti-Cas antibody commercially available from Santa Cruz Biotechnology. The Transduction Laboratories antibody readily cross-reacts with HEF1/Cas-L, while the C-20 antibody is Cas-specific. In most cell types, the apparent molecular weight of HEF1/Cas-L is around 110 kDa, although several splice variants and/or degradation products of both Cas and HEF1/Cas-L exist in several cell lines (our unpublished observations).[15,31] In addition to the antibodies that are commercially available, we have used rabbit polyclonal antibodies to Cas generated in our laboratory and by Dr. Hisamaru Hirai (University of Tokyo, Japan). Other scientists, including Drs. Thomas Parsons and Amy Bouton (University of Virginia), have also described generation of anti-Cas antibodies.[3,32,33] Horseradish peroxidase-conjugated py20 monoclonal antibody against phosphotyrosine is from Transduction Laboratories.
3. **Reagents for cell culture, immunoprecipitation, and immunoblotting** — DMEM is from Gibco-BRL (Gaithersburg, MD) supplemented with 10% fetal calf serum (Tissue Culture Biologicals, Tulare, CA), 50 units/ml penicillin and 50 μg/ml streptomycin (Irvine Scientific, Santa Ana, CA). Trypsin/EDTA and soybean trypsin inhibitor are from Sigma. BCA protein assay kit and the ECL detection kit are from Pierce (Rockford, IL). Gammabind Sepharose is from Pharmacia (Uppsala, Sweden) and Protein A-Sepharose from Sigma. Precast SDS-PAGE gels are from Novex (San Diego, CA). Immobilon-P membrane is from Millipore (Bedford, MA).

D. Buffers

1. 1xPBS: 137 mM NaCl, 2.7 mM KCl, 4.3 mM Na_2HPO_4, 1.4 mM KH_2PO_4 pH 7.3.
2. Modified RIPA buffer: 50 mM Tris, pH 7.5, 150 mM NaCl, 5 mM EDTA, 1% Triton X-100, 0.1% SDS, 1% deoxycholate, 50 mM NaF, 0.5 mM Na_3VO_4, 0.1 U/ml aprotinin, 10 μg/ml leupeptin and 4 μg/ml pepstatin A.
3. Wash buffer for immunoprecipitations: modified RIPA buffer without SDS and deoxycholate.
4. Blocking buffer for immunoblotting: 2% BSA/0.1% Tween-20 in PBS.
5. Washing buffer for immunoblotting: PBS/0.1% Tween-20.
6. Stripping buffer for immunoblots: 62.5 mM Tris-HCl, pH 6.7, 100 mM β-mercaptoethanol, 2% SDS.

III. Protein-Protein Interactions Mediated by Tyrosine-Phosphorylated Cas

A. Overview

Tyrosine-phosphorylated Cas is known to bind to multiple SH2-domain containing signaling molecules following adhesion-induced tyrosine phosphorylation (see above). Several of these interactions were originally characterized by *in vitro* association experiments using glutathione S-transferase (GST)-fusion proteins of various SH2-domains for "pull-down" experiments or for blotting in an overlay ("Far Western") analysis; these experiments are described below. These analyses require prior knowledge of the putative interactive partners for Cas, and are therefore useful for studying the capability of known proteins to interact with Cas. The *in vitro* findings need to be confirmed at the level of endogenous proteins by coimmunoprecipitation experiments (for an example of Cas-Crk immunoprecipitation, see Reference 10). A more quantitative analysis of known protein-protein interactions can be done, e.g., by using surface plasmon resonance technology (BiaCore, Pharmacia; for a review, see, e.g., Reference 34). Songyang and co-workers have identified and characterized specific sequence motifs that are recognized by individual signaling domains, including various SH2-domains.[35] These predictions have greatly facilitated the identification of protein complex formation by signaling molecules, such as Cas. As an alternative approach, known and novel protein-protein interactions mediated by Cas and other signaling molecules can be studied by using various cDNA library screening methods, such as the yeast two-hybrid screening or λgt11 expression cloning. Both of these screening methods have been modified to specifically facilitate the detection of tyrosine phosphorylation-dependent interactions; in these assays, tyrosine-phosphorylated Cas can be used as a "bait" (two-hybrid system) or as a "probe" (λgt11 expression cloning) to search for and directly clone known and novel molecules interacting with Cas in a phosphorylation-dependent manner. The reader is referred to several recent articles for further information on these techniques.[36-39]

B. Protocols

1. **Purification of GST-fusion proteins** — Both the *in vitro* "pull-down" experimentation and "Far Western" analysis utilize GST-fusion proteins. A protein domain of interest, in our case various SH2-domains that are candidates for interacting with tyrosine-phosphorylated Cas based on the binding motif prediction by Songyang et al.,[35] are cloned in frame to one of the pGEX-series bacterial expression vectors. The protocol described below for fusion protein production has been used successfully for expression of multiple different GST-fusion proteins, including various Cas domains. Expression of individual GST-fusion proteins may require adjustment of the growth or induction conditions.

   Day 1 — *E. coli* DH5α bacteria (or any other bacterial strain suitable for isopropyl-β-D-thiogalactopyranoside [IPTG] induction) previously transformed with the pGEX vector of interest are inoculated in 25 ml of LB broth, containing

100 μg/ml ampicillin and incubated at 37°C overnight on a horizontal shaker at 300 rpm.

Day 2 — The 25 ml culture is diluted in 250 ml of LB broth containing 100 μg/ml ampicillin in 1 L Erlenmeyer flask and grown at 37°C on a horizontal shaker until $O.D._{600}$ = 0.4 to 0.6 (cells reach the desired optical density in about 1 h). Fusion protein expression is induced by adding IPTG to a final concentration of 0.1 mM, and the bacteria are grown at room temperature for 3 h on a horizontal shaker. Cells are collected by centrifugation at 5000 rpm in a Sorval RC-5B centrifuge, using a GSA rotor, the supernatant is removed and the pellet can be stored at –20°C. Alternatively, one can proceed directly to the next step (Day 3).

Day 3 — After being thawed on ice, cells are thoroughly resuspended in 5 ml of 0.9% NaCl, and spun at 5000 rpm in a Sorval RC-5B centrifuge using a SS-34 rotor. Pellet is then resuspended (be sure that no clumps are left behind) in 2.5 ml of Buffer A containing 1 mM PMSF, 5 μg/ml leupeptin, and incubated on ice for 30 min. After addition of 10 ml of Buffer B (containing PMSF and leupeptin as above) and addition of 15 to 20 mg of lysozyme, the suspension is incubated on ice for 1 h. 0.1% (weight/volume) of sodium deoxycholate (approximately 12.5 mg for a 250 ml starting culture volume), 10 mM $MgCl_2$ (125 μl of a 1 M solution), and 200 μg DNase I are then added, and the suspension is incubated on ice for 15 min (a significant decrease in viscosity of the suspension should be clearly evident after 15 min incubation; incubation on ice can be continued up to 45 min, if necessary). The suspension is centrifuged at 15,000 rpm for 1 h at 4°C in a SS-34 rotor. Supernatant (approximately 12.5 ml) is collected in a 15 ml Falcon tube, 0.2 to 0.3 ml of a 50% suspension of Glutathione-Sepharose in Buffer C is added to the supernatant, and the mixture is rotated for 45 to 120 min at 4°C. After the incubation, beads are washed four times with 10 ml of Buffer C (with proteinase inhibitors as above), transferred to an Eppendorf tube and washed four more times with 1 ml of the same buffer. Finally, beads are resuspended in an equal volume of Buffer C and stored at 4°C. For the "pull-down" experiments, the fusion proteins can be left coupled to the Glutathione Sepharose. For the "Far Western" analysis, 1.0. ml of Glutathione Elution Buffer is added and gently mixed per ml bed volume of Glutathione Sepharose. The suspension is incubated at room temperature for 10 min to elute the fusion protein from the matrix. A 5 min centrifugation at 5000 rpm in a microcentrifuge is carried out to sediment the matrix, and the supernatant is removed and stored in a fresh tube. The elution and centrifugation steps are repeated twice more, and the eluates of the purified fusion proteins are pooled and can be stored in aliquots at 4°C or –20°C. Expression and purity of the GST-fusion protein preparation are analyzed using precast SDS-PAGE gels stained with Coomassie Blue; the molecular weight of the GST portion is 26 kDa. The yield can be also estimated by measuring the absorbance at 280; for the GST-affinity tag, 1 A280 = 0.5 mg/ml. The protein yield may also be determined by standard chromogenic methods. If a Lowry or BCA-type method is used, the eluted sample must first be dialyzed against PBS to remove glutathione, which interferes with the assay. The Bradford method can be performed in the presence of glutathione.

2. ***In vitro* association ("pull-down") experiments with GST-SH2-domains**— In this assay, bacterial GST-SH2-fusion proteins are used to identify tyrosine-phosphorylated SH2-binding proteins, in our case Cas, in a cell lysate. RIPA cell lysates (250 μg,

1 mg/ml) prepared from suspended and ECM-adherent cells (as described above for studies on tyrosine phosphorylation of Cas) are first precleared by incubation with GST immobilized on glutathione-Sepharose beads for 30 min with rotation at 4°C. The lysates are then incubated with 5 μg of GST alone or of the GST-SH2 fusion proteins coupled to Glutathione Sepharose, for 2 h rotating at 4°C. The beads are collected and washed twice with RIPA buffer and twice with Tris-buffered saline, and the bound proteins are released by boiling in Laemmli sample buffer followed by SDS-PAGE and immunoblot analysis for Cas as described above. The GST alone or an irrelevant GST-fusion protein serves as a negative control. In general, a positive signal with a relevant SH2-fusion protein is expected to be obtained in lysates prepared from adherent, but not from suspended, cells. To date, the SH2-domains of phosphatidylinositol 3′-kinase, PLC-γ, Src and the adapter proteins Grb2, Nck and Crk have been shown to interact with Cas in cell lysates prepared from adherent cells, and many of these interactions have been confirmed with endogenous, full-length proteins by coimmunoprecipitation experiments.[9,10,25] This experiment can also be carried out in reverse: a fusion protein encoding for a fragment of Cas of interest, such as the substrate domain, can be produced in *E. coli*, and the fusion protein is phosphorylated by Src *in vitro*. This fusion protein, and the corresponding non-phosphorylated fusion protein, can be used as an affinity reagent to "pull-down" proteins from cell lysates. Immunoblot analysis with antibodies specific for various SH2-domain containing signaling molecules are used to detect the putative interaction (for a more detailed description of the protocol, see Reference 40).

3. **Blotting ("Far Western") analysis with GST-fusion proteins**— When performing a "Far Western" analysis for a direct SH2-domain interaction with tyrosine-phosphorylated Cas, Cas is immunoprecipitated, the precipitate is resolved by SDS-PAGE gel, transferred to a membrane, and the blots are probed with GST-SH2 fusion proteins of interest. Optimal immunoprecipitation conditions should be used to assure that a strong signal will be obtained. We typically prepare lysates from 1×10^6 cells that have been transiently transfected to overexpress Cas and plated on ECM proteins (see the next section for a transfection protocol), or start with a RIPA cell lysate prepared from 5×10^6 ECM-adherent fibroblast cells. This equals to about 1 mg of protein (see above for lysate preparation and immunoprecipitation), the final concentration of a lysate to be used is 1 mg/ml). Following immunoprecipitation, the precipitate is resolved on SDS-PAGE and transferred on nitrocellulose membrane or Immobilon with a trans-blot apparatus as above. Blots are blocked overnight at room temperature in PBS/1% BSA-blocking buffer. Alternatively, we have used a 0.05% Tween-20-containing blocking buffer; this buffer results in better specificity, but lower sensitivity. The blots are probed with 5 μg/ml (approximately 100 nM) of purified GST-SH2-fusion protein of interest in the blocking solution for 1 to 2 h at room temperature, then washed with PBS/1% BSA/0.1% Tween-20 three times for 5 min. To detect bound GST-SH2 fusion proteins, the blots are first incubated with a monoclonal anti-GST antibody (1/1000 dilution), washed and incubated with horseradish peroxidase anti-mouse Ig, washed and detection performed with enhanced chemiluminescence, as above. A positive signal with a relevant SH2-fusion protein is expected to be obtained in immunoprecipitates prepared from adherent, but not from suspended, cells. This assay can also be done in reverse: the endogenous SH2-containing proteins of interest are immunoprecipitated and separated on an SDS-PAGE, transferred on a membrane, and blotting is carried out with an *in vitro* phosphorylated GST-fusion protein of Cas (for a detailed protocol, see Reference 40). "Far Western" analysis can also be utilized to detect previously unidentified interactions mediated by Cas. In this case, total cell lysates, rather than immu-

noprecipitates, are separated on SDS-PAGE, and blotting is carried out with a Cas fusion protein. Molecular weights of the proteins that bind Cas will give indications as to the identity of the proteins. "Far Western" analysis with immunoprecipitates of the candidate proteins will confirm the identity (for a detailed protocol, see Reference 34).

C. Materials

1. **Vectors and antibodies** — pGEX-vectors for the generation of GST-fusion protein constructs are from Pharmacia. Monoclonal anti-GST antibody is from Sigma (Cat. #G-1160).
2. **Reagents and bacterial culture for the fusion protein preparation** — *E. coli* DH5α are grown in LB broth supplemented with 100 μg/ml ampicillin. Glutathione-Sepharose 4B is from Pharmacia. Lysozyme and DNAse I are from Sigma.

D. Buffers

1. Buffer A: 50 mM Tris base, pH 7.5, 1.62 M Sucrose, 10 mM EDTA (supplemented with 1 mM PMSF and 5 μg/ml leupeptin).
2. Buffer B: 50 mM Tris-HCl, pH 7.5, 1 mM DTT, 100 mM KCl, 1 mM EDTA (supplemented with proteinase inhibitors as above).
3. Buffer C: 10 mM HEPES, pH 8.00, 1 mM DTT, 10 mM $MgCl_2$, 150 mM NaCl (supplemented with proteinase inhibitors as above).
4. Glutathione elution buffer: 10 mM Glutathione, 50 mM Tris-HCl, pH. 8.0.
5. Blocking buffer for "Far Western:" PBS/1% BSA, or PBS/1% BSA/0.05% Tween-20.
6. Wash buffer for "Far Western:" PBS/1% BSA/0.1% Tween-20.

IV. Functional Studies on the Cas-Crk Signaling Complex

A. Overview

As discussed above, Cas binds to several SH2-domain containing signaling molecules in tyrosine phosphorylation- and integrin ligand binding-dependent manner. Stoichiometrically, the most abundant Cas-SH2-protein interaction has been found to be the interaction between Cas and c-Crk.[10] Crk, which was originally cloned as the avian retroviral oncogene v-Crk,[41] is the first identified member of the adaptor protein family consisting mostly of the SH2- and SH3-domains. Two cellular homologs for v-Crk, c-CrkI and c-CrkII, are produced from the same gene by alternative splicing.[42] The SH3-domain of c-Crk can bind to multiple target molecules, including two guanine nucleotide exchange proteins, C3G[43] and Sos,[44] and

DOCK 180, a 180-kDa protein of unknown function;[45] we have found that these proteins are efficiently recruited by c-Crk to the Cas-Crk signaling complex upon cellular stimulation.[10] Taken together, Crk is likely to be involved in integrin-mediated signaling pathways, but a physiological role for c-Crk has remained largely unknown. It has been shown before that integrin ligand binding stimulates activation of the various family members of the mitogen-activated protein kinases (MAPKs), including c-Jun kinases (JNKs) and extracellular-regulated kinases (Erks).[46-48] We found recently that both Cas and Crk, when overexpressed, induce activation of the JNK pathway but not of the Erk pathway. Furthermore, dominant-negative forms of Crk prevent Cas-induced JNK activation, and dominant-negative forms of the two proteins prevent integrin-mediated JNK activation.[57] The studies on JNK activation are described in this section, and demonstrate the utility of the Cas and Crk reagents in studies concerning other cellular functions potentially mediated by the two signaling molecules.

In order to detect the activation of the JNK pathway in response to extracellular stimuli (integrin ligand binding, growth factors, etc.) or intracellular stimulation (ectopic expression of p130Cas, c-CrkII, etc.), we have employed two different techniques: (i) *in vitro* kinase assays and (ii) a luciferase trans-acting reporter system, which are described below.

In vitro kinase assay represents a rapid and quantitative method for detection of JNK activity, and is based on the ability of the immunoprecipitated JNK protein to phosphorylate specific substrates *in vitro*.[49] The immunoprecipitated JNK is either the endogenous JNK protein or an ectopically expressed, tagged form of JNK. The latter approach is used when putative intracellular JNK regulators, such as Cas and Crk, are transiently expressed in cells and tested for their ability to interfere positively or negatively with the JNK pathway. A truncated c-Jun fusion protein containing amino acids 1 to 79 is the substrate of choice for the JNK kinase assay. This truncated form of c-Jun, expressed as a GST-fusion protein, contains the JNK binding region (aas 30–60) and two serine residues (Ser 63 and 73), which are specifically phosphorylated by JNK. The truncated c-Jun fusion protein lacks the DNA binding region that contains three additional serine residues phosphorylated by other kinases.[50] The immunoprecipitated JNK is mixed together with the substrate in the presence of [γ^{32}P]ATP, and the kinase reaction is allowed to proceed, after which the sample is analyzed by SDS-PAGE and autoradiography.

We have also used a luciferase trans-acting reporter system that allows a rapid, quantitative and reliable *in vivo* assessment of the activation of the JNK kinase in response to the ectopic expression of a gene of interest or following extracellular stimuli (Stratagene, PathDetect — *in vitro* pathway reporting systems). The application of this technique to study the induction of the JNK pathway by expression of Cas and the modulation of this pathway by the coexpression of dominant negative mutants of Crk is described below. A similar approach can be utilized to study the participation of any other protein in the JNK activation pathway.

The procedure involves transient cotransfection of several plasmids into mammalian cells: (i) the construct that contains the gene(s) of interest (i.e., Cas, Crk constructs); (ii) the pathway-specific fusion transactivator plasmid, Gal4-Jun (1-223); and (iii) the reporter plasmid (5xGal4-Luciferase). Following transfection, the cells are serum starved and subsequently assayed for the activity of the reporter gene. If the expression of the gene of interest (Cas) is either directly or indirectly involved in the activation of the JNK signal transduction pathway, the activity of the reporter gene (luciferase) will be readily detected and quantified. A similar procedure can also be applied to study the activation of the JNK pathway by extracellular stimuli.

The trans-activator plasmid encodes a recombinant protein, in which the yeast Gal4 DNA binding domain (aas 1 to 147) is fused in frame with the activation domain of the mammalian c-Jun transcription factor (aas 1 to 223), a cellular substrate of the JNK kinases. This truncation of c-Jun encompasses the JNK-docking site and the JNK specific phosphoserine acceptor sites (see above). The phosphorylation of these residues enhances the ability of c-Jun to activate transcription. As a result, the Gal4-Jun fusion protein is efficiently and specifically phosphorylated by JNK kinases, and — most importantly — its transcriptional activity is greatly enhanced upon phosphorylation.

The reporter plasmid contains the entire sequence of the *Photinus pylaris* (firefly) luciferase gene downstream of a basic promoter element (TATA box) joined to five tandem repeats of the Gal4 binding element. The binding of the phosphorylated (activated) Gal4-Jun transcription factor to the activating sequence in the reporter plasmid results in luciferase expression. Therefore, the relative luciferase activity in the cell population can be directly correlated with the activation/inhibition of the JNK signal transduction pathway.

B. Protocols

1. *In vitro* kinase assay

a. Transient transfection, JNK immunoprecipitation, and kinase assay — In order to study the capability of Cas and Crk to activate JNK, transient transfection approach is utilized. We have chosen to utilize COS cells for our experiments. These cells are very easy to transfect using Lipofectamine or other methods, and usually are able to express high levels of ectopic proteins in transient transfection experiments. Similar protocol for transfection can be used for other cell lines, such as HeLa, NIH 3T3, or HEK293, but typical expression levels are lower in these cells than in COS cells.

Day 1 — Confluent cells, exponentially growing in DMEM/10% fetal calf serum are detached by trypsin and plated in 6-well plates in 2 ml of DMEM/10% fetal calf serum and antibiotics (250,000 to 300,000 cells/well).

Day 2 — Cells in each well are transfected by the Lipofectamine method, using 1 μg total of DNA (containing equal parts of an expression vector carrying a tagged form of JNK, such as pSRα3-(HA)JNK1[49,51] or pCMV5-(M2)JNK1,[52] and the gene of interest to be studied, such as the pCAGGS-(Myc)CrkII-vector for CrkII[42,53] and either a non-tagged (pSSRα-p130Cas)[5] or a GST-tagged version (pEBG-Cas) of Cas.[54] The transfection is carried out according to the instructions provided by the manufacturer (Gibco-BRL). We recommend mixing a total amount of 1 μg DNA and 4 μl of Lipofectamine for each well, and incubating the mixture for 45 min at room temperature before adding it to the cells. 5 h following incubation at 37°C/5% CO_2, an equal volume of DMEM containing

20% fetal calf serum, but no antibiotics, is added to the cells. Lipofectamine may be toxic for COS cells; if a problem is observed, we recommend using a slightly modified transfection protocol described below for the luciferase assay.

Day 3 — After approximately 24 h from the transfection, the medium is removed, cells are rinsed twice with PBS and serum-starved for 24 h in DMEM without serum, but with antibiotics.

Day 4 — After 24 h, the medium is removed from the plate, cells are rinsed twice with ice-cold PBS and then immediately lysed in 250 µl of Lysis Buffer for each well. Cells are then collected in Eppendorf tubes, centrifuged at 14,000 rpm for 10 min at 4°C; the total amount of protein in a lysate is determined by the BCA method. 2 µl of rabbit polyclonal anti-HA or monoclonal anti-Flag (M2) antibodies together with 15 µl of a 50% suspension of Protein A-Sepharose are added to supernatants containing equal amounts of proteins to immunoprecipitate the transfected JNK. After 3 h of rotation at 4°C, the beads are washed twice with Lysis Buffer and twice with Kinase Buffer. Following a complete removal of the last wash buffer, kinase assay is then performed by adding a total volume of 20 µl of Kinase Buffer containing 10 µM ATP, 1 µCi [γ-^{32}P] ATP and 10 µg GST-c-Jun(1-79) as substrate. After incubation at 30°C for 20 min, the reaction is stopped by adding Laemmli sample buffer and by boiling the mixture for 5 min. Samples are then separated by 4 to 20% precast SDS-PAGE gels. The gels are stained with Coomassie Blue to verify that equal amounts of antibodies and substrate were loaded in each reaction. The gels are dried, visualized by autoradiography, and the radioactivity incorporated to the substrate (which is about 38 kDa in molecular weight) is quantified by Phosphoimager.

b. **Detection of the protein expression levels by immunoblotting**— In order to determine that all the cDNA transiently transfected were equally expressed in the different transfections, 10 to 20 µg of each lysate prepared for JNK kinase assays (see above) are separated on precast Novex gels and immunoblotted with specific antibodies as described in the first section of this chapter. Anti-HA or anti-Flag (M2) antibodies for JNK detection are diluted 1/250 and 1/500, respectively, in PBS containing 0.5% BSA and 0.1% Tween-20. Anti-Cas and anti-GST antibodies for Cas detection are diluted 1/1000. Expression levels of c-CrkII (Myc-tagged) may be detected using commercial antibodies against the Myc-epitope (although this appears to be of low sensitivity) or against the Crk protein; the Myc-tagged form of Crk is up-shifted with respect to the endogenous Crk protein.

c. **Preparation of c-Jun substrate for the kinase assay**— GST-c-Jun(1-79) is prepared as described above. After the final washes in Buffer C, the protein is eluted from the beads using 1 ml of 5 to 10 mM Glutathione in Kinase Buffer containing 1 mM DTT (per 200 µl of the beads) at 4°C overnight. Beads are washed twice with 0.5 ml of the same buffer. Proteins are then concentrated up to 10 mg/ml using Centricon 25, aliquotted and stored at –20°C.

d. **Detection of JNK activation in response to cell adhesion and use of dominant-negative forms of Cas and Crk**— As described above, Cas and Crk form a molecular complex upon integrin ligand binding, and both integrin ligation and ectopic expression of Cas or Crk in eukaryotic cells induce JNK activation. In order to study whether the integrin-induced activation of JNK activation is mediated via Cas and Crk, dominant-interfering mutants of Crk and Cas were utilized. The following mutants, which lack the domains necessary for the Cas-Crk interaction were used: a mutant form of c-CrkII in which the functionality of the SH2-domain is knocked out by a point mutation (R38V, from hereon referred as the "Crk-SH2" construct) and a form of Cas in which

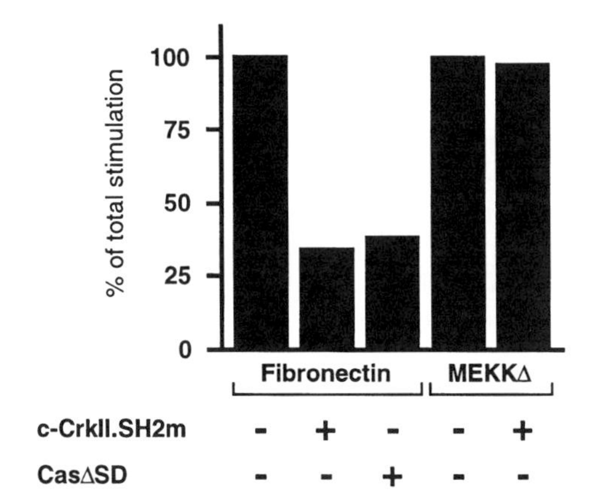

FIGURE 2
Effect of CasΔSD and Crk-SH2 on adhesion-induced JNK activation. COS-7 cells were grown in DMEM supplemented with 10% fetal calf serum and antibiotics, and transfected as described in the text using pCMV5-(M2)JNK1 and either pSSRα-CasΔSD ("CasΔSD"), pCAGGS-(Myc)c-CrkIIR38V ("c-CrkII.SH2m") or the empty vector as control. After 24 h starvation, cells were detached and replated on fibronectin-coated cell dishes for 15 min. The specificity of the dominant negative forms activity on JNK signaling pathway was analyzed in cells transfected with a constitutive active form of MEKK [using the plasmid pSRα3-(HA)MEKKΔ] (results for the Crk-SH2 are shown). JNK activity was analyzed in immunocomplex kinase assays using c-Jun (1-79) as the substrate. The expression levels of the different proteins were analyzed by immunoblotting as described in the text.

the substrate domain (CasΔSD) is deleted. When expressed in COS cells, neither CasΔSD nor Crk-SH2 are able to coimmunoprecipitate with the endogenous Crk or Cas proteins, respectively, confirming that the interaction between the two proteins is due to the domains described above. Similarly, a form of c-CrkII in which the SH3-domain function has been disrupted (W169L, the "Crk-SH3"-construct) is dominant-inhibitory with respect to upstream molecules, such as Cas.

For the experiment, COS cells are transfected by the Lipofectamine method using either an empty vector or CasΔSD or Crk-SH2 constructs, together with a tagged form of JNK (see above). Two days after the transfection, cells are detached by brief trypsinization, washed with soybean trypsin inhibitor, and either lysed as above in the lysis buffer or replated on fibronectin for 15 min and lysed. JNK immunocomplex kinase assay is carried out as described above. The results of the experiment are summarized in Figure 2. Simple short-term detachment of COS cells is not sufficient to activate JNK (compare, however, to, e.g., MDCK-cells[55]). In contrast, JNK is readily activated when cells were plated on fibronectin. Cas and Crk mutants were able to completely inhibit JNK activation induced by the cell attachment on fibronectin; in COS-cells, this adhesion is mediated by the $\alpha5\beta1$ integrin. The specificity of the two mutant constructs can be analyzed by transfecting cells with a truncated form of MEKK (MEKKΔ, the MAPKKK of JNK),[56] HA-tagged JNK and either CasΔSD or Crk-SH2. MEKK is a kinase upstream of JNK in the kinase cascade, and is able to activate JNK independent of further upstream stimuli; accordingly, the dominant-negative forms of Cas and Crk do not interfere with MEKK-induced JNK activation (Figure 2).

2. Luciferase trans-activator reporter system

a. **Cell transfection** — A day prior to transfection, in a 6-well or 35 mm tissue culture plates, 3 to 4 × 10^5 cells/well are seeded in 2 ml of complete growth medium (DMEM/10% fetal calf serum supplemented with penicillin and streptomycin). All the assays should be performed in triplicate, thereby three independent wells per condition assayed are required. Cells are transfected by the Lipofectamine method as described above, with slight modifications: a total of 2 µg DNA and 3 and 6 µl of Lipofectamine for COS cells and HeLa cells, respectively, is used per well. After incubating the cells with the DNA/Lipofectamine mixture for 5 h at 37°C in a 5% CO_2 atmosphere, we followed distinct procedures for HeLa cells and COS cells. For HeLa cells, an equal volume (1 ml) of DMEM with 20% serum and without antibiotics is added on top of the DNA/lipid mixture and the cells are incubated overnight (12 to 18 h) in 37°C/5% CO_2. The DNA/lipid mixture is completely removed from the cells after the incubation and the wells are rinsed twice with DMEM without serum. Fresh DMEM without serum is added to the cells, and the cells are serum-starved for 24 h in 37°C/5% CO_2, after which the luciferase assay is carried out.

As mentioned earlier, Lipofectamine is fairly toxic for COS-1 and COS-7 cells, and we have therefore optimized the transfection protocol for these cells, especially when carrying out the luciferase assay. We recommend removing thoroughly the DNA/lipid sample after the initial 5 h incubation. Cells are then covered with 2 ml of DMEM containing 10% serum and antibiotics and incubated overnight as above. Next day, the complete medium is replaced with DMEM without serum to starve the cultures; 24 h later the cells are processed for the luciferase expression analysis.

b. **Luciferase assay** — To assess the firefly luciferase activity, we essentially follow the manufacturer's recommendations (Promega). After 24 h of serum starvation, the media is carefully removed from the wells and the cells are rinsed twice with PBS at room temperature. 200 µl of Cell Lysis Buffer is added to the cells. The cells are scraped from the plates with a rubber polishman, and the lysates are transferred to Eppendorf tubes. Care should be taken that entire cell lysate is recovered. Cell debris is removed by centrifugation for 5 min at 10,000 rpm in a microcentrifuge and the supernatants are transferred to new Eppendorf tubes. The lysates can be assayed immediately for firefly luciferase activity. If required, the extracts can be stored at 4°C for 1 to 2 h or at –70°C for long-term storage (all the samples will be uniformly affected by the exposure to a freeze/thaw cycle).

We assay the firefly luciferase enzymatic activity using a luminometer. 20 µl of cell extract is mixed with 100 µl of the Luciferase Assay Reagent (provided with the Luciferase Assay Kit, Promega) in a 96-well plate; both the extract and the Assay Reagent should be at room temperature. The mixture should be set up immediately prior to the measurement. To avoid any cross-reflection of light between adjacent samples, plates opaque to light should be used. The most suitable ones are white-colored (a black plate will yield light emission values that are >10 times smaller than the ones achieved with the white plate). The light luminescence is measured for a period of 10 to 20 sec. The light intensity is constant for a period of ~20 sec and then decays slowly, with a half-life of about 5 min. If an automated injection device is used, an injection volume of 100 µl should be set for the Luciferase Assay Reagent and an initial delay of 2 sec should be allowed before measuring luminescence.

c. **Specific amounts of DNA used for the cell transfections** — The total amount of plasmid DNA used for the luciferase experiments is always 2 µg per sample. This includes the luciferase reporter vector, the plasmid encoding the Gal4-Jun (aa 1-223)

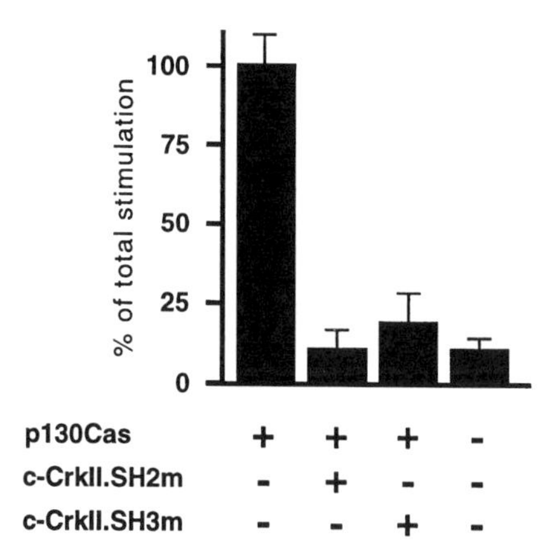

FIGURE 3
Gal4/Jun transcriptional activity induced by the expression of Cas in the absence or presence of c-CrkII dominant-negative mutants and determined using the trans-acting luciferase reporter system outlined in this chapter. COS-7 cells were cotransfected with the reporter plasmid pFR-Luc (1 μg) and the trans-activator plasmid pFA-Jun (Gal4/Jun, 50 ng) together with (+) or without (–) plasmids expressing p130Cas (pSSRα-p130Cas, 300 ng), c-CrkII.SH2m [pCAGGS-(Myc) c-CrkII (R38V), 300 ng] and/or c-CrkII.SH3m mutants [pCAGGS-(Myc) c-CrkII (W169L), 300 ng]. The total amount of DNA per sample was normalized to 2 μg using pSSRα vector. Two days after transfection the activity of firefly luciferase was assessed as indicated in the text. Data are shown as percent of activation — average of three independent determinations — normalized to the value obtained for the cells transfected with Cas. Error bars represent standard deviation.

transactivator protein, the gene(s) of interest (Cas, Crk constructs) in a mammalian expression vector and a plasmid to adjust the final amount of DNA to 2 μg (an empty mammalian expression vector). For the pFA-c-Jun, we always use 50 ng/sample. Increasing the amount of pFA-c-Jun increases the background without significantly improving the signal/noise ratio. For the pFR-Luc, we typically use 1 μg/sample. For the plasmid pSSRα-p130Cas, amounts of 100 to 300 ng of plasmid DNA will yield ~2- to 5-fold enhancement of luciferase enzymatic activity (Figure 3). Note that the system is saturable, and increasing the amount of pSSRα-p130Cas results in a reduction of the firefly luciferase expression for Crk-SH2 and Crk-SH3 mutants; amounts ranging between 250 to 500 μg DNA per sample have been used and will reduce more than 90% the Gal4/Jun transcriptional activity induced by expressing Cas (Figure 3).

C. Materials

(1 to 4 are for the *in vitro* kinase assay experiments.)

1. **Cells** — We have obtained COS, HeLa and HEK293 cells from American Type Culture Collection.
2. **Antibodies** — The antibodies mentioned in the protocol are obtained from commercial sources. Monoclonal anti-Crk antibody is from Transduction Laboratories. Anti-HA rabbit polyclonal antibodies Y-11 are from Santa Cruz Biotechnology and mouse monoclonal anti-Flag (M2) antibodies are from Kodak/IBI (Rochester, NY). Anti-Myc antibodies can be obtained from Calbiochem (San Diego, CA).

3. **Reagents for cell culture, immunoprecipitation and kinase assays** — Cells were grown in DMEM, supplemented with 10% fetal calf serum and antibiotics as described in the first section of this chapter. Protein A-Sepharose CL-4B is from Pharmacia. [γ–^{32}P] ATP (6000 Ci/mmol) is from Du Pont/NEN (Boston, MA). All the other reagents mentioned in the protocols are purchased from Sigma unless otherwise specified. Lipofectamine is from Gibco (Cat. #18324-012).
4. **Plasmids** — The following plasmids were used in the studies: pSRα3-(HA)JNK1,[51] pCMV5-(M2)JNK1,[52] pGEX3X-c-Jun(1-79),[49,51] pSSRα-p130Cas, pSSRα-CasΔSD,[5] pCAGGS-(Myc)c-CrkII, pCAGGS-(Myc)c-CrkII-R38V, pCAGGS-(Myc)c-CrkII-W169L,[42,53] pSRα3-(HA)MEKKΔ.[51]

(5 to 8 are for the luciferase assay.)

5. **Mammalian cell lines** — Several cell lines have been successfully used to study the activation of JNK in the luciferase assay by Cas including COS-1, COS-7, and HeLa cells. Some other cell lines, e.g., HEK293 cells, demonstrate a very high constitutive activation of the reporter vector; therefore, we do not recommend their use.
6. **Plasmids** — The following plasmids are used for luciferase assay: pFA-c-Jun (Stratagene, Cat. #219053), encoding the fusion transactivator protein Gal4-Jun(1-223); pFR-Luc (Stratagene, Cat. #219050), reporter plasmid containing the firefly luciferase gene joined to the 5 × Gal4/TATA promoter; gene(s) of interest cloned into an appropriate mammalian expression vector; for Cas- and Crk-constructs, see above.
7. **Luciferase assay reagents** — We recommend the use of a system that provides a constant emission of light mediated by the activity of firefly luciferase. We use the Luciferase Assay System from Promega (Cat. #E1500), but similar systems are available from other commercial sources.
8. **Instruments** — With the Luciferase Assay System (Promega), either a scintillation counter or luminometer can be used (see manufacturer's recommendations for details). A luminometer can measure as little as 10^{-2} moles (0.001 pg) of luciferase, whereas a scintillation counter has a less sensitive detection limit. Because of the constant light output with the Luciferase Assay System, the luminometer does not require an automated injection device.

D. Buffers

1. Lysis Buffer for the kinase assay: 50 mM HEPES, pH 7.6, 250 mM NaCl, 3 mM EDTA, 3 mM EGTA, 0.5% NP-40, 100 mM Na_3VO_4, supplemented with proteinase inhibitors (1 mM PMSF; 10 μg/ml leupeptin; 10 μg/ml aprotinin).
2. Kinase Buffer: 50 mM HEPES, pH 7.6, 10 mM $MgCl_2$.
3. Cell Lysis Buffer for luciferase assay: 25 mM Tris-phosphate pH, 7.8, 2 mM DTT, 2 mM 1,2-diaminocyclohexane-N,N,N′,N′-tetraacetic acid, 10% glycerol, 1% Triton X-100.

Acknowledgments

The original work in the authors' laboratory was supported by NIH grants CA72560 and CA76037. F.D. and M.G.-G. were supported by fellowships from the American-Italian Cancer Foundation and Human Frontier Science Program, respectively. K. V. is a PEW scholar in biomedical sciences.

References

1. Birge, R. B., Fajardo, J. E., Mayer, B. J., and Hanafusa, H., Tyrosine-phosphorylated epidermal growth factor receptor and cellular p130 provide high affinity binding substrates to analyze Crk-phosphotyrosine-dependent interactions *in vitro*, *J. Biol. Chem.*, 267, 10588, 1992.
2. Matsuda, M., Mayer, B. J., Fukui, Y., and Hanafusa, H., Binding of transforming protein, P47gag-crk, to a broad range of phosphotyrosine-containing proteins, *Science,* 248, 1537, 1990.
3. Kanner, S. B., Reynolds, A. B., Vines, R. R., and Parsons, J. T., Monoclonal antibodies to individual tyrosine-phosphorylated protein substrates of oncogene-encoded tyrosine kinases, *Proc. Natl. Acad. Sci. USA,* 87, 3328, 1990.
4. Kanner, S. B., Reynolds, A. B., Wang, H. C., Vines, R. R., and Parsons, J. T., The SH2 and SH3 domains of pp60^{src} direct stable association with tyrosine phosphorylated proteins p130 and p110, *EMBO J.,* 10, 1689, 1991.
5. Sakai, R., Iwamatsu, A., Hirano, N., Ogawa, S., Tanaka, T., Mano, H., Yazaki, Y., and Hirai, H., A novel signaling molecule, p130, forms stable complexes *in vivo* with v-Crk and v-Src in a tyrosine phosphorylation-dependent manner, *EMBO J.,* 13, 3748, 1994.
6. Polte, T. R. and Hanks, S. K., Interaction between focal adhesion kinase and Crk-associated tyrosine kinase substrate p130^{Cas}, *Proc. Natl. Acad. Sci. USA,* 92, 10678, 1995.
7. Garton, A. J., Burnham, M. R., Bouton, A. H., and Tonks, N. K., Association of PTP-PEST with the SH3 domain of p130^{Cas}: a novel mechanism of protein tyrosine phosphatase substrate recognition, *Oncogene,* 15, 877, 1997.
8. Liu, F., Hill, D. E., and Chernoff, J., Direct binding of the proline-rich region of protein tyrosine phosphatase 1B to the Src homology 3 domain of p130^{Cas}, *J. Biol. Chem.,* 271, 31290, 1996.
9. Nakamoto, T., Sakai, R., Ozawa, K., Yazaki, Y., and Hirai, H., Direct binding of C-terminal region of p130^{Cas} to SH2 and SH3 domains of Src kinase, *J. Biol. Chem.,* 271, 8959, 1996.
10. Vuori, K., Hirai, H., Aizawa, S., and Ruoslahti, E., Induction of p130^{Cas} signaling complex formation upon integrin-mediated cell adhesion: a role for Src family kinases, *Mol. Cell. Biol.,* 16, 2606, 1996.
11. Tachibana, K., Urano, T., Fujita, H., Ohashi, Y., Kamiguchi, K., Iwata, S., Hirai, H., and Morimoto, C., Tyrosine phosphorylation of Crk-associated substrates by focal adhesion kinase, *J. Biol. Chem.,* 272, 29083, 1997.

12. Nakamoto, T., Sakai, R., Honda, H., Ogawa, S., Ueno, H., Suzuki, T., Aizawa, S., Yazaki, Y., and Hirai, H., Requirements for localization of p130^{Cas}to focal adhesions, *Mol. Cell. Biol.,* 17, 3884, 1997.
13. Ishino, M., Ohba, T., Sasaki, H., and Sasaki, T., Molecular cloning of a cDNA encoding a phosphoprotein, Efs, which contains a Src homology 3 domain and associates with Fyn, *Oncogene,* 11, 2331, 1995.
14. Alexandropoulos, K. and Baltimore, D., Coordinate activation of c-Src by SH3- and SH2-binding sites on a novel p130^{Cas}-related protein, Sin, *Genes Dev.,* 10, 1341, 1996.
15. Law, S. F., Estojak, J., Wang, B., Mysliwiec, T., Kruh, G., and Golemis, E. A., Human enhancer of filamentation 1, a novel p130^{Cas}-like docking protein, associates with focal adhesion kinase and induces pseudohyphal growth in Saccharomyces cerevisiae, *Mol. Cell. Biol.,* 16, 3327, 1996.
16. Minegishi, M., Tachibana, K., Sato, T., Iwata, S., Nojima, Y., and Morimoto, C., Structure and function of Cas-L, a 105-kD Crk-associated substrate-related protein that is involved in beta 1 integrin-mediated signaling in lymphocytes, *J. Exp. Med.,* 184, 1365, 1996.
17. Petch, L. A., Bockholt, S. M., Bouton, A., Parsons, J. T., and Burridge, K., Adhesion-induced tyrosine phosphorylation of the p130 src substrate, *J. Cell Sci.,* 108, 1371, 1995.
18. Vuori, K. and Ruoslahti, E., Tyrosine phosphorylation of p130^{Cas}and cortactin accompanies integrin-mediated cell adhesion to extracellular matrix, *J. Biol. Chem.,* 270, 22259, 1995.
19. Nojima, Y., Morino, N., Mimura, T., Hamasaki, K., Furuya, H., Sakai, R., Sato, T., Tachibana, K., Morimoto, C., Yazaki, Y., et al., Integrin-mediated cell adhesion promotes tyrosine phosphorylation of p130^{Cas}, a Src homology 3-containing molecule having multiple Src homology 2-binding motifs, *J. Biol. Chem.,* 270, 15398, 1995.
20. Ingham, R. J., Krebs, D. L., Barbazuk, S. M., Turck, C. W., Hirai, H., Matsuda, M., and Gold, M. R., B cell antigen receptor signaling induces the formation of complexes containing the Crk adapter proteins, *J. Biol. Chem.,* 271, 32306, 1996.
21. Schraw, W. and Richmond, A., Melanoma growth stimulatory activity signaling through the class II interleukin-8 receptor enhances the tyrosine phosphorylation of Crk-associated substrate, p130, and a 70-kilodalton protein, *Biochemistry,* 34, 13760, 1995.
22. Ribon, V. and Saltiel, A. R., Nerve growth factor stimulates the tyrosine phosphorylation of endogenous Crk-II and augments its association with p130^{Cas}in PC-12 cells, *J. Biol. Chem.,* 271, 7375, 1996.
23. Casamassima, A. and Rozengurt, E., Tyrosine phosphorylation of p130^{Cas} by bombesin, lysophosphatidic acid, phorbol esters, and platelet-derived growth factor. Signaling pathways and formation of a p130(cas)-Crk complex, *J. Biol. Chem.,* 272, 9363, 1997.
24. Ojaniemi, M. and Vuori, K., Epidermal growth factor modulates tyrosine phosphorylation of p130^{Cas}. Involvement of phosphatidylinositol 3′-kinase and actin cytoskeleton, *J. Biol. Chem.,* 272, 25993, 1997.
25. Schlaepfer, D. D., Broome, M. A., and Hunter, T., Fibronectin-stimulated signaling from a focal adhesion kinase-c-Src complex: involvement of the Grb2, p130^{Cas}, and Nck adaptor proteins, *Mol. Cell. Biol.,* 17, 1702, 1997.

26. Avraham, S., London, R., Fu, Y., Ota, S., Hiregowdara, D., Li, J., Jiang, S., Pasztor, L. M., White, R. A., Groopman, J. E., et al., Identification and characterization of a novel related adhesion focal tyrosine kinase (RAFTK) from megakaryocytes and brain, *J. Biol. Chem.,* 270, 27742, 1995.
27. Lev, S., Moreno, H., Martinez, R., Canoll, P., Peles, E., Musacchio, J. M., Plowman, G. D., Rudy, B., and Schlessinger, J., Protein tyrosine kinase PYK2 involved in Ca(2+)-induced regulation of ion channel and MAP kinase functions, *Nature,* 376, 737, 1995.
28. Sasaki, H., Nagura, K., Ishino, M., Tobioka, H., Kotani, K., and Sasaki, T., Cloning and characterization of cell adhesion kinase beta, a novel protein-tyrosine kinase of the focal adhesion kinase subfamily, *J. Biol. Chem.,* 270, 21206, 1995.
29. Engvall, E. and Ruoslahti, E., Binding of soluble form of fibroblast surface protein, fibronectin, to collagen, *Int. J. Cancer,* 20, 1, 1977.
30. Yatohgo, T., Izumi, M., Kashiwagi, H., and Hayashi, M., Novel purification of vitronectin from human plasma by heparin affinity chromatography, *Cell. Struct. Funct.,* 13, 281, 1988.
31. Law, S. F., Zhang, Y. Z., Kleinszanto, A. J. P., and Golemis, E. A., Cell cycle-regulated processing of Hef1 to multiple protein forms differentially targeted to multiple subcellular compartments, *Mol. Cell. Biol.,* 18, 3540, 1998.
32. Harte, M. T., Hildebrand, J. D., Burnham, M. R., Bouton, A. H., and Parsons, J. T., p130^{Cas}, a substrate associated with v-Src and v-Crk, localizes to focal adhesions and binds to focal adhesion kinase, *J. Biol. Chem.,* 271, 13649, 1996.
33. Bouton, A. H. and Burnham, M. R., Detection of distinct pools of the adapter protein p130^{Cas}using a panel of monoclonal antibodies, *Hybridoma,* 16, 403, 1997.
34. Gish, G., Larose, L., Shen, R., and Pawson, T., Biochemical analysis of SH2 domain-mediated protein interactions, *Methods Enzymol.,* 254, 503, 1995.
35. Songyang, Z., Shoelson, S. E., Chaudhuri, M., Gish, G., Pawson, T., Haser, W. G., King, F., Roberts, T., Ratnofsky, S., Lechleider, R. J., Neel, B. G., Birge, R. B., Fajardo, J. E., Chou, M. M., Hanafusa, H., Schaffhausen, B., and Cantley, L. C., SH2 domains recognize specific phosphopeptide sequences, *Cell,* 72, 767, 1993.
36. Vojtek, A. B. and Hollenberg, S. M., Ras-Raf interaction: two-hybrid analysis, *Methods Enzymol.,* 255, 331, 1995.
37. Keegan, K. and Cooper, J. A., Use of the two hybrid system to detect the association of the protein-tyrosine-phosphatase, SHPTP2, with another SH2-containing protein, Grb7., *Oncogene,* 12, 1537, 1996.
38. Skolnik, E. Y., Margolis, B., Mohammadi, M., Lowenstein, E., Fischer, R., Drepps, A., Ullrich, A., and Schlessinger, J., Cloning of PI3 kinase-associated p85 utilizing a novel method for expression/cloning of target proteins for receptor tyrosine kinases, *Cell,* 65, 83, 1991.
39. Blackwood, E. M. and Eisenman, R. N., Identification of protein-protein interactions by lambda gt11 expression cloning, *Methods Enzymol.,* 254, 229, 1995.
40. Burnham, M. R., Harte, M. T., Richardson, A., Parsons, J. T., and Bouton, A. H., The identification of p130^{Cas}-binding proteins and their role in cellular transformation, *Oncogene,* 12, 2467, 1996.
41. Mayer, B. J., Hamaguchi, M., and Hanafusa, H., A novel viral oncogene with structural similarity to phospholipase C, *Nature,* 332, 272, 1988.

42. Matsuda, M., Tanaka, S., Nagata, S., Kojima, A., Kurata, T., and Shibuya, M., Two species of human CRK cDNA encode proteins with distinct biological activities, *Mol. Cell. Biol.,* 12, 3482, 1992.
43. Tanaka, S., Morishita, T., Hashimoto, Y., Hattori, S., Nakamura, S., Shibuya, M., Matuoka, K., Takenawa, T., Kurata, T., Nagashima, K., and Matsuda, M., C3G, a guanine nucleotide-releasing protein expressed ubiquitously, binds to the Src homology 3 domains of CRK and GRB2/ASH proteins, *Proc. Natl. Acad. Sci. USA,* 91, 3443, 1994.
44. Matsuda, M., Hashimoto, Y., Muroya, K., Hasegawa, H., Kurata, T., Tanaka, S., Nakamura, S., and Hattori, S., CRK protein binds to two guanine nucleotide-releasing proteins for the Ras family and modulates nerve growth factor-induced activation of Ras in PC12 cells, *Mol. Cell. Biol.,* 14, 5495, 1994.
45. Hasegawa, H., Kiyokawa, E., Tanaka, S., Nagashima, K., Gotoh, N., Shibuya, M., Kurata, T., and Matsuda, M., DOCK180, a major CRK-binding protein, alters cell morphology upon translocation to the cell membrane, *Mol. Cell. Biol.,* 16, 1770, 1996.
46. Schwartz, M. A., Schaller, M. D., and Ginsberg, M. H., Integrins: emerging paradigms of signal transduction, *Annu. Rev. Cell Dev. Biol.,* 11, 549, 1995.
47. Mainiero, F., Murgia, C., Wary, K. K., Curatola, A. M., Pepe, A., Blumemberg, M., Westwick, J. K., Der, C. J., and Giancotti, F. G., The coupling of alpha6beta4 integrin to Ras-MAP kinase pathways mediated by Shc controls keratinocyte proliferation, *EMBO J.,* 16, 2365, 1997.
48. Miyamoto, S., Teramoto, H., Coso, O. A., Gutkind, J. S., Burbelo, P. D., Akiyama, S. K., and Yamada, K. M., Integrin function: molecular hierarchies of cytoskeletal and signaling molecules, *J. Cell Biol.,* 131, 791, 1995.
49. Cavigelli, M., Dolfi, F., Claret, F. X., and Karin, M., Induction of c-fos expression through JNK-mediated TCF/Elk-1 phosphorylation, *EMBO J.,* 14, 5957, 1995.
50. Lin, A., Frost, J., Deng, T., Smeal, T., Al-Alawi, N., Kikkawa, U., Hunter, T., Brenner, D., and Karin, M., Casein kinase II is a negative regulator of c-Jun DNA binding and AP-1 activity, *Cell,* 70, 777, 1992.
51. Minden, A., Lin, A., McMahon, M., Lange-Carter, C., Derijard, B., Davis, R. J., Johnson, G. L., and Karin, M., Differential activation of ERK and JNK mitogen-activated protein kinases by Raf-1 and MEKK, *Science,* 266, 1719, 1994.
52. Derijard, B., Hibi, M., Wu, I. H., Barrett, T., Su, B., Deng, T., Karin, M., and Davis, R. J., JNK1: a protein kinase stimulated by UV light and Ha-Ras that binds and phosphorylates the c-Jun activation domain, *Cell,* 76, 1025, 1994.
53. Tanaka, S., Hattori, S., Kurata, T., Nagashima, K., Fukui, Y., Nakamura, S., and Matsuda, M., Both the SH2 and SH3 domains of human CRK protein are required for neuronal differentiation of PC12 cells, *Mol. Cell. Biol.,* 13, 4409, 1993.
54. Mayer, B. J., Hirai, H., and Sakai, R., Evidence that SH2 domains promote processive phosphorylation by protein-tyrosine kinases, *Curr. Biol.,* 5, 296, 1995.
55. Frisch, S. M., Vuori, K., Kelaita, D., and Sicks, S., A role for Jun-N-terminal kinase in anoikis; suppression by bcl-2 and crmA, *J. Cell Biol.,* 135, 1377, 1996.
56. Minden, A. and Karin, M., Regulation and function of the JNK subgroup of MAP kinases, *Biochim. Biophys. Acta,* 1333, F85, 1997.
57. Dolfi, F., Garcia-Guzman, M., Ojaniemil, M., Nakamura, H., Matsuda, M., and Vuori, K., The adaptor protein Crk connects multiple cellular stimuli to the JNK signaling pathway, *Proc. Natl. Acad. Sci. USA*, 95, 15394, 1998.

Chapter 7

Specificity of Integrin Signaling

Kishore K. Wary, Agnese Mariotti, and Filippo G. Giancotti

Contents

I. Introduction 102
II. Experimental Overview 103
III. Coimmunoprecipitation of Integrins with Caveolin-1 and Caveolin-1 with Fyn 104
  A. Protocol 104
  B. Reagents 105
IV. Integrin Ligation 106
  A. Crosslinking of Integrins with Soluble Antibodies 106
  B. Ligation of Integrins with Antibody-Coated Polystyrene Beads 107
  C. Adhesion to Anti-Integrin Antibody- or ECM-Coated Dishes 107
  D. Reagents 108
V. Coimmunoprecipitation of Integrins and Caveolin-1 with Shc 109
  A. Protocol 109
  B. Reagents 109
VI. Caveolin-1 Associated c-Fyn Tyrosine Kinase Assay 110
  A. Protocol 110
  B. Reagents 110
VII. Coimmunoprecipitation of Fyn with Shc Following Integrin Ligation 111
  A. Protocol 111
  B. Reagents 111
VIII. Tyrosine Phosphorylation of Shc and Recruitment of Grb-2 112
  A. Protocol 112
  B. Reagents 112

0-8493-3385-7/99/$0.00+$.50

IX. Association of Shc GST-Fusion Proteins with Tyrosine Phosphorylated β4 113
A. Protocol 113
B. Reagents 113
Acknowledgments 114
References 114

I. Introduction

The binding of integrins to extracellular matrix (ECM) proteins promotes their aggregation on the plane of the plasma membrane and their interaction with cytoskeletal elements as well as signaling molecules. These events lead to the assembly of focal adhesions and the activation of signaling pathways which regulate gene expression.[1,2] Ligation of all β1 and αv-containing integrins results in the recruitment and activation of Focal Adhesion Kinase (FAK). This process appears to involve the interaction of FAK with talin, since it requires the portion of β1 cytoplasmic domain which binds talin. Upon activation and autophosphorylation, FAK combines with a Src-family kinase, c-Src or c-Fyn, which phosphorylates its C-terminal domain promoting the recruitment of other signaling molecules. The signaling pathways activated by the FAK-Src family kinase complex appear to play a role in regulating the assembly/disassembly of focal adhesions during cell migration, regulating cell proliferation, and protecting cells from programmed cell death.[3]

We have recently provided evidence that ligation of α1β1, α5β1, αvβ3, and α6β4, but not α2β1, α3β1, and α6β1, causes recruitment and tyrosine phosphorylation of Shc.[4,5] Shc is an adaptor protein containing a Src Homology-2 (SH2), a phosphotyrosine binding (PTB), and a collagen homology domain. Shc is known to link various tyrosine kinases to Ras. Upon recruitment to activated tyrosine phosphorylated signal transducers via the SH2 and/or PTB domain, Shc is phosphorylated on tyrosine and binds the Grb2/mSOS complex. This process results in the recruitment of mSOS to the plasma membrane and consequent activation of Ras. We have observed that the α6β4 integrin is associated with a cytoplasmic tyrosine kinase that phosphorylates the cytoplasmic domain of β4. Since the SH2 and the PTB domain of Shc can interact directly *in vitro* with distinct tyrosine phosphorylated residues on the β4 tail, it is likely that *in vivo* the recruitment of Shc to activated α6β4 is mediated by the cooperative binding of both Shc domains to tyrosine phosphorylated β4 tail.[4] By contrast, the recruitment of Shc to activated β1 and αv integrins is indirect and mediated by the interaction of the integrin α subunit with caveolin-1.[6] Recent studies have indicated that caveolin-1, a protein implicated in the biogenesis of caveolae, also functions as a membrane adaptor to link integrins to the tyrosine kinase Fyn. Upon integrin-mediated activation, Fyn undergoes a conformational transition that allows its SH3 domain to interact with Shc.[6] Upon recruitment to activated integrins, Shc becomes phosphorylated on tyrosine-317 and binds to the Grb2/mSOS complex. These events result in the activation of the Ras-Extracellular signal Regulated Kinase (ERK) signaling pathway and appear to be required to promote progression through the G1 phase of the cell cycle in response to soluble mitogens. Accordingly, ligation of integrins which are not capable of recruiting Shc results in exit from the cell cycle and apoptotic death even in the presence of otherwise mitogenic concentrations of soluble growth factors.[5-7] Taken together, our observations indicate that integrins activate not only common (FAK) but also subgroup-specific (Shc) signaling pathways.

II. Experimental Overview

Here, we describe methods used in our laboratory to examine integrin-mediated Shc signaling. Specifically, we provide methods to analyze the constitutive association of integrins with caveolin-1 and that of caveolin-1 with Fyn, as well as methods to examine the activation of the fraction of Fyn associated with caveolin-1, the recruitment and tyrosine phosphorylation of Shc, and the association of Shc with Grb2 in response to integrin ligation. The recruitment of Shc to activated integrins can be detected by immunoprecipitating the integrins and probing the resulting samples by immunoblotting with anti-Shc antibodies. The tyrosine phosphorylation of Shc and the association of Shc with Grb2 can be studied simultaneously by immunoprecipitating Shc and, after transfer to nitrocellulose, probing the top portion of the blot with anti-phosphotyrosine (anti-P-Tyr) and the bottom portion with anti-Grb2 antibodies. Finally, we describe a GST-pull down assay which can be used to detect the binding of tyrosine phosphorylated β4 to the SH2 or the PTB domain of Shc.

Integrin-mediated signaling can be activated: 1) by plating the cells onto dishes coated with ECM components or anti-integrin antibodies; 2) by incubating the cells in suspension with polystyrene beads coated with anti-integrin antibodies; or 3) by incubating the cells in suspension with soluble anti-integrin antibodies followed by appropriate secondary antibodies. The use of antibodies offers the advantage of activating a specific integrin heterodimer, thus allowing the examination of the signaling pathways activated by that integrin. This specificity can be more difficult to obtain with ECM proteins, since the same ECM protein is often recognized by several integrins.[4-7] Several affinity-purified monoclonal antibodies to individual integrin subunits are now commercially available, and the American Type Culture Collection (ATCC, Rockville, MD) provides various hybridomas producing anti-integrin monoclonal antibodies. The majority of anti-integrin antibodies are able to cluster integrins at the cell surface, whether or not they bind to the ligand binding region and inhibit adhesion. Since integrin aggregation is sufficient to stimulate signaling events such as the activation of FAK and the recruitment and tyrosine phosphorylation of Shc, most anti-integrin antibodies are suitable for studies of integrin signaling.[4-9]

In our studies on integrin-mediated Shc signaling we have employed untransformed (and often primary, nonimmortalized) fibroblasts, endothelial cells, and epithelial cells. There is reason to believe that integrin-mediated Shc signaling may be defective in neoplastic cells. In fact, caveolin-1 and at least two of the Shc-linked integrins, $\alpha1\beta1$ and $\alpha5\beta1$, are often downregulated in oncogenically transformed cells.[10, 11] Thus, unless one desires to compare integrin-mediated signaling in normal and neoplastic cells, we recommend the use of primary cells.

The selective activation of the fraction of Fyn associated with caveolin-1, the association of Fyn with Shc, the tyrosine phosphorylation of Shc, and the association of Shc with Grb2 are readily examined after ECM ligand- or antibody-mediated ligation of integrins. In contrast, we have so far been unable to coimmunoprecipitate integrins with Shc upon plating of the cells on extracellular matrix ligand. This is probably due to the fact that integrins are ligated in an asynchronous manner during physiological adhesion to the extracellular matrix. In addition, it is possible that a fraction of ligated integrins becomes insoluble in the non-ionic detergent used to preserve the association of integrins with Shc. Finally, the coimmunoprecipitation of $\beta1$ and αv integrins with Shc is by its very nature a complicated experiment because it requires the successful extraction and preservation of a quaternary complex, which includes the relevant integrin, caveolin-1, Fyn, and Shc. Accordingly, it is easier to coimmunoprecipitate caveolin-1 and Fyn with Shc than it is to coisolate integrins with Shc. The association of integrins with caveolin-1 and that of caveolin-1 with Fyn are constitutive, and rather easily detectable.

III. Coimmunoprecipitation of Integrins with Caveolin-1 and Caveolin-1 with Fyn

A. Protocol

1. Use two 150 mm diameter dishes of cells for each immunoprecipitation. Wash with cold PBS and lyse in 1 ml complete cell lysis buffer. Centrifuge lysate at 13,000 *g* for 20 min at 4°C. Protein assays can be used to verify that lysate contains approximately 4 mg proteins.
2. To coimmunoprecipitate integrins with caveolin-1, pre-clear lysates with 100 µl packed-prewashed* agarose mouse IgG beads for 1 h at 4°C. For coimmunoprecipitation of caveolin with Fyn, pre-clear lysates with 100 µl packed-prewashed* agarose rabbit IgG beads for 1 h at 4°C. After preclearing, spin at 10,000 *g* for 2 min and transfer the supernatant to a new tube.
3. To coimmunoprecipitate integrins with caveolin-1, add 10 µg of purified monoclonal anti-integrin or control anti-MHC antibody and 25 µl packed-prewashed* sepharose protein-G beads. To coimmunoprecipitate caveolin-1 and Fyn, add 5 µg of affinity purified anti-caveolin-1 (C13630) rabbit antibodies or control normal rabbit IgG and 25 µl of packed-prewashed* Protein-A beads. Incubate at 4°C for 2 to 3 h.
4. Wash the beads five times with cold lysis buffer.

* Sepharose Protein-A/G and agarose rabbit/mouse IgG beads are provided as 40 to 50% slurry. Wash appropriate volume of agarose/sepharose slurry (beads) with lysis buffer three times before use. The beads will henceforth be referred to as packed-prewashed Protein-A/G or agarose rabbit/mouse IgG beads.

5. Resolve samples on a 10% SDS-PAGE. Run 50 to 75 μg of total lysate as a positive control. Transfer proteins to nitrocellulose by electroblotting.
6. Saturate the nitrocellulose membrane in blocking buffer A for 1 h at room temperature and then rinse briefly twice in blotting wash buffer.
7. Incubate the blot containing the integrin immunoprecipitates with anti-caveolin-1 antibody (N-20) and the blot containing the caveolin-1 immunoprecipitates and relative negative and positive controls with anti-Fyn antibody (Fyn-3G) for 2 h at room temperature.
8. Wash the membranes with wash buffer for 30 min at room temperature, changing the solution every 10 min.
9. Incubate the membranes with HRP-conjugated Protein-A diluted 1:3000 in secondary reagent dilution buffer for 45 min at room temperature.
10. Wash membranes with wash buffer 30 min at room temperature, changing the solution every 10 min.
11. Remove excess wash buffer and subject to ECL following standard protocol.
12. Expose to X-ray film for 1 min. Adjust exposure time according to the signal intensity. The intensity of chemiluminescence decreases over time. No signal is generally detected after 30 min.

B. Reagents

1. Stock solutions of protease and phosphatase inhibitor: Prepare 100 mM Phenylmethanesulfonyl fluoride (PMSF) stock solution in isopropanol. Prepare 1 mg/ml Aprotinin, 1 mg/ml Leupeptin, 0.5 M NaF, and 0.5 M Na_3VO_4 in double distilled water (ddw). Prepare Pepstatin-A 1 mg/ml in methanol. Store at –20°C.
2. Lysis buffer: 50 mM HEPES, pH 7.4, 150 mM NaCl, 1% Triton X-100, 1 mM $CaCl_2$, 1 mM $MgCl_2$, and 10% glycerol. Just before use, add 1 mM PMSF, 10 μg/ml Aprotinin, 10 μg/ml Leupeptin, 10 μg/ml of Pepstatin A, 2 mM Na_3VO_4, 10 mM $Na_4P_2O_7$, and 25 mM NaF.
3. Mouse IgG Agarose (Sigma) and Protein-G Sepharose (Pharmacia Biotech Inc., Piscataway, NJ).
4. Rabbit IgG Agarose (Sigma, Chemical Co., St. Louis, MO) and Protein A Sepharose (Pharmacia).
5. Purified anti-integrin and control anti-MHC Class I monoclonal antibodies (Sigma).
6. Affinity purified rabbit anti-caveolin-1 (Cat. #C13630; Transduction Labs, Lexington, KY) and anti-Fyn polyclonal antibodies.
7. Control normal rabbit IgGs (Sigma).
8. Reagents for SDS-Polyacrylamide Gel Electrophoresis (PAGE) and electroblotting to nitrocellulose membranes.
9. Blocking buffer A: 5.0% BSA in 10 mM Tris, pH 7.5, 100 mM NaCl, pH 7.4 (TBS). Adjust pH to 7.4 with NaOH after dissolving BSA.
10. Wash buffer: 10 mM Tris pH 7.5, 100 mM sodium chloride, 0.1% Tween-20 (Sigma).
11. Affinity purified goat antibodies to a synthetic peptide modeled after amino acid residues 28 to 48 of human Fyn (Fyn-3G) can be purchased from Santa Cruz

Biotechnology (Santa Cruz, CA; Cat. #sc-16G). These antibodies recognize human, mouse, rat, and chicken c-Fyn.

12. Affinity purified antibodies to a synthetic peptide modeled after amino acid residues 2 to 20 of caveolin-1 (N-20) can be purchased from Santa Cruz Biotechnology (Cat. #894). These antibodies immunoblot human, mouse, and rat caveolin-1.
13. Horseradish Peroxidase (HRP)-conjugated Protein A and anti-goat IgGs (Amersham, Arlington Heights, IL).
14. Secondary reagent dilution buffer: 5% nonfat dry milk in Tris-buffered saline (TBS) containing 0.1% Tween-20 (Sigma), pH 7.4.
15. Enhanced chemiluminescence (ECL) reagents (Pierce, Rockford, IL).

IV. Integrin Ligation

Culture cells according to standard methods, and deprive cells of growth factors when they reach 80% confluency. Starve NIH-3T3 fibroblasts for 18 h, human umbilical vein endothelial cells, human HaCat keratinocytes, and Fisher Rat Thyroid (FRT) for 36 h. Avoid using confluent cells, because it is difficult to obtain single cell suspension following EDTA treatment. Do not use trypsin, since some integrins are sensitive to trypsin treatment, but detach the cells with 10 mM EDTA in PBS. Determine cell surface expression of integrins by FACS analysis, or immunoprecipitation of surface labeled cells. It should be noted here that it may be preferable to examine Shc signaling after crosslinking with anti-β1 integrin antibodies only in cells which express the Shc-linked integrins α1β1 and α5β1 at levels higher than those of other β1 integrins.

A. Crosslinking of Integrins with Soluble Antibodies

1. Detach cells from culture dish with EDTA solution. Wash and resuspend at 2×10^7/150 µl density in serum-free DMEM supplemented with 0.1% heat-inactivated BSA.
2. Keep cells in suspension at room temperature for 30 min. This step is required to fully deactivate integrin signaling.
3. Add 10 µg of anti-integrin monoclonal antibody, or control anti-MHC class I monoclonal antibody to 2×10^7 cells and incubate on ice for 30 min; mix the cell suspension gently from time to time.
4. Wash twice with cold DMEM by centrifugation at 1,000 *g* for 5 min, to remove the unbound antibodies.
5. Resuspend the cell pellet in 150 µl of DMEM containing 5 µg of affinity purified rabbit anti-mouse IgGs.
6. Incubate at 37°C for desired periods of time (e.g., 2, 5, 10, 15, and 30 min). Mix gently every few minutes.
7. At the end of the incubation, add 1 ml of cold PBS containing 0.5 mM Na_3VO_4 and centrifuge at 1,000 *g* for 5 min to remove the unbound antibodies.

8. Lyse the cell pellet.

B. Ligation of Integrins with Antibody-Coated Polystyrene Beads

1. Coating of the beads: the polystyrene beads are usually provided as a 8% slurry. Pipette the volume needed to obtain 50 μl packed beads per sample. Wash the beads three times with 1 ml of MES buffer pH 5.5 by centrifugation at 10,000 *g* for 2 min at room temperature. Resuspend the beads in 300 μl of MES buffer pH 5.5, and add 20 μg of purified anti-integrin monoclonal antibody or anti-MHC class I monoclonal antibody for control beads. Rock the tubes for 1 h at room temperature. At the end of incubation, spin the tubes at 10,000 *g* for 2 min at room temperature. Wash the beads four times with 1 ml of PBS (without Ca^{++} and Mg^{++}), and resuspend them in 150 μl of DMEM.
2. Prepare a cell suspension as described in Section IV.A. Keep cells in suspension at room temperature for 30 min. This step is required to fully deactivate integrin signaling.
3. Gently mix 150 μl of cell suspension with 150 μl of anti-integrin antibody-coated beads in an Eppendorf tube.
4. Incubate at 37°C for desired periods of time (e.g., 2, 5, 10, 15, and 30 min). Mix gently every few minutes.
5. At the end of the incubation, add 1 ml of cold PBS containing 0.5 mM Na_3VO_4, and centrifuge at 1,000 *g* for 5 min to remove the unbound antibodies.
6. Lyse the cell pellet as required.

C. Adhesion to Anti-Integrin Antibody- or ECM-Coated Dishes

1. Coating of the dishes: dilute purified ECM proteins in PBS to a concentration of 10 μg/ml (poly-L-lysine to 3 μg/ml as a control); for coating with antibodies, dilute the affinity purified rabbit anti-mouse IgGs in PBS to a concentration of 10 μg/ml. Incubate 10 cm tissue culture dishes with 5 ml of the ECM protein or antibody solution for 2 h at room temperature. Wash twice with PBS. Saturate with 0.1% BSA in PBS for 1 h at room temperature. Incubate the anti-mouse IgG-coated dishes with anti-integrin monoclonal antibody (20 μg/ml in PBS) for 2 h at room temperature. Wash all plates three times with DMEM and maintain wet until use.
2. Preparation of Laminin-5 coated dishes: Laminin 5-enriched matrices are obtained from the murine cell line RAC-11P/SD.[12] Culture cells to confluency, wash them three times with PBS, and then incubate overnight at 4°C in PBS containing 20 mM EDTA, 10 μg/ml leupeptin, 1 mM phenylmethanesulfonyl fluoride (PMSF), and 10 μg/ml soybean trypsin inhibitor. Detach the cells from the underlying matrix by forceful pipetting, removing them as a single continuous layer. Saturate the plates with 0.1% BSA in PBS and either use immediately or store at –20°C in 10% DMSO in PBS.

3. Detach cells from culture dish with EDTA solution. Wash and resuspend to a concentration of 10^7 cells in 6 ml of serum-free DMEM supplemented with 0.1% heat-inactivated BSA.
4. Keep cells in suspension at room temperature for 30 min.
5. Plate the cells on ECM- or anti-integrin antibody-coated dishes and incubate at 37°C in a 5% CO_2 incubator for desired periods of time (e.g., 15, 30, 45, 60, 90, and 120 min). Plate the cells on poly-L-lysine or anti-MHC class I antibody-coated dishes as a control. The number of cells to be plated on a 10 cm dish may vary according to the cell type (10^7 cells is optimal for fibroblasts). The ideal number of cells has to be determined considering that the cells should barely touch each other after spreading.
6. At the end of each incubation, wash the cells gently with cold PBS in order to remove unattached cells and lyse the attached cells.

D. Reagents

1. The ECM proteins fibronectin, collagen I, collagen IV, laminin 1, laminin 4, and vitronectin can be purchased from Gibco-BRL (Gaithersburgh, MD), Upstate Biotechnology Inc. (Lake Placid, NY), and Collaborative Research (Lexington, MA). The control polypeptide poly-L-lysine is purchased from Sigma (Cat #P9155). Dissolve ECMs and poly-L-lysine in PBS. Quality and quantity is judged by running aliquots of these samples on a SDS-PAGE. Store in aliquots at –80°C. Preparation of Laminin 5-enriched matrices is described above (Step C.2).
2. Purified anti-human β4 integrin (3E1), α1 integrin (TS 2/7), α2 integrin (P1E6), α3 integrin (P1B5), and α5 integrin (P1D6) monoclonal antibodies are bought from Gibco-BRL; anti-mouse α5 integrin (5H10-27) monoclonal antibody is obtained from Pharmingen (San Diego, CA); anti-mouse and human α6 integrin (GoH3) monoclonal antibody is purchased from Immunotech (Westbrook, ME); anti-human β1 integrin (4B4) monoclonal antibody is bought from Coulter Inc. (Hialeah, FL); control anti-MHC Class I monoclonal antibody from Sigma.
3. Polystyrene beads, mean diameter approximately 2.7 μm (Interfacial Dynamics Corporation, Portland, OR).
4. 100 mM methan-sulphonic acid ethyl ester (MES), pH 5.5.
5. Affinity purified rabbit anti-mouse IgGs (Amersham).
6. Dulbecco's Modified Eagle Medium (without serum).
7. PBS without Ca^{++} and Mg^{++}.
8. 0.5 M sodium orthovanadate in ddw.
9. 0.1% BSA in PBS, sterile filtered through 0.22 μm.
10. 10 mM EDTA in PBS, pH 7.4, autoclaved.

V. Coimmunoprecipitation of Integrins and Caveolin-1 with Shc

A. Protocol

1. Perform integrin ligation for various periods of time (as described in Section IV. A or B), add 1 ml of complete lysis buffer per sample and incubate on ice for 30 min. Clarify the extracts by centrifugation at 13,000 *g* for 30 min at 4°C and measure protein concentration. Use 4.0 mg of total proteins for each immunoprecipitation.
2. Add 10 μg of purified anti-integrin monoclonal antibody and 50 μl packed Protein-G Sepharose beads, and rotate the tubes for 2 to 3 h at 4°C. At the end of incubation, pellet the beads by centrifugation at 13,000 *g* for two min at 4°C.
3. Wash the beads with cell lysis buffer five times, boil the immunocomplexes in reducing sample buffer, and resolve on a 10% SDS-PAGE; load prestained molecular weight markers. Transfer proteins onto nitrocellulose membrane and block with blocking reagent.
4. Cut the membrane approximately at the level of the 80 kDa molecular weight marker, and at the level of the 40 kDa molecular weight marker.
5. Incubate the blots with affinity purified rabbit anti-β1 integrin (top portion of the blot), anti-Shc (middle portion), and anti-caveolin-1 (bottom portion) polyclonal antibodies for 1 h at room temperature. (Recommended dilution of antibodies: prepare 1:500 dilution of anti-β1 integrin, 0.5 to 1.0 μg/ml of anti-Shc, and 0.5 to 1.0 μg/ml of anti-caveolin-1 rabbit polyclonal antibodies.)
6. Wash the blot with washing buffer for 30 min, incubate with HRP-conjugated Protein A, and do the ECL reaction as described in III.A, Steps 8 through 12.

B. Reagents

1. Lysis buffer: 50 mM HEPES, pH 7.4, 150 mM NaCl, 1% Triton X-100 (Sigma), 1 mM $CaCl_2$, 1 mM $MgCl_2$, 10% glycerol (Gibco-BRL). Add protease and phosphatase inhibitors before use.
2. Items 4, 6, and 9 through 11 from Section III.B.
3. Purified monoclonal antibody to the N-terminal cytoplasmic tail of caveolin-1 (CO60) from Transduction Laboratories.
4. Affinity purified rabbit anti-Shc antibodies (Upstate Biotechnology).
5. HRP-conjugated Protein A (Amersham).
6. ECL reagents.

VI. Caveolin-1 Associated c-Fyn Tyrosine Kinase Assay

A. Protocol

1. Carry out integrin ligation for various periods of time as described in Section IV.A or B, add 1 ml of modified RIPA buffer containing proper inhibitors and lyse for 30 min on ice. Sonicate briefly and centrifuge at 13,000 *g* for 30 min. Measure protein concentration. Use 2 mg protein for each immunoprecipitation.
2. Preclear the lysates with 100 µl packed-prewashed agarose rabbit IgG beads for 1 h at 4°C, as described in Section V.B.
3. Add 5 µg of affinity purified rabbit anti-Fyn or anti-caveolin-1 antibodies and 30 µl of packed-prewashed Sepharose Protein-A beads and rotate for 3 h at 4°C.
4. Pellet the beads, wash them twice with cold Triton wash buffer and twice with kinase wash buffer.
5. Resuspend the beads in 30 µl of kinase assay buffer containing 20 µCi of [γ-^{32}P] ATP.*
6. Shake tightly sealed Eppendorf tubes containing radioactive ATP in a vortex shaker for 15 min at room temperature.
7. At the end of the kinase assay, spin the tubes at low speed (1000 RPM) for 10 sec to pellet the beads, remove the radioactive supernatant and discard it carefully in a ^{32}P radioactive disposal container. (This step will significantly reduce the amount of free unbound ^{32}P ATP loaded onto the gel, thus eliminating unnecessary background.)
8. Resolve immunocomplexes in a 10% SDS-PAGE.
9. Fix gel in 8% acetic acid and 15% methanol for 1 h.
10. Incubate the gel in 1 M KOH at 60°C for 2 h to remove radioactive phosphates bound to serine and threonine amino acid residues. Tyrosine-bound phosphate groups are largely alkali resistant.
11. Repeat Step 9.
12. Dry gel at 80°C for 2 h.
13. Expose it to X-ray film for 15 min at –80°C. Develop the film and adjust the exposure time to obtain the desired intensity of the bands.

B. Reagents

1. Items 2, 5, 7 through 9, and 12 from Section III.B.
2. RIPA buffer: 50 mM HEPES, pH 7.4, 150 mM NaCl, 2.5 mM $MgCl_2$, 1 mM EGTA, 1% Triton X-100 (Sigma), 0.5% Sodium Deoxycholate (Sigma), 0.1% Sodium Dodecyl Sulphate (SDS) (Sigma), 10% glycerol (Gibco-BRL). Store at 4°C. Add protease and phosphatase inhibitors to the lysis buffer before use.

---

* While working with radioactive isotopes, proper shielding and disposal methods should be employed. Wear gloves, apron, and use other protective devices.

3. Affinity purified rabbit antibodies to Fyn can be obtained from Santa Cruz Biotechnology, Inc. (Fyn-3; Cat. #SC-16).
4. Triton wash buffer: 25 mM HEPES, pH 7.4, 150 mM NaCl, 1.5 mM $MgCl_2$, 1 mM EGTA, 0.1% Triton X-100 (Sigma), and 10% glycerol (Gibco-BRL).
5. Kinase wash buffer: 25 mM HEPES, pH 7.4, 100 mM NaCl, 10 mM $MgCl_2$.
6. Kinase assay buffer: 25 mM HEPES, pH 7.5, 100 mM NaCl, 10 mM $MgCl_2$, 20 μM p-nitrophenyl phosphate (PNPP) and 10 μM ATP. Store in aliquots at –20°C.
7. [γ-^{32}P]-ATP (at least 4500 Ci/mmol, ICN Pharmaceuticals Inc., Costa Mesa).

VII. Coimmunoprecipitation of Fyn with Shc Following Integrin Ligation

A. Protocol

1. Perform integrin ligation for various periods of time as described in Section IV, add 1 ml of complete lysis buffer per sample, and incubate on ice for 30 min. Clarify the extracts by centrifugation at 13,000 *g* for 30 min at 4°C and measure protein concentration. Use 4.0 mg of total proteins for each immunoprecipitation.
2. Preclear lysates using washed agarose mouse IgG for 1 h at 4°C, as described in Section III.A.
3. Transfer the supernatants to new tubes and add 10 μg of purified monoclonal anti-Fyn (Mab 15) antibody or control anti-MHC (W6.32) antibody together with 50 μl packed prewashed Protein-G Sepharose beads; rotate the tubes for 3 to 4 h at 4°C. At the end of the incubation, pellet the beads by centrifugation at 13,000 *g* for 5 min at 4°C.
4. Wash the beads with cell lysis buffer five times, boil the immunocomplexes in reducing sample buffer, and resolve on a 10% SDS-PAGE. Transfer proteins onto nitrocellulose membrane and block with blocking reagent.
5. Incubate the blot with anti-Shc polyclonal antibody.
6. Wash the blot with washing buffer for 30 min, changing the wash solution every 10 min.
7. Incubate the blots with HRP-conjugated Protein A diluted in secondary antibody dilution buffer (1:3000) for 1 h at room temperature.
8. Wash the blot for 45 min to 1 h with wash buffer, then perform the ECL reaction.

B. Reagents

1. Items 1 through 3, 9, and 10 from Section III.B.
2. Purified monoclonal antibody to•amino acids 85 to 206 of human Fyn (Mab15) (Santa Cruz Biotech, Cat. #sc-434); anti-Shc polyclonal antibodies (Santa Cruz Biotechnology).
3. Reagents required for 10% SDS-PAGE, nitrocellulose membrane, HRP-conjugated protein A, and ECL reagent.

VIII. Tyrosine Phosphorylation of Shc and Recruitment of Grb-2

A. Protocol

1. Perform integrin ligation as described in Section IV.A, B, or C, wash cells and lyse in 1 ml of complete lysis buffer. Estimate the amount of protein, and use 2 mg of total proteins for each immunoprecipitation.
2. Preclear lysates using washed agarose mouse IgG for 1 h at 4°C, as described in Section III.A.
3. Add to the supernatants 5 μg of anti-Shc (PG-797) or control anti-MHC (W6.32) monoclonal antibodies together with 25 μl of packed Protein-G Sepharose, and carry out the immunoprecipitation at 4°C for 3 h.
4. Wash the immunoprecipitates five times with lysis buffer, resolve the samples on a 10% SDS-PAGE; load prestained molecular weight markers.
5. Transfer the proteins onto a nitrocellulose membrane. Cut the blot approximately at the level of the 39 kDa molecular weight marker.
6. Incubate the top portion of the membrane with blocking buffer B, and the bottom portion with blocking buffer A for 1 h at room temperature. Rinse twice with wash buffer.
7. Incubate the top portion of the blot with 50 to 100 ng/ml HRP-conjugated recombinant anti-P-Tyr monoclonal antibody RC-20.* for 30 min at 37°C. Incubate the bottom portion with 0.5 μg/ml affinity purified rabbit anti-Grb2 antibodies in TBS-1% BSA for 1 h at room temperature. Wash the blots with wash buffer for 30 min at room temperature.
8. Carry out ECL reaction for the top portion of the blot. Incubate the bottom blot with HRP-conjugated protein A (1:3000) in secondary dilution reagent for 45 min, wash, and perform the ECL reaction. Tyrosine phosphorylated Shc is predominantly the p52 isoform.** Grb-2 runs at 25 Kd molecular size.

B. Reagents

1. Items 1 through 3, 9, and 10 from Section III.B.
2. Monoclonal antibody to the SH2 domain of human Shc (PG-797) (Santa Cruz Biotechnology, Cat. #sc-967).

---

* The HRP-conjugated recombinant PY20 monoclonal antibody RC-20 can detect very small quantities of phosphotyrosine, and outperforms all the other anti-P-Tyr reagents that we used. Since it is directly conjugated to HRP, this reagent should not be reused.

** The Shc protein isoforms migrate at 46, 52, and 66 kD molecular sizes. The p46 and p52 Shc proteins are products of two alternate start sites, while p66 originates from a distinct mRNA generated by alternative splicing. Hemopoietic cells generally do not express p66. The major isoform recruited to the activated integrins and thereby phosphorylated on tyrosine is p52.

3. Blocking buffer B: 1% BSA in TBS+0.1% Tween-20, pH 7.5.
4. HRP-conjugated recombinant anti-phosphotyrosine (RC-20) monoclonal antibody (Transduction Laboratories).
5. Affinity purified rabbit anti-Grb2 antibodies (Cat. #sc-255) (Santa Cruz Biotechnology).

IX. Association of Shc GST-Fusion Proteins with Tyrosine Phosphorylated β4

A. Protocol

1. Perform α6β4 ligation by one of the methods described in Section IV.A. or IV.B. using the monoclonal antibody 3E1 (or W6.32 as a control) and lyse the cells in SDS buffer. Heat lysate to 80°C for 5 min, sonicate briefly, clarify by centrifuging at 13,000 *g* for 10 min, and dilute with 9 volumes of lysis buffer. Use 5 mg of total proteins for each pull-down assay.
2. Incubate glutathione-agarose beads carrying 10 to 50 μg of the GST-fusion protein containing the PTB or SH2 domain of Shc with the denatured lysate for 2 h at 4°C. Incubate control samples with glutathione beads carrying GST only.
3. Wash the beads five times with lysis buffer by centrifugation at 10,000 *g* for 2 min at 4°C.
4. Resolve the samples on a 7.5% SDS-PAGE gel and transfer to nitrocellulose.
5. Incubate the membrane in blocking buffer (5% milk, 1% BSA in TBS, pH 7.4) for 1 h at room temperature.
6. Blot with rabbit anti-β4 cytoplasmic peptide serum (1:500 in TBS, pH 7.4 containing 3% BSA) for 1 h at room temperature.
7. Wash the blot, incubate in secondary HRP-conjugated protein A (1:3000) reagent for 1 h, and do the ECL reaction.

B. Reagents

1. Items 1, 2 and 8 through 10 from Section III.B.
2. SDS buffer: 50 mM Tris, pH 7.4, 150 mM NaCl, 1% SDS. Store at room temperature.
3. Glutathione Agarose beads (Sigma).
4. Rabbit anti-β4 cytoplasmic peptide serum (Chemicon, Temecula, CA), anti-β4 integrin monoclonal antibody (3E1) and anti-MHC class I monoclonal antibody (W6.32).
5. Gst-SH2 and Gst-PTB fusion proteins of Shc are produced and purified according to established protocols.[13]
6. Items 13 through 15 from Section III.B.

Acknowledgments

Research in the author's laboratory is supported by grants from the National Institutes of Health, the Department of Defense Breast Cancer Program, and the American Heart Association.

References

1. Yamada, K. M. and Miyamoto, S., Integrin transmembrane signaling and cytoskeletal control, *Curr. Biol.*, 7, 681, 1995.
2. Giancotti, F.G., Integrin signaling: specificity and control of cell survival and cell cycle progression, *Curr. Opin. Cell Biol.*, 9, 691, 1997.
3. Parsons, J. T., Integrin-mediated signaling: regulation by protein tyrosine kinases and small GTP-binding proteins, *Curr. Opin. Cell Biol.*, 8, 146, 1996.
4. Mainiero, F., Pepe, A., Wary, K. K., Spinardi, L., Mohammadi, M., Schlessinger, J., and Giancotti, F. G., Signal transduction by the $\alpha6\beta4$ integrin: distinct $\beta4$ subunit sites mediate recruitment of Shc/Grb2 and association with the cytoskeleton of hemidesmosomes, *EMBO J.*, 14, 4470, 1995.
5. Wary, K. K., Mainiero, F., Isakoff, S. J., Marcantonio, E. E., and Giancotti, F. G., The adaptor protein Shc couples a class of integrins to the control of cell cycle progression, *Cell*, 87, 733, 1996.
6. Wary, K. K., Mariotti, A., Zurzolo, C., and Giancotti, F. G., A requirement for Caveolin-1 and associated tyrosine kinase Fyn in integrin signaling and anchorage-dependent cell growth, *Cell*, 94, 625, 1998.
7. Mainiero, F., Murgia, C., Wary, K. K., Curatola, A. M., Pepe, A., Blumemberg, M., Westwick, J. K., Der, C. J., and Giancotti, F. G., The coupling of $\alpha6\beta4$ integrin to Ras-MAP kinase pathways mediated by Shc controls keratinocyte proliferation, *EMBO J.*, 16, 2365, 1997.
8. Miyamoto, S., Akiyama, S. K., and Yamada, K. M., Synergistic roles for receptor occupancy and aggregation in integrin transmembrane function, *Science*, 267, 883, 1995.
9. Miyamoto, S., Teramoto, H., Coso, O. A., Gutkind, J. S., Burbelo, P. D., Akiyama, S. K., and Yamada, K. M., Integrin Function: Molecular hierarchies of cytoskeletal and signaling molecules, *J. Cell Biol.*, 131, 791, 1995.
10. Koleske, A. J., Baltimore, D., and Lisanti, M., Reduction of caveolin and caveolae in oncogenically transformed cells, *Proc. Natl. Acad. Sci. USA*, 92, 1381, 1995.
11. Plantefaber, L.C. and Hynes, R.O., Changes in integrin expression in oncogenically transformed cells, *Cell*, 56, 281, 1989.
12. Sonnenberg, A., de Melker, A. A., Martinez de Velasco, A. M., Janssen, H., Calafat, J., and Niessen, C. M., Formation of hemidesmosomes in cells of a transformed murine mammary tumor cell line and mechanisms involved in adherence of these cells to laminin and kalinin, *J. Cell Sci.*, 106, 1083, 1993.
13. Smith, S. B. and Johnson, K. S., Single-step purification of polypeptides expressed in Escherichia coli as fusions with glutathione X-transferase, *Gene*, 67, 31, 1988.

SECTION II

Late Events and Biological Functions of Integrin Signaling

Chapter 8

Methods for Study of Integrin Regulation of MAP Kinase Cascades

Rudy L. Juliano, Andrew E. Aplin, Alan Howe, Tsung H. Lin, and Qiming Chen

Contents

I. Introduction 117
II. Measurement of MAP Kinase by Immunoprecipitation and *in vitro* Kinase Assay 118
III. Use of Phosphospecific Antibodies to Analyze Activation of Signal Transduction Pathways 119
IV. Coupled Two-Stage Raf Immunoprecipitation and *in vitro* Kinase Assay 120
V. Ras GTP/GDP Loading Assay 122
VI. Coimmunoprecipitation of Raf and Ras or Rap 123
VII. Simultaneous Examination of Morphology and Signaling by Transient Cotransfections 124
References 126

I. Introduction

Our understanding of the role of integrins in biology has changed substantially, as it has become clear that integrins are involved not only in cell adhesion but also in signal transduction. There are really two modes of integrin-related signal transduction. The first is direct signaling where stimulation of integrins by extracellular matrix proteins or other ligands triggers intracellular signaling events. The second

0-8493-3385-7/99/$0.00+$.50

mode is integrin regulation of mitogen signaling, where integrin-mediated cell anchorage influences signaling pathways activated by growth factors. In both of these cases integrin signaling is intimately linked with activation of elements of the Ras-MAP kinase cascade.

The Ras-MAP kinase pathway usually starts with the activation of a tyrosine kinase (for example, a growth factor receptor). Adaptor proteins with SH2 domains, such as Grb2, can bind to the phosphate groups on tyrosine residues in the activated kinase. The SH3 domain of Grb2 bind to proline-rich domains of SOS, a guanine nucleotide exchange factor for Ras, thus bringing Ras into its active GTP-loaded form. On the other hand, inactivation is mediated by GTPase activating proteins (GAPs), that stimulate the intrinsic GTPase activity of Ras. Once activated, Ras has effects on cytoplasmic serine/threonine kinase cascades. The principal targets of the activated form of Ras seem to be the various isoforms of Raf kinases. Activated Raf-1, in turn, phosphorylates and activates a second kinase, MEK (aka MAP kinase kinase) which is an activator of the ERK sub-family of MAP kinases. MAP kinases (ERKs) can migrate from the cytosol into the nucleus after activation, thus finally delivering the signal into the nucleus by phosphorylating transcription factors involved in growth regulation. In studying integrin regulation of MAP kinase cascades, it is important to be able to address the level of activation of different elements of the cascade.

Our laboratory has explored several aspects of the linkage between integrin and the Ras-MAP kinase cascade, working mainly in mouse fibroblasts. Some of the assays that have worked well for us in this context are described in the paragraphs below. We have tried to provide detailed step-by-step protocols with indications for specific reagents. Naturally, some of these protocols may need modification if the assays are done on other types of cells or tissue samples. Utilization of most of these protocols to address experimental issues can be found in the articles from our laboratory mentioned in the bibliography.

II. Measurement of MAP Kinase by Immunoprecipitation and *in vitro* Kinase Assay

This is the basic assay for evaluating the activity of the ERK1, 2 forms of MAP kinase. Although there are now commercial antibodies available that detect activated forms of MAP kinases (see below), it is still very desirable to use a quantitative biochemical measurement of this type. The assay involves phosphorylation of myelin basic protein (MBP), a commonly used substrate for ERKs. This protocol is similar to ones we have reported previously.[1,2]

1. Wash cells twice with ice-cold PBS.
2. Add modified RIPA lysis buffer (0.5% deoxycholate, 1.0% NP40, 50 mM Tris, pH 7.4, 150 mM NaCl, 1 mM EDTA) containing phosphatase and protease inhibitors (for example, use PMSF, aprotinin, pepstatin, NaF and Na-orthovanadate). Use 0.5 to 1.0 ml lysis buffer for 10 cm plate (depending on cell density), and scale up or down accordingly.

3. Scrape lysate and transfer to microfuge tube.
4. Vortex (10 sec) and incubate on ice for 10 min.
5. Spin in a microfuge for 10 min at 4°C at 14,000 rpm.
6. Transfer supernatant lysate to new microfuge tube. Freeze at –80°C or proceed directly to IP. At this point you can pre-clear the lysate by incubating with 30 µl protein G sepharose for 30 min at 4°C with rocking, then centrifuging in the cold for 1 min at 14,000 rpm and transferring the supernatant lysate to a fresh tube. Depending on the cell type you are using, this may or may not be necessary — however, it never hurts to do it.
7. Determine protein concentration by your preferred method.
8. Typically, 150 to 200 µg protein is used per immunoprecipitation. You can use less (as little as 50) but the higher amount gives a good, easily detectable amount of activity.
9. To the lysate, add 1.5 µg of anti-ERK antibody (Santa Cruz, SC154 [Erk2] or SC93 [Erk1]), and incubate, with rocking, at 4°C for 1 h.
10. Add 20 to 30 µl of protein G sepharose (50% slurry) and incubate at 4°C for 30 min to 1 h.
11. Prepare kinase buffer (50 mM HEPES or Tris pH 7.4, 10 mM $MgCl_2$, 10 mM $MnCl_2$, 1 mM DTT) and reaction mixture, which is 15 µM ATP, 500 µg/ml MBP in kinase buffer.
12. Wash the immunoprecipitate three times with RIPA buffer and once with kinase buffer.
13. Add 40 µl of the reaction mixture, along with 15 µCi of γ-^{32}P-ATP (3000 Ci/mmol, NEN) to the washed beads, and mix very gently by tapping the tube (try to avoid getting beads stuck to the side of the tube, above the level of the reaction mixture).
14. Incubate at 25°C (room temperature) for 25 min, mixing occasionally.
15. Add sample buffer to 1X (usually add 20 µl of 3X), and boil samples for 3 to 5 min.
16. Run samples on 12% or 15% gel. Stain gel, dry, and expose to film or Phosphorimager screen. Usually between 5 and 10 µl is loaded, providing a strong signal overnight. The sample can also be stored, after boiling, at –20°C.

III. Use of Phosphospecific Antibodies to Analyze Activation of Signal Transduction Pathways

Activation of separate protein kinase cascades can now be analyzed rapidly and without the necessity for radioactivity, by utilizing phosphospecific antibodies. These reagents recognize phosphorylated and hence, theoretically, either active or inactive components of pathways. Phosphospecific antibodies are now available from several commercial sources to many of the signal proteins involved in growth control, development, and apoptotic pathways. We have utilized[3] an anti-active ERK antibody (Promega Corp.) that preferentially recognizes the dually phosphorylated pT-E-pY motif within the mitogen-activated proteins kinases, ERK1 and ERK2 (see Figure 1). This antibody can be used at relatively low dilutions (1:3000 to 1:5000) for Western blot analysis. Total levels (phosphorylated and non-phosphorylated) of ERK1 and ERK2 in samples should also be determined by Western blotting (using, for example, Santa Cruz antibody Sc094).

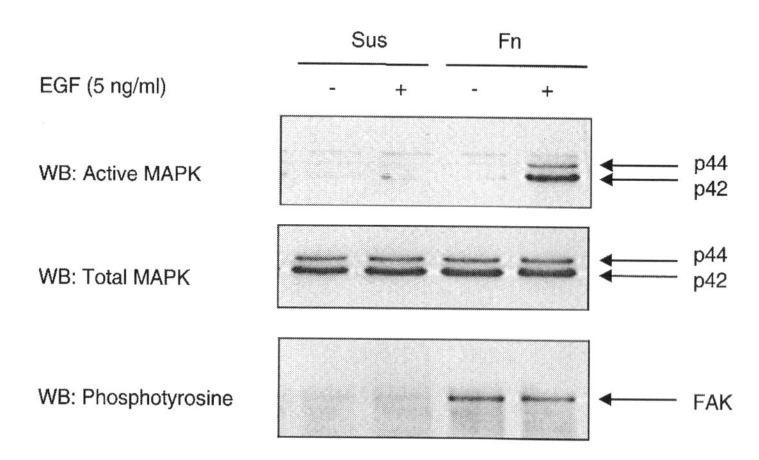

FIGURE 1
Analysis of integrin-mediated adhesion on EGF signaling to MAPK with anti-active MAPK antibody. Serum-starved NIH 3T3 cells were either maintained in suspension or allowed to adhere to fibronectin for 2 h, and then treated ± with 5 ng/ml EGF. Cells were washed and lysed in modified RIPA buffer. Samples (20 μg of lysate) were analyzed by Western blotting using anti-active MAPK/ERK antibody (upper panel), total MAPK/ERK antibody (middle panel), or antiphosphotyrosine (lower panel).

1. Separate protein samples (20 μg) by 10% SDS-PAGE.
2. Electroblot proteins onto PVDF membrane.
3. Block non-specific sites with 1% BSA overnight at 4°C.
4. Transfer membrane into primary antibody (1:3000 dilution of anti-active Erk antibody in 1% BSA/PBS/0.1% Tween) for 1 h at room temperature.
5. Wash three times for 20 min with PBS/0.1% Tween.
6. Transfer membrane into secondary antibody (1:5000, goat anti-rabbit-horse radish peroxidase conjugate) for 1 h at room temperature.
7. Wash three times for 20 min with PBS/0.1% Tween.
8. Wash once for 10 min with PBS.
9. Develop using ECL reagents (Amersham).

IV. Coupled Two-Stage Raf Immunoprecipitation and *in vitro* Kinase Assay

This assay involves immunoprecipitation of the Raf kinase and then a test of its activity by its ability to phosphorylate and activate MEK, which in turn phosphorylates and activates MAP kinase (ERK). The readout is the phosphorylation of myelin basic protein by the ERKs. The amount of protein required is generally 500 μg lysate per IP. This method is adapted from that developed by C. Marshall and colleagues[4,5] and has been used extensively by us.[2,6]* This particular protocol

* Howe & Juliano, submitted

uses commercially available recombinant MEK and MAPK. However, we have noticed that reagents from some suppliers often give very high background activity, i.e., the cpm in a reaction-only mix are very high, and increases with time. Thus, we would recommend using lab-prepared reagents; alternatively, reagents from UBI have often been satisfactory. Expressing and purifying your own batches of MEK and MAPK is highly recommended: it is far more economical, the results are generally much more reproducible, and quality control is in the hands of the investigator. Expression plasmids for GST- or His_6-tagged MEK and MAPK have been generously provided to the research community by several investigators.[4,7]

1. Cells are lysed with modified RIPA and Raf-1 is immunoprecipitated with anti-Raf-1 (Santa Cruz, c12), followed by protein G sepharose.
2. Spin down immunoconjugates after immunoprecipitation and aspirate excess.
3. Wash three times with 0.5 ml Buffer A and twice with 0.5 ml Buffer B. These are termed the 'IPs.'
4. Add 30 µl First Reaction Mix to each IP, to an empty tube labeled 'Rxn Only,' and to one containing 30 µl 2X sample buffer (SDS containing sample buffer for PAGE) labeled 'Rxn Stop.'
5. Mix gently and incubate at room temperature for 30 min or at 30° for 15 min, with occasional, gentle mixing.
6. During incubation, prepare MBP Reaction Mix and aliquot 35 µl into the appropriate number of microfuge tubes, then add 1 µl ^{32}P-ATP to each tube; also, if desired, add 10 µl of 2X sample buffer to another set of microfuge tubes.
7. Briefly spin all First Reaction tubes.
8. Remove 10 µl and add to MBP Reaction Mix tubes; remove another 10 µl and add to 2X SDS-sample buffer tubes ('FS') (this is to monitor background in the absence of substrate phsophorylation); incubate MBP Reaction tubes at room temperature for 15 min.
9. Remove and discard remaining supernatant (~10 µl) from IP tubes and add 30 to 50 µl 1X SDS sample buffer ('IP blot').
10. Add 45 µl 2X SDS-sample buffer to MBP reaction tubes ('MBP').
11. Boil all tubes for 5 min.
12. Run 10 µl 'FS' on 10% gels and blot with anti-MAPK (Santa Cruz Sc094) and active-MAPK antibodies (Promega).
13. Run 15 to 25 µl 'IP blot' on 7.5% gel and blot with anti-Raf antibody.
14. Run 20 to 30 µl 'MBP' on 15% gel, stain, dry, and expose to film or Phosphoimager plate.

Materials

Buffer A: 10 mM Tris 7.4, 5 mM EDTA, 50 mM NaF, 50 mM NaCl, 1% Triton X-100, 0.1% BSA

Buffer B: 50 mM Tris 7.4, 0.1 mM EGTA, 0.5 mM Na vanadate, 0.1% β mercaptoethanol

Buffer C: 60 mM Tris 7.4, 0.2 mM EGTA, 0.2% β mercaptoethanol, 0.6% Brij-35, 20 mM $MgCl_2$

First Reaction Mix (per 100 μl) = 50 μl 2X Buffer C

5 μl 1 mM ATP (Final = 50 μM)
1 μl GST-MEK (UBI) (Final = 2.5 μg/ml)
1.25 μl MAPK (Santa Cruz) (Final = ~20 μg/ml)
42.75 μl H_2O

MBP Reaction Mix (1 ml) = 50 μl Tris 7.4 (Final = 50 μM)

0.5 μl EGTA (Final = 0.1 mM)
12.5 μl $MgCl_2$ (Final = 12.5 μM)
66.7 μl MBP (Stock = 2.99 mg/ml; Final = ~0.2 mg/ml)
870 μl H_2O

V. Ras GTP/GDP Loading Assay

This is a tricky assay; however, it really is the only way to actually measure the amount of activated (GTP loaded) Ras. If done carefully[2,6] this assay can be quite revealing. It is important to use the right antibody in this assay.

1. Culture cells in 100 mm tissue culture dishes.
2. Serum starve cells in phosphate-free DMEM (8 ml) containing 2.5 mCi [P-32] orthophosphate overnight (10 to 12 h).
3. Stimulate cells (e.g., with growth factor or by integrin mediated adhesion).
4. Wash cells once with cold PBS.
5. Lyse cells with 1 ml of lysis buffer consisting of 12.5 μg of anti-Ras antibody (Oncogene Science, v-H-ras (Ab-1), clone Y13-259) for each 100 mm dish of cells.
6. Collect cell lysate, vortex, and incubate on ice for 15 min.
7. Spin to remove debris and incubate on ice for another 45 min.
8. Add 40 μl of goat anti-rat IgG bound to protein G agarose beads and incubate on a rotator at 4°C for 1 h.
9. Wash the immunocomplex three times with lysis buffer and a final wash with cold PBS.
10. Elute GTP/GDP by boiling the immunocomplex for 3 min with 20 μl of elution buffer.
11. Spot 3 μl of eluates on TLC plate (Cellulose PEI TLC plate, J. T. Baker).
12. Analyze the nucleotides by thin-layer chromatography and quantitate by Phosphorimaging.
13. Present results as 100X GTP/[GTP + (1.5 GDP)] (this corrects for the relative amounts of phosphate in GTP and GDP).

Materials

Lysis buffer: 50 mM HEPES, pH 7.5; 500 mM NaCl; 5 mM $MgCl_2$; 1% Triton X-100; 0.5% DOC; 0.05% SDS; 1 mM EGTA (before use, add to 10 ml, Aprotinin (1% V/V), Soybean-trypsin inhibitor (10 μg/ml), Leupeptin (10 μg/ml), Benzamidine 10 mM).

Elution buffer and TLC running buffer: 1 M KH_2PO_4, pH 3.4.

Immuno reagents: -H-ras (Ab-1), clone Y13-259, Calbiochem, Cat # OPO1L; Goat anti-rat IgG (Cappel); Protein G agarose beads (Calbiochem).

VI. Coimmunoprecipitation of Raf and Ras or Rap

The assay is very straightforward, the only drawback being the requirement for a rather large amount of total protein from which to immunoprecipitate. It offers a very direct means for documenting Ras/Raf and Rap/Raf interactions. The assay is adapted, with minor changes, from References 8 and 9.

1. For each experimental point, the equivalent of a confluent 150 mm dish (2.5 to 5 mg total protein) is required. For experiments using transient transfections, if your transfection efficiency and your levels of expression are high enough, you can probably get away with the equivalent of a 100 mm dish (1 to 2 mg total protein).
2. Wash the cells twice with ice-cold PBS.
3. Lyse the cells in NP-40 lysis buffer [50 mM HEPES (pH 7.5), 150 mM NaCl, 10% glycerol, 1.5 mM $MgCl_2$, 1 mM EGTA, 1.0% (v/v) NP-40, 100 mM NaF, 10 μM aprotinin, 10 μM leupeptin, 1 mM PMSF], using 1 ml lysis buffer per 5×10^7 cells, or 2 ml per 150 mm dish equivalent.
4. Scrape lysate into a 15 ml conical, screw-cap centrifuge tube and incubate on ice for 10 min.
5. Briefly (15 sec) sonicate each lysate once, at a setting of 7 on a Branson Sonifier.
6. Preclear the lysates with 1/40th the volume of protein A/G beads (50% slurry, washed twice in lysis buffer) for 1 h at 4°C with gentle mixing.
7. During the pre-clearing, prepare the antibody-bead conjugate (this can also be started the day before performing the immunoprecipitation, and let go overnight):
   a) For every 2 mg extract, use 1 μg of anti-Ras antibody (Y13-238, available from Oncogene Science or Santa Cruz Biotechnology). Make an amount of immunoconjugate sufficient for all of the immunoprecipitations, i.e., if you have 4 experimental points (i.e., 4X 2 mg extracts), prepare the immunoconjugate using 4 μg of antibody. The amount of antibody necessary depends on the total amount of Ras and Raf in the cells you're using, as well as the amount of activated Ras in your experimental point; thus, the amount should be titrated. However, the above ratio is a good general starting point.
   b) Combine the antibody with protein A/G beads (10 μl of a 50% slurry for every 1 μg of antibody) and add enough lysis buffer to allow efficient mixing (total volume in a microfuge tube should be no less than 400 μl).
   c) Incubate the mixture at 4°C for at least 1 h. Spin out beads.

d) Wash the beads three times with lysis buffer, resuspend in a suitable volume, and aliquot into the appropriate number of tubes.

8. Centrifuge the lysates (2000 × *g*, 4°C, 5 min) and carefully remove the supernatants and place into the tubes containing the immunoconjugate.
9. Incubate, with gentle mixing, 4 to 12 h at 4°C.
10. Wash the immunoprecipitates three times with 3 to 5 ml lysis buffer, vortexing gently after the buffer is added. Carefully aspirate the last wash, trying to remove as much buffer as possible without losing any beads.
11. Add 50 to 100 µl 1X SDS sample buffer (it is usually easiest to add the sample buffer, resuspend the beads by trituration, and transfer to a microfuge tube), and boil for 5 min.
12. If desired, centrifuge the tubes in a microfuge for 1 min at 14,000 rpm, then carefully remove the supernatant to a new tube. This will minimize the chance of loading beads into your wells.
13. Run 10 to 15% of the sample on a 15% gel (for Western blotting with anti-Ras antibody) and the rest on a 7.5% gel (for blotting with anti-Raf antibody).

Notes:

a. An excellent control for specificity in this experiment is the use of Y13-259, another anti-Ras antibody, which cannot efficiently precipitate Ras-Raf complexes.
b. When doing the anti-Ras Western, try to blot back with antibodies from a different species, as cross-reactivity of the secondary antibody with the light chain of the immunoprecipitating antibody may obscure the Ras signal.
c. This protocol also works for analysis of Raf-Rap interaction, using anti-Rap1 antibody from Santa Cruz Biotechnology.

VII. Simultaneous Examination of Morphology and Signaling by Transient Cotransfections

One of the major issues in integrin signaling is the role of the cytoskeleton. The assays below allow simultaneous visualization of effects on the cytoskeleton and detection of effects on the MAP kinase cascade. For example, one might test whether Rho family GTPases, which are known to affect actin filament organization, also affect the activity of MAP kinase. Active or dominant negative forms of other GTPases, or kinases, or other proteins, can just as easily be substituted for the Rho GTPases. Variations on this procedure could be used to look at the relationship between cytoskeletal effects and other signaling pathways.

To analyze the effects of Rho GTPases in NIH 3T3 fibroblasts, we have transiently cotransfected cells with constitutively active forms of these constructs along with cDNA encoding the green fluorescent protein (GFP; pGreen Lantern-1, GIBCO BRL). After allowing transfected cells to re-adhere to fibronectin for a given time the changes on the actin cytoskeleton and focal adhesion structures are analyzed with TRITC-phalloidin (Sigma) or anti-phosphotyrosine (Upstate Biotech. Incorp.), anti-vinculin (Sigma) or anti-paxillin (Transduction Labs.). Transfected cells can be readily identified in the fluorescein channel by their expression of GFP (see Figure 2). Effects on MAP kinase can be evaluated by cotransfection of an epitope-tagged MAPK construct.[10]

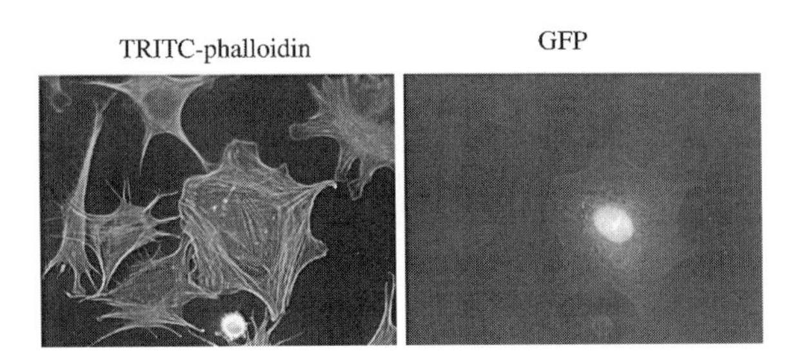

FIGURE 2

Transient cotransfection of cDNA construct of interest with GFP to observe morphological changes. NIH 3T3 cells were co-transfected with cDNA encoding the constitutively active form of the small GTPase, Rac1 and green fluorescent protein (GFP). Cells were serum-starved, detached, and allowed to reattach to fibronectin for 2 h. Cells were fixed and stained for actin (TRITC-phalloidin). GFP-positive cells were identified in the fluorescein channel. Cells expressing a constitutively-active form of Rac appear more spread and contain prominent actin-containing membrane ruffles.

1. Prepare coated coverslips by flaming coverslips (Corning Cover Glass No. 1 22 mm sq., Cat #12524B) and inserting in wells of 6-well dish. Coat overnight with fibronectin (10 μg/ml) at 4°C.
2. Wash coverslips three times with PBS, block for 1 h at 37°C with DMEM/2% BSA.
3. Cotransfect cells with plasmid of interest with pGreen Lantern1 (GIBCO BRL) encoding GFP at a plasmid ratio of 1:4.
4. After recovery period (40 h), seed cotransfected cells to give approximately 50% density; allow to adhere and spread for 2 to 3 h.
5. After the adherence period, wash cells twice with cold PBS, fix for 10 min at room temperature in 3.7% formaldehyde/PBS in wells.
6. Wash three times in PBS (5 min washes each).
7. Permeabilize for 5 min at room temperature with 0.5% Triton X-100 in PBS.
8. Wash three times in PBS.
9. Block in 2% BSA/PBS for 1 h at room temperature.
10. Remove coverslip and incubate upside down on primary antibody drop (50 to 75 μl) on level parafilm surface in a humidified chamber either overnight at 4°C or 1 h at room temperature. Antibody diluted in 2% BSA/PBS (for anti-PY 4G10 (UBI) use 1:100 dilution).
11. Remove and place coverslip face up in 6-well dish. Wash three times for 10 min with PBS.

12. Secondary antibody incubation (1:100 dilution in 2% BSA/PBS) in humidified chamber 1 h at room temperature. Use a TRITC-labeled secondary, either goat anti-mouse TRITC (Sigma T7657) or goat anti-rabbit TRITC (Sigma T6778).
13. Wash twice for 10 min with PBS, once with water.
14. Drain off excess water, mount in Permafluor (Thomas), and visualize under microscope.

The alterations in the actin cytoskeleton and focal adhesion complexes induced by Rho GTPases, as well as the ability of cells to signal, can be analyzed under the same conditions. NIH 3T3 cells are cotransfected with the construct of interest and cDNA encoding a hemagglutinin (HA)-tagged version of ERK1 (pcDNAI-ERK1-HA) (originally obtained from Dr. J. Pouysseguer, Nice France).

1. Transfect cells with GFP and pcDNAI-ERK1-HA plasmids.
2. Transfected cells are allowed to readhere to fibronectin-coated dishes for a given time and then stimulate with growth factor.
3. Wash cells twice with ice-cold PBS and then lyse in modified RIPA buffer (50 mM HEPES, pH 7.5, 1% NP-40, 0.5% sodium deoxycholate, 150 mM NaCl, 50 mM NaF, 1 mM sodium vanadate, 1 mM nitrophenylphosphate, 5 mM benzamidine, 0.2 μM calyculin A, 2 mM PMSF, and 10 μg/ml aprotinin).
4. Lyse for 20 min on ice and clarify lysates by centrifugation at 16,000 *g* for 10 min at 4°C.
5. Preclear lysates with 30 μl of a 1:1 slurry of protein G-sepharose fast flow (Pharmacia Biotech, Cat #17-0618-01) on a rotator at 4°C for 30 min.
6. Incubate precleared lysates with anti-HA antibody on ice for 2 h.
7. Add 30 μl Protein G beads and incubate on a rotator at 4°C for 2 h.
8. Wash immunocomplex once with cold lysis buffer.
9. Wash twice with cold wash buffer (0.1 M NaCl, 0.25 M Tris-HCl, pH 7.5).
10. To washed immunocomplex, add 40 μl of reaction mixture (10 mM Tris-HCl, pH 7.5, 10 mM $MgCl_2$, 1 mM DTT, 25 μM ATP, 5 μCi ^{32}P-ATP (Dupont NEN Cat #NEG-002A) and 10 μg myelin basic protein [UBI or GIBCO]) per assay and incubate on a shaking platform at RT for 30 min.
11. Add 13 μl 4X SDS sample buffer and boil for 5 min to stop the reaction.
12. Use SDS PAGE (15% gel) to analyze incorporation of ^{32}P into myelin basic protein and (10% gel) for Western blotting to obtain levels of expressed HA-ERK1.

References

1. Chen, Q., Kinch, M. S., Lin, T. H., Burridge, K., and Juliano, R. L., Integrin mediated cell adhesion activates MAP kinases, *J. Biol. Chem.*, 269, 26602, 1994.
2. Chen, Q., Lin, T. H., Der, C. J., and Juliano, R. L., Integrin-mediated activation of MEK and mitogen-activated protein kinase is independent of Ras, *J. Biol. Chem.*, 271, 18122, 1996.
3. Short, S., Talbot, G., and Juliano, R. L., Integrin mediated signaling events in human endothelial cells, *Mol. Biol. Cell.*, in press.

4. Alessi, D. R., Cohen, P., Ashworth, A., Cowley, S., Leevers, S. J., and Marshall, C. J., Assay and expression of mitogen-activated protein kinase, MAP kinase kinase, and Raf, *Methods Enzymol.*, 255, 279, 1995.
5. Bogoyevitch, M. A., Marshall, C. J., and Sugden, P. H., Hypertrophic agonists stimulate the activities of the protein kinases c-Raf and A-Raf in cultured ventricular myocytes, *J. Biol. Chem.*, 270, 26303, 1995.
6. Lin, T. H., Chen, Q., Howe, A., and Juliano, R. L., Cell anchorage permits efficient signal transduction between Ras and its downstream kinases, *J. Biol. Chem.*, 272, 8849, 1997b.
7. Lange-Carter, C. A. and Johnson, G. L, Assay of MEK kinases, *Methods Enzymol.*, 255, 290, 1995.
8. Finney, R. and Herrera, D., Ras-Raf complexes: analyses of complexes formed *in vivo*, *Methods Enzymol.*, 255, 310, 1995.
9. Hallberg, B., Rayter, S. I., and Downward, J., Interaction of Ras and Raf in intact mammalian cells upon extracellular stimulation, *J. Biol. Chem.*, 269, 3913, 1994.
10. Lin, T. H., Aplin, E., Shen, Y., Chen, Q., Schaller, M., Romer, L., Aukhil, I., and Juliano, R. L., Integrin-mediated activation of MAP kinase is independent of FAK: evidence for dual integrin signaling pathways in fibroblasts, *J. Cell Biol.*, 136, 1385, 1997a.

Chapter

Methods for Analysis of Adhesion-Dependent Cell Cycle Progression

Xiaoyun Zhu, Kristin Roovers, Gabriela Davey, and Richard K. Assoian

Contents

I. Introduction 130
II. Cell Culture 130
 A. Maintenance 130
 B. Synchronization 131
 C. Induction of Cell Cycle Progression 131
 D. Collection and Extraction of Cells 132
 E. Special Issues Regarding the Analysis of ERK Phosphorylation in Monolayer and Suspension 133
III. SDS-Polyacrylamide Gel Electrophoresis and Western Blotting 133
 A. Quantification of Protein in Cell Lysates 133
 B. Electrophoresis 134
 C. Electrophoretic Transfer 134
 D. Immunoblotting 134
IV. Fluorescent Staining of Cells 136
 A. Immunofluorescence 136
 B. Staining of Focal Contacts and Actin Stress Fibers with Fluorescein-Phalloidin 137
V. *In vitro* Analysis of Cyclin-cdk2 Kinase Activity 137
VI. Analysis of Cyclin-cdk Complexes and Association of cdk Inhibitors by Immunoprecipitation/Western Blotting 138

0-8493-3385-7/99/$0.00+$.50

VII. Preparation of Recombinant Proteins 138
VIII. General Proliferation Assays for Anchorage-Dependent Growth 139
Acknowledgments 140
References 140

I. Introduction

Cell adhesion and mitogenic growth factors are jointly required for cell proliferation of almost all cell types. Cell proliferation reflects continued progression through the cell cycle, and progression through the cell cycle is mediated by the sequential activation of cyclin-dependent kinases (cdks). These enzymes are expressed constitutively, but they are inactive in the absence of their cyclin partners. Additionally, cyclin-cdk complexes are regulated by stimulatory and inhibitory phosphorylations (on the cdk) as well as by the binding of cdk inhibitors (CKIs).

We and others have shown that the adhesion requirement for cell cycle progression maps to G1 phase of the cell cycle. For example, if anchorage-dependent cells are cultured in the absence of a substratum or in the presence of cytochalasin D (which disrupts cytoskeletal spreading), they fail to phosphorylate the retinoblastoma protein (pRb) and fail to induce the expression of cyclin A (a hallmark of entry into S phase). Consistent with these molecular events, flow cytometric analyses reveal that cells cultured in the absence of substratum are arrested in G1 phase. Detailed analyses by us and others have shown that the lack of pRb phosphorylation and cyclin A induction in suspended cells reflects the fact that cell adhesion cooperates with growth factors to regulate the activities of the G1 phase cyclin-cdks.

Cell cycle progression through G1 phase is mediated by a subset of the cyclin-dependent kinases: cyclin D-cdk4 or cyclin D-cdk6, cyclin E-cdk2, and cyclin A-cdk2. Adhesion is required for the induction of cyclin D1 and D3 (the major D type cyclins in nonhematopoietic cells) and the downregulation of the cdk inhibitors, $p21^{cip1}$ and $p27^{kip1}$. The detailed description of these effects and their consequences for cell cycle progression can be found in Assoian[1] and Zhu and Assoian,[2] as well as references therein. In this chapter, we describe our methods for analyzing the regulation of G1 phase cyclins, cdk inhibitors, and cdk activities.

II. Cell Culture

A. Maintenance

We use the following conditions for culturing fibroblast cell lines. NIH 3T3 cells are maintained in 5% calf serum in Dulbecco's modified Eagle's medium with low glucose (DMEM). NRK cells are maintained in 5% newborn calf serum-DMEM and established cultures of mouse embryo fibroblasts (MEFs), and early passage cultures of freshly isolated human foreskin fibroblasts are maintained in 10% fetal calf serum-DMEM. Because overgrowth of NIH 3T3 cells usually leads to a transformed phenotype, the cells need to be passaged as they reach 50% confluence. In our hands, MEFs, NRK cells, and human fibroblasts can be brought to confluence prior to trypsinization without detrimental effect on cell morphology or the phenotypes of growth factor or anchorage-dependence. Eventually, all cell lines in culture lose their "normal" phenotype. These cells are discarded and replaced with early passage freeze-downs.

B. Synchronization

To render cells quiescent in G0, cultures are brought to confluence and then incubated in serum-free DMEM or DMEM with 1 to 2 mg/ml BSA. In our hands, NIH 3T3 cells require a 1-day starvation, MEFs require 2 days, NRK cells require 4 days, and human fibroblasts require 6 days (the starvation medium for human fibroblasts also contains a 1:500 dilution of ITS+, Collaborative Research, Bedford, MA). Starved cells show essentially no mitotic figures and have a very well-spread and flattened morphology. At the molecular level, we define good quiescence as the lack of cyclin D1, $p21^{cip1}$, and cyclin A expression, and predominance of the unphosphorylated forms of pRb and the ERKs. In most fibroblastic cells, high expression of $p27^{kip1}$ is also a valuable marker of the quiescent state.

C. Induction of Cell Cycle Progression

In order to examine cell cycle progression through G1 phase, G0-synchronized cells are typically trypsinized, suspended in a small volume of DMEM-5% FCS, counted, and quickly reseeded in monolayer and suspension (see below). A duplicate plate of quiescent cells is collected by scraping, lysed, and analyzed (all as described below) to assess the degree of quiescence. Human fibroblasts are treated similarly but stimulated with 10% FCS-DMEM. In most experiments, the mitogenic activity of FCS is supplemented by use of EGF (2 nM final concentration). If cells are to be plated on specific matrix proteins in serum-free medium, or if an aliquot of the cell suspension (rather than an entire plate) is to be used to monitor the state of quiescence, it is important to suspend the trypsinized cells in DMEM with heat-inactivated BSA (1 mg/ml) rather than DMEM with serum. In general, we seed 1 to 2×10^6 cells per 100 mm plate and 2 to 4×10^6 cells per 150 mm plate. Because different cells are different sizes, the most important consideration is to seed at a cell density that results in sub-confluent monolayer cultures (so that there will be no contact-inhibition of cell cycle progression) and suspension cultures having a minimal amount of cell-cell aggregation.

When we stimulate cell cycle progression in monolayer, the trypsinized cells are seeded in untreated tissue culture dishes in serum-containing medium or in matrix-protein coated dishes (either tissue culture or bacteriologic grade). The dishes are pre-coated by incubation with selected amounts of purified matrix proteins (usually 10 to 20 μg in 2 ml PBS for 35 mm dishes and 50 to 100 μg in 10 ml PBS for 100 mm dishes). Coating is performed for 2 h at room temperature or overnight at 4°C. After careful removal of the PBS, the dishes are "blocked" with a 2 mg/ml heat-inactivated (30 min, 68°C), fatty-acid free BSA in PBS (for 2 h at room temperature or 30 min at 37°C). The final amounts of matrix protein used to coat dishes typically depends on the cell line being studied. We finalize matrix protein concentrations based on rates of attachment and spreading. To ensure that cell adhesion is mediated by integrins, we also seed cells in dishes coated with BSA alone (the cells should remain in suspension) and/or poly-L-lysine (50 μg in 2 ml PBS for 35 mm dishes and 250 μg in 10 ml PBS for 100 mm dishes). Cells attach to poly-L-lysine in a non-integrin dependent manner, so they should not spread. Cell spreading on poly-L-lysine-coated dishes is a strong indication that significant amounts of matrix protein have been secreted by the cells during the experimental incubation.

To block adhesion-dependent signaling, we culture cells in suspension or in monolayer in the presence of cytochalasin D. Cells seeded in suspension are added to petri dishes that have been coated with a thin layer of molecular biology-grade agarose. The agarose is mixed in cell culture-quality water to a final concentration of 1%. The agarose is dissolved and the solution is sterilized by boiling in a water bath for 10 to 20 min. After the solution cools to 70°C, about 5 to 10 ml of melted agarose is quickly added to petri dishes. As soon as the dish is completely covered, the agarose is aspirated. Usually, 1.5 to 2 ml of agarose remains in the 150 mm dish as a thin clear film which solidifies over the next 5 to 10 min. When we use cytochalasin D to block adhesion signaling, cells are suspended in medium containing 2 μg/ml cytochalasin D (Calbiochem; a 1000-fold dilution of a 2 mg/ml stock solution prepared in DMSO), and either attached to tissue culture dishes in the presence of serum or matrix-protein coated dishes in the absence of serum. Alternatively, serum-starved cells can be directly stimulated with mitogens in the presence and absence of 2 μg/ml cytochalasin D. In the former procedure, cells attach but fail to spread. In the latter procedure, the spread cells usually become round after a 30 to 120 min incubation in cytochalasin D.

D. Collection and Extraction of Cells

Monolayer cells are washed 2 to 3 times in PBS prior to trypsinization or scraping. The cells are then suspended in a small volume (about 1 ml PBS) and transferred to an Eppendorf tube for a final wash and collection in an Eppendorf micro-centrifuge (centrifugation is performed at half-speed for 2 to 3 min at 4°C). For suspension cultures, the cells are collected by centrifugation (1000 to 1200 rpm in a Beckman JS5.2 rotor for 5 to 10 min at 4°C) in either 15 or 50 ml conical tubes. The pellet is suspended in 1 ml PBS, transferred to an Eppendorf centrifuge tube and then washed 2 to 3 times by repeated centrifugation (half speed, 2 to 3 min at 4°C) with PBS. For both the monolayer and suspension cultures, the final washed cell pellets are lysed by moderate vortexing in 50 mM Tris-HCl, pH 8.0, 250 mM NaCl, 2 mM EDTA, 1% NP-40, 10 μg/ml leupeptin, 10 μg/ml aprotinin, 1 mM phenylmethyl sulfonyl flouride, 50 mM sodium fluoride, and 1 mM sodium orthovanadate using a volume of approximately 50 μl/10^6 cells. After a 5 to 10 min incubation on ice, lysates were clarified by centrifugation at full speed in an Eppendorf micro-centrifuge for 10 min at 4°C. The supernatants are collected for immediate use or quick-frozen on dry ice and stored at –70°C. Alternatively, the washed cell pellets can be quick-frozen in dry ice and stored at –70°C for later lysis and analysis. We find that cyclin-dependent kinase activity will be stable for about two weeks when cells are stored as frozen pellets, but only for a few days when cells are lysed and stored as clarified extracts.

Note for collection of G0-synchronzed cells as reference: because certain enzymes associated with G1 phase cell cycle progression, e.g., the ERK subfamily of MAP kinases, are phosphorylated in response to trypsin, it is preferable to avoid trypsinization for collection of cells that will be used to define the degree of quiescence. We usually have one dish of serum-starved cells set aside for scraping and use as the "G0-control." Otherwise, it is imperative that the trypsinized cells are suspended in DMEM with BSA rather than serum.

E. Special Issues Regarding the Analysis of ERK Phosphorylation in Monolayer and Suspension

Trypsin-dependent activation of ERKs should be considered when setting up experiments designed to examine time- or anchorage-dependent ERK activation. Trypsinized cells must be suspended in serum-free DMEM, 1 mg/ml BSA or 50 μg/ml soybean trypsin inhibitor, and incubated in suspension in 50 ml conical tubes for 30 to 60 min at room temperature (to allow for dephosphorylation of trypsin-activated MAP kinase) prior to seeding in the presence of serum, purified growth factor, or purified matrix protein. Alternatively, cells can be detached with versene or directly stimulated in monolayer without trypsinization. Cells destined for MAP kinase analysis should be rinsed twice with PBS containing 1 mM orthovanadate (to prevent dephosphorylation of MAP kinase) and collected by scraping instead of by trypsinization. Lysates are prepared and stored as described above, except the buffer is 50 mM Tris-HCl, pH 7.4, 150 mM NaCl, 2 mM EDTA, 1% NP-40, 1 mM phenylmethylsulfonyl fluoride, 10 μg/ml aprotinin, 10 μg/ml leupeptin, 50 mM sodium fluoride, and 10 mM sodium orthovanadate.

III. SDS-Polyacrylamide Gel Electrophoresis and Western Blotting

A. Quantification of Protein in Cell Lysates

We use Coomassie staining (Bio-Rad Protein Assay) to quantify the amount of total protein in cell lysates. A 1 to 2 μl aliquot of each cell lysate is added to 0.8 ml of distilled water or PBS, and then 0.2 ml of the Bio-Rad reagent is added. After mixing well and waiting about 5 min, 0.2 ml of the solutions are added in duplicate to a 96-well plate. A protein standard (usually 0-5 μg of BSA) is prepared in parallel;

the absorbance of standards and samples are read at 595 nm in a Bio-Rad Microplate reader (Model 550).

B. Electrophoresis

SDS-gels are prepared using a bisacrylamide/acrylamide ratio of 1/37.5 (Bio-Rad Laboratories) and run with reductant using standard procedures. Gels containing 7.5% give best separation of unphosphorylated and phosphorylated pRb and also allow for ready separation of cyclins A, E, and D, the cdks, and the ERKs. We use SDS-gels with 15% acrylamide for fractionating cdk inhibitors. Alternatively, a 5 to 15% gradient gel can be used for analysis of pRb, the G1 cyclins, cdks, and cdk inhibitors. We typically fractionate 100 μg cell extract for analysis of pRb, cyclins, cdks, and cdk inhibitors. For MAP kinase and MEK-1, 15 to 25 μg cell extract is sufficient for analysis; significantly higher amounts of protein will obscure resolution of the phosphorylated and nonphosphorylated ERKs by gel-shift. We usually run 1.5 mm × 12 cm × 14 cm SDS-gels overnight at approximately 40 volts or for 3 h at 200 volts.

C. Electrophoretic Transfer

After electrophoresis, the fractionated proteins are transferred to nitrocellulose membranes electrophoretically at 42 volts (initial amperage is 0.25 to 0.35A) for 2 h in 1x transfer buffer (20x transfer buffer = 0.25 M Tris base, 2M glycine). After the transfer, the transferred proteins are stained with Ponceau S (0.2% in 3% trichloroacetic acid) for 2 to 5 min, and then destained in about 2 to 5 min by washing the membrane in distilled water. Ponceau S staining does not interfere with subsequent immunoblotting, and allows for the visualization of transferred proteins and molecular weight markers. The positions of the molecular weight markers are scored, and the membrane is subjected to the immunoblotting procedure outlined below.

D. Immunoblotting

The stained nitrocellulose membranes are blocked for either 1 to 2 h at room temperature, or 16 h overnight at 4°C. The blocking solution contains either 3 to 5% BSA or 3 to 5% non-fat powdered milk (Carnation), 1x washing buffer with 0.1% Tween-20 [20x washing buffer = 0.4 M Tris base, 1 M NaCl, 50 mM EDTA, adjusted to pH 7.5 with HCl]. The choice of BSA or milk is made empirically. We find that blocking in "milk" usually results in lower backgrounds, but milk can contain phosphatases which prevent the analysis of some, but not all, phosphoproteins. After blocking, the membranes are incubated with first antibody (typically 10 ml of a 100 to 1000-fold dilution of the first antibody stock in the appropriate blocking solution with 0.1% sodium azide) for 1 to 2 h at room temperature or overnight at 4°C, again determined empirically. The membranes are washed 2 to 3

TABLE 1
Commercially available antibodies presently in use in our laboratory

Antibody	Source	Use	WB dilution	IF dilution
cyclin D1	UBI #06-137	WB/IP	1:1000	
cyclin E	Santa Cruz sc-481	WB	1:500	
cdk2	UBI #06-505	WB	1:1000	
cdk4	Santa Cruz sc-260	WB/IP	1:1000	
p21 (for human p21)	Oncogene Sci. 39940203	WB	1:500	
p21 (for rat p21)	Santa Cruz sc-397	WB	1:500	
p27	Transduction Lab. K25020	WB	1:2500	
p107	Santa Cruz sc-318	WB	1:500	
phosphotyrosine	UBI #05-321	WB[a]	1:1000	
ERK	Transduction Lab. M12320	WB	1:1000	
ERK1	Santa Cruz sc-94	IF		1:100
phosho-ERK	Promega V667A	WB	1:5000	1:200
MEK1	Transduction Lab. M17020	WB/IF	1:5000	1:100

Abbreviations used: WB, Western blot; IP, immunoprecipitation; IF, immunofluorescence; UBI, Upstate Biotechnologies.

[a] Western blots are blocked with BSA.

times (30 to 45 min). Horseradish peroxidase (HRP)-labeled secondary antibodies or HRP-protein A (a 10,000-fold dilution in wash buffer) is incubated with the membrane at room temperature for 1 h. After washing as described above, the immunoreactive proteins are detected by enhanced chemiluminesence. Note that the primary antibody solutions can often be reused several times, and the membrane can be stored at 4°C for several months for future probing with other antibodies or even the same antibodies. The commercially available antibodies we presently use are listed in Table 1. When necessary, membranes were incubated in 0.1M Tris-HCl, ph 6.7, 2% SDS, 1% 2-mercaptoethanol (15 min at room temperature) to release bound antibodies.

IV. Fluorescent Staining of Cells

A. Immunofluorescence

For studies in monolayer, G0-synchronized cells are trypsinized and seeded (2 to 3 $\times 10^5$ cells/2ml DMEM with serum) into 35-mm dishes with autoclaved cover slips (Fisher, #12-545-82D, 0.15 mm). For studies in suspension culture, cells (1 to 1.5 $\times 10^6$ cells/10 ml DMEM with serum) are seeded in 100-mm agarose-coated dishes. To examine the effects of specific matrix proteins, cells are suspended in serum-free medium and added to 35-mm dishes containing cover slips that have been coated with purified matrix protein or poly-L-lysine (see above). As control, cells are also added to BSA-coated 35-mm dishes using the procedures outlined above. Cells cultured on BSA-coated dishes should remain suspended in serum-free medium. In all cases, adherent cultures should be 70 to 80% confulent so that several cells will be present in each microscope field during the subsequent analysis.

At selected times spanning G1 phase, monolayer cultures are washed once with PBS and fixed with 3.7% formaldehyde in PBS for 10 to 20 min at room temperature. Cells in suspension are collected and 2 ml aliquots are transferred to 35-mm dishes containing coverslips that have been precoated with poly-L-lysine and incubated for 2 min at room temperature. (This procedure results in cells that are seeded at a high cell density such that several cells can be seen in each microscope field during analysis.) As soon as the cells attach, they are gently washed and fixed as described for monolayer cells. After fixation, the formaldehyde solution is aspirated and replaced with 2 ml PBS. (The fixed cells can be stored at 4°C in PBS for at least a month.) For staining, fixed cells are permeabilized with 0.2% Triton X-100 in PBS (2 ml for 5 min at room temperature). Background fluorescence is often reduced by pre-incubation of the fixed cells with 50 mM NH_4Cl in PBS for 10 min (followed by two 5 min washes with PBS) prior to permeabilizing cells with Triton.

After washing twice with PBS, the permeabilized cells are incubated in 1 ml of first antibody solution, typically a 1:100 dilution in PBS, 2% BSA. Double staining is similarly performed, except that the two first antibodies must be from immunologically distinguishable antibodies. After washing three times with PBS, cells are incubated for 1 h at room temperature with secondary antibody: usually fluorescein-labeled goat anti-mouse or anti-rabbit IgG (100 to 300 fold dilution, Life Technologies or Sigma Cat. numbers F0511 or F5262). For double staining, we replace the fluorescein labeled anti-rabbit IgG with a "sandwich" detection using biotin-labeled goat anti-rabbit IgG (100-fold dilution, PharMingen or Life Technologies) and Texas Red-labeled streptavidin (200-fold dilution, Life Technologies). In each case, incubation is for 1 h at room temperature with reagents diluted in 2% BSA-PBS. For double staining, the anti-mouse and anti-rabbit primary antibodies are added sequentially rather than at the same time. Cells are washed three times with PBS between each antibody incubation, and the incubations with fluorescein or Texas Red-conjugated proteins are done in the dark. Stained cells are washed, in the dark, three times with PBS and once (for 5 to 10 min) with 0.5 ml Slow-Fade equilibration buffer (Molecular Probes, S2828). The cover slips are then removed from the wells, placed face down on slides containing a 40 μl drop of Slow-Fade antifade reagent in glycerol/PBS (Molecular Probes), and sealed with a clear nail polish. Staining patterns can be assessed by either epifluoresent or confocal fluorescent microscopy.

B. Staining of Focal Contacts and Actin Stress Fibers with Fluorescein-Phalloidin

Autoclaved cover slips are added to 35 mm dishes and coated with purified matrix protein. Cells (0.5 to 1 × 10^5) are seeded and incubated in 5% FCS-DMEM. After 24 h or a particular experimental treatment, the cells are fixed, permeabilized, and washed with PBS as described above. In the meantime, an aliquot of fluorescein-phalloidin in methanol (2 units, Molecular Probes F-432) is placed under nitrogen to remove the solvent, and the dried fluorescein-phalloidin is dissolved in 1 ml of PBS. Droplets (40 μl) of fluorescein-phalloidin solution are placed on parafilm. Cover slips are removed from wells, placed cell-side down on the droplets, and incubated in the dark for 20 to 30 min. After staining, the cover slips are returned to the wells and washed four times for 5 min with PBS and once for 5 min with water, in the dark. Slides are mounted in Slow-Fade, as described above. Staining patterns are visualized by epifluorescent microscopy, using 40× magnification. For long-term storage and analysis, stained cells can be mounted in anti-fade solutions containing polyvinyl alcohol.[3] In this case, the cover slips are placed cell-side down onto slides containing a 40 μl drop of mounting medium, and the mounted slides are allowed to dry for at least 2 h in the dark at room temperature.

V. *In vitro* Analysis of Cyclin-cdk2 Kinase Activity

For cyclin E/cdk2 and cyclin A/cdk2 kinase assays, cells are lysed in NP-40 based lysis buffer (described above for immunoblotting of cyclins). Protein concentration of the post-nuclear supernatants is determined (see above) and 200 μg aliquots of lysate are used for the kinase assay. The cell lysates are adjusted to a volume of 50 μl by addition of NP-40 lysis buffer prior to the addition of 1 to 5 μl of cyclin or cdk2 antibodies. After incubation at 4°C for 1 h, the solution is diluted to 0.5 ml. Protein A-agarose (50 μl, Life Technologies) is added to antibody-kinase complexes at 4°C with rocking for 1 h. The collected immunoprecipitates are washed three times with cold NP-40 lysis buffer and then twice with cold kinase reaction buffer (50 mM Tris-HCl, pH 8.0, 10 mM $MgCl_2$). The washed pellet is resuspended in 30 μl of kinase reaction solution including kinase buffer, 5 μg histone H1 (Life Technologies), 10 uCi ^{32}P-ATP (3000 Ci/mmol) and 20 μMATP. The kinase reaction is mixed well by vortexing and incubated at room temperature for 30 min. We resuspend the beads on the vortex mixer once or twice during the 30 min reaction. The reaction is stopped by adding 30 μl of 2x SDS-PAGE sample buffer (with reductant and 20% glycerol or 8 M urea) and heating the solution for 5 min at 90 to 100°C. Centrifugation (full speed for 10 sec in an Eppendorf microcentrifuge) removes the protein A agarose, and the supernatant is fractionated on a 12% acrylamide gel. The next day, the gel is stained for 1 h with Coomassie Brilliant Blue, destained, dried in a vacuum gel drier, and exposed to X-ray film. These gels can also be Western blotted to examine cyclin, cdk, and cdk inhibitor levels (see Section VI).

VI. Analysis of Cyclin-cdk Complexes and Association of cdk Inhibitors by Immunoprecipitation/Western Blotting

When interpreting results from kinase assays, it is usually beneficial to know the degree to which expressed cyclins and cdks are present as the cyclin-cdk complex as well as the degree to which cdk inhibitors are associating with the complex (or free cdk in the case of INK4s). Each of these issues can be addressed by immunoprecipitation/Western blotting (IP/Westerns). The cyclin-cdk complex is immunoprecipitated by use of either cyclin or cdk antibodies (depending on the particular complex and the availability of suitable antibodies), the immunoprecipitate is fractionated on SDS-gels, and the gels are subjected to Western blotting (with anti-cyclin, cdk, or cdk inhibitor antibodies). An alternative procedure is to immunoprecipitate with antibodies to particular cdk inhibitors and then Western blot with cyclin or cdk antibodies. We typically use 500 μg or more of cell extracts and excess antibody (determined empirically) for the initial immunoprecipitation.

Because this procedure is fundamentally similar to that used for *in vitro* kinase reactions, the SDS-gels used to assess kinase activity can also be used for Western blotting. In principle, this procedure allows for the assessment of cyclin-cdk kinase activity, complex formation, and association of cdk inhibitors within the same sample. When doing this combined analysis, the SDS gel (containing ^{32}P-labeled substrate) can be wrapped in plastic and directly exposed to X-ray film without staining, destaining, or drying. Exposure times, at least for histone H1 kinase assays, are about 10 to 60 min. If longer exposure times are needed, the fractionated proteins will diffuse in the gel. In these cases, the fractionated samples are first electrophoretically transferred to nitrocellulose as described above and the nitrocellulose is exposed to X-ray film for assessment of kinase activity. After determining kinase activity, the filters are blocked and subjected to immunoblotting as described above, with the caveat that proteins co-migrating with the heavy and light chains of the immunoglobulin (used in the immunoprecipitation) will not be detectable. Affinity purification of antibodies may also be helpful to reduce background noise in the Western blotting.

VII. Preparation of Recombinant Proteins

We prepare recombinant cyclins from pET vectors for use as standards in SDS gels, competitors in immunoprecipitations, and immunogens for preparation of antibodies.

Recombinant cyclins are prepared by transforming competent *E. coli* BL21(DE3) (Stratagene 200131) with 5 ng of pET-cyclin vector using a heat pulse of 15 to 30 sec. Colonies grown on LB-ampicillin agar plates are picked and cultured in 5 ml LB-ampicillin overnight. The overnight culture is added to 100 ml of LB-ampicillin (the resulting Ab^{600} is 0.1 to 0.2) and incubated for 1 h at 37°C with shaking. IPTG is then added to the culture to a final concentration of 1 mM, and the incubation is continued for another 2.5 to 3 h. The bacteria are harvested by centrifugation at 4°C in a Sorvall RC-5 centrifuge (6000 rpm, 10 min) and resuspended in 5 ml lysis buffer (25 mM Tris-HCl, pH 7.4, 100 mM NaCl, 1 mM $MgCl_2$, 10 mM EDTA) containing 1 mg/ml lysozyme samples are incubated on ice for 30 min, and then subjected to ultrasonication using a Virsonic 475 sonicator fitted with a microprobe. Ultrasonication is performed at 20% maximal output power (near maximal power for this probe) for a total of 3 to 5 min with repeated 20 sec pulses followed by 20 sec on ice. During this time, the bacterial suspension becomes somewhat more translucent.

The lysate is collected by centrifugation (8000 to 10,000 rpm in an Sorvall SS-34 rotor for 10 min at 40°C) and washed in 5 ml of wash buffer (0.2 M Tris-HCl, pH 8.2, 0.5 M NaCl). To recover recombinant cyclins from the inclusion bodies, the washed pellet is resuspended in 2 ml 8 M urea with 5 mM dithiothreitol (DTT) and incubated for 1 h at 4°C with rocking. The supernatant is carefully collected after centrifugation and dialyzed three times against 1 l of wash buffer containing 5 mM DTT. The dialysate is centrifuged, and the supernatant is collected and stored at –70°C. Recombinant protein is quantified by Coomassie staining (Bio-Rad Protein Assay); a typical yield is about 5 mg. The recombinant cyclin is the only predominant protein on Coomassie-stained SDS gels and comprises about 50% of the total protein in deliberately overloaded SDS-gels. Methods for further purification depend on the biochemical properties of the specific recombinant cyclin and the contaminants.

VIII. General Proliferation Assays for Anchorage-Dependent Growth

In addition to the subcellular analyses described above, we qualitatively monitor the degree of anchorage-dependent growth by culturing cells in soft agar. We quantify the degree of anchorage-dependence by ^{3}H-thymidine labeling with cells in preparative suspension culture.[4] For the soft agar assay, 1 g of Difco Noble agar is added to 20 ml water, dissolved, and sterilized by autoclaving. The melted agar is diluted tenfold in DMEM, and the medium is brought to approximately 70°C prior to the addition of serum (to a final concentration as needed). This soft agar solution is maintained at 45 to 48°C; 1 ml volumes are added to 35 mm dishes and allowed to harden to form the bottom layer for the assay. For the top layer, 0.8 ml of soft agar (prepared with serum as described above) is inoculated with cells (5×10^3 in 0.2 ml FCS-DMEM), gently mixed, and then added to the hardened bottom layer. If purified growth factors are to be included, they are added to the soft agar before adding the cells. [We typically prepare a growth factor cocktail in 0.2 ml of 4 mM HCl, 1 mg/ml BSA, and then add 0.6 ml of soft agar and 0.2 ml of the cell suspension. When calculating the amount of growth factors to be added, we assume that the growth factors will freely diffuse throughout the entire volume of the top and bottom layer.] After the top layer hardens, the dishes are incubated at 37°C for 3 to 14 days to detect strong and weak colony formation, respectively. Colonies are readily visible in a phase contrast microscope at 4× or 10× magnification. The working concentrations of soft agar may need to be varied somewhat, and agarose can be used instead of Difco agar. Agar/agarose concentrations that are too high will dehydrate the cells with time, while concentrations that are too low will result in loose top layers that fail to completely immobilize the cells.

For preparative suspension cultures, bottom layers are poured as described above. Cells (asynchronous or synchronized; 2×10^4) are suspended in 2 ml of FCS-DMEM and gently added to the hardened bottom layers. Growth factors are added to the cell suspension as needed. At selected times from 1 to 7 days, the cultures are pulsed with 1 to 3 uCi of ^{3}H-thymidine (80 Ci/mmol) to assess the degree of cell cycling by incorporation of radiolabel into DNA. Radiolabel is added in 25 μl of serum-free DMEM. After 24 h, the cells are collected quantitatively by multiple washes of the bottom layer with 1 to 2 ml cold PBS. The cells are transferred to 12 to 15 ml conical tubes and brought to a final volume of 9 ml with cold PBS. Low molecular weight DNA (0.25 mg) is added to each sample as carrier for the radiolabeled DNA. TCA (1 ml of a 50% solution) is added to the cell suspension and mixed; cells lyse immediately, and the DNA is precipitated by overnight incubation at 4°C. The DNA precipitate is collected by centrifugation (Beckman JS5.2 rotor, 2500 rpm, 10 to 20 min at 4°C) and washed three times with ice-cold 5% TCA. The final DNA pellet is dissolved in 0.2 ml formic acid, and the associated radioactivity is determined by counting in a beta scintillation counter.

Acknowledgments

Work in our laboratory is supported by grants from the National Institute of General Medical Sciences and the National Cancer Institute. K.R. is supported by a predoctoral fellowship from the American Heart Association, Florida Affiliate.

References

1. Assoian, R. K., Anchorage-dependent cell cycle progression, *J. Cell Biol.,* 136, 1, 1997.
2. Assoian, R. K. and Zhu, X., Cell anchorage and the cytoskeleton as partners in growth factor-dependent cell cycle progression, *Curr. Opin. Cell Biol.,* 9, 93, 1997.
3. Salas, P. J. I., Rodriguiz, M. L., Viciana, A., Vega-Salas, D. E., and Hauri, H. P., The apical sub-membrane cytoskeleton participates in the organization of the apical pole in epithelial cells, *J. Cell Biol.,* 137, 359, 1997.
4. Assoian, R. K., Boardman, L. A., and Drosinos, S., A preparative suspension culture system permitting quantitation of anchorage-independent growth by direct radiolabeling of cellular DNA, *Anal. Biochem.,* 177, 95, 1989.

Chapter 10

Integrin Modulation of Mitogenic Pathways Involved in Muscle Differentiation

Tho Q. Truong, Sarita K. Sastry, Margot Lakonishok, and Alan F. Horwitz

Contents

I. Introduction 142
II. General Methods 144
A. Overview 144
B. Isolation of Primary Quail Myoblasts 144
C. Cell Culture 145
D. Expression Vectors 145
III. Modulation of Mitogenic Signaling by Integrin Subunits 146
A. Overview 146
B. Protocols 148
IV. Downstream Effectors of Integrin Signaling 151
A. Overview 151
B. Protocols 152
V. Integrins and Growth Factors Cooperatively Regulate the Onset of Terminal Differentiation in Myoblasts 154
A. Overview 154
B. Protocols 156
VI. Discussion and Future Directions 156
References 157

0-8493-3385-7/99/$0.00+$.50

I. Introduction

Skeletal muscle development relies on complex regulatory mechanisms which must coordinate cell cycle withdrawal, cell alignment, and membrane fusion with the synthesis, organization, and innervation of a contractile apparatus. Cell adhesion through the integrins regulates many of the molecular events which underlie each of these processes by modulating a variety of intracellular signaling molecules. Some of these molecules constitute signaling pathways mediated by soluble peptide mitogens, and may provide points for synergy between adhesion- and growth factor-mediated signals. Such effectors include the mitogen activated kinases (MAPKs), e.g., extracellular related kinase-1 and -2 (ERK-1 and -2), as well as the proto-oncogenes, which include the cell cycle regulating transcription factor, c-myc, the cell membrane-associated src family of kinases, and the GTP binding protein, H-ras.[1-3]

The extensive study of muscle development itself has also resulted in the establishment of a primary cell culture model with which to investigate the contribution of integrin signaling and its synergy with growth factors to important cellular events, such as cell survival and proliferation, potentially relevant to other tissue types. This chapter will provide methods and discuss experimental examples which use primary quail myoblast differentiation as a model to study the roles of various integrin subunits and the signaling events which these proteins mediate.

A large body of literature established early cell culture models, particularly in embryonic chicken and quail, that greatly enhanced the study of muscle differentiation. Konisberg described the culture conditions necessary for muscle precursor cells, isolated from nine-day-old quail embryo, to proliferate and subsequently differentiate into myotubes *in vitro*.[4] During this stage of embryonic development, committed myoblasts have migrated from the somite, are localized to specific regions, and form the muscles along the body wall as well as the limb muscles. During the first 48 h after plating in culture, these cells undergo a period of rapid proliferation before withdrawing from the cell cycle and terminally differentiating, typically within the next 24 h.[5] Muscle-specific genes are then expressed which encode transcription factors or characteristic structural proteins that lead to terminal myogenesis. The nuclear binding protein, myoblast determination factor (MyoD), commits post-migratory somitic cells to become myoblasts.[6] At this stage, the structural protein desmin and antigenic myoblast marker L4 are also expressed. After the terminal differentiation program has been initiated, muscle α-actinin and titin appear; both may be detected immunochemically as early markers for terminal differentiation. At this point in quail, myogenin is also activated, and can serve as an early nuclear marker for cells which have entered the terminal step in myogenesis; its constitutive inhibition delays further differentiation.[7] Muscle LIM protein (MLP) expression then marks cell fusion into myotubes, and the expression of myosin heavy chains and organization of α-actinin into striations follows (Table 1). Later, other muscle-specific proteins appear and give rise to various specialized membrane junctions, including the costameres, and the myotendinous and neuromuscular junctions.[8,9]

TABLE 1
Markers for Early Terminal Differentiation

Activity	Replicating Myoblast	Committed Myoblast	Fusion	Myofibrillogenesis
Regulatory Gene Activated	MyoD →			→
	MRF4 →			→
		Myogenin →		→
	MYF-5, -6 →			→
Cytoskeletal Proteins	Desmin	Titin	Muscle LIM Protein	Myosin heavy chain
Expressed	L4	Muscle α-actinin		Muscle α-actinin in striations

Compliments of Stanley Wu.[7,9] Previously unpublished table.

Myoblast differentiation provides a particularly productive model in which to investigate the role of integrins in regulating the cell's decision to proliferate or withdraw from the cycle cell and initiate terminal differentiation. For instance, muscle-specific gene expression requires concomitant cell cycle withdrawal; consequently, proliferating vs. differentiating cells may be clearly distinguished morphologically and biochemically. In addition, a relatively comprehensive knowledge of the transcriptional activities and sequence of protein expression exists.[10] Hence, molecules required for the muscle phenotype, e.g., α-actinin, may be detected to determine whether cells have undergone the transition from proliferation to terminal differentiation. Assays to detect and characterize the extent of terminal myoblast differentiation are relatively straightforward and typically yield unambiguous, biological phenotypes. Finally, the effects of several soluble peptide mitogens on muscle differentiation is known, and hence, this allows for the study of the molecular basis of integrin and growth factor signal integration.[11]

Primary quail myoblasts in particular provide an auspicious cell culture model for elucidating the signaling mechanisms through which integrins can modulate the cell cycle and gene activation. These cells undergo more complete and faster differentiation in culture than do other existing primary myoblasts such as chick. In contrast to chick, quail myoblasts are easily cloned and can also be carried to a much higher passage. In fact, healthy cells can typically be carried through at least a dozen passages and still terminally differentiate after 72 h of plating without further passage, or upon serum withdrawal.[12] Chick myoblasts do not proliferate well in culture; hence, the more proliferative fibroblasts will predominate in culture dish after only a few passages. Further, the selective removal or termination of fibroblasts from chick primary myoblast preparations inhibits myoblast proliferation. As a result, more preparations of chick myoblasts must be made for each series of experiments. Quail primary muscle precursor cells also reproduce *in vivo* muscle structures, both morphologically and biochemically, with greater fidelity and expedience than myoblastic cell lines, which may also have abnormal or unnatural signaling profiles as a consequence of transformation.[13] Finally, primary quail myoblasts may be efficiently and stably transfected by lipofection (with typically a 50 to 70% transfection efficiency), in contrast to chick myoblasts, which tend to expel transfected plasmids before chromosomal incorporation can occur.[14].

II. General Methods

A. Overview

Despite intense interest and rapid progress in integrin signaling research, mechanistic models which link integrin-dependent signal transduction from the cell membrane to downstream effector molecules that coordinate myogenesis are still nascent. Numerous experiments, for instance, show biochemical correlations between integrin ligation and the activation of intracellular components in growth factor receptor tyrosine kinase pathways; but considerably less data exist which demonstrate the phenotypic significance of integrin and growth factor synergy. In particular, the roles of specific integrin subunits and integrin-mediated downstream effectors in regulating the cellular decision to terminally differentiate need to be elucidated. To this end, we ectopically expressed various integrin subunits and putative effectors of integrin signaling in primary quail myoblasts and studied any subsequent effects on myogenesis.

B. Isolation of Primary Quail Myoblasts

Primary myoblasts are isolated from pectoralis muscle of nine-day Japanese quail embryos; fertilized eggs may be obtained from the University of California (Davis, CA) or from the GQF Manufacturing Co. (Savannah, GA). Briefly, as described earlier,[12,15] breast muscles are dissected from the embryo and myoblasts are dissociated from muscle tissue with 0.1% dispase (Sigma, St. Louis, MO) in calcium-magnesium free phosphate buffered saline (CMF-PBS). The cell suspension is then filtered through a Sweeney filter; cells are seeded onto gelatin-coated tissue culture (TC) plates (0.1% gelatin in CMF-PBS), typically at a density of 3 × 10^4 cells/cm^2. Cells at passage 1 may be stored frozen (–80°C) for future use (up to 6 mos.) in 95% fetal bovine serum + 5% DMSO at a density of 2 × 10^6 cells/mL in a cyrovial, insulated in a foam rack.

Considerations: *Embryo extraction from the egg and dissection should be done in a dissection hood. It is critical to maintain antiseptic conditions. We highly recommend that an aliquot of freshly isolated myoblasts be examined for contamination by microorganisms, especially Mycoplasma, as well as by other cell types before using in experiments. Symptoms of contamination by Mycoplasma include slow cell division after plating in culture and the appearance of speckled nuclei by DAPI staining. The myoblast-specific antigenic marker L4 (Iowa Developmental Hybridoma Bank) can be used to distinguish myoblasts from other cell types. Also important, slight but noticeable variability between batches of myoblast preparations may affect* *the precise onset of differentiation after plating in culture and the rate of proliferation. It is therefore necessary for control cells to be from the same myoblast preparation as experimental cells.*

C. Cell Culture

Myoblast cultures are maintained in complete myoblast medium [DMEM (Sigma) containing 15% horse serum, 5% chick embryo extract, 1% pen/strep, and 1.25 mg/ml fungizone (Gibco)]. Myoblasts are subcultured using trypsin-EDTA (0.06% trypsin, 0.02% EDTA) and used between passages 1 and 10.[15] Cells which will be carried through several passages should be passed at approximately 3 to 4 $\times 10^5$ cells per 6 cm TC plate coated with sterile 0.1% gelatin or 10 μg/mL fibronectin in PBS-CMF for at least 1 h at 37°C (aspirate gelatin and wash plates once with PBS-CMF before plating cells). Note that these myoblasts always need to be plated on a substrate; in the absence of a matrix, myoblasts neither attach well nor fully differentiate *in vitro*. Gelatin is least expensive and facilitates the adherence of fibronectin secreted from the myoblasts and present in the serum of complete medium to TC plastic.

Considerations: *It is highly recommended that cultures which will be carried (rather than be seeded for assays) be passaged at least every 48 h. Cells should therefore be plated such that approximately 70% confluency will be reached in time for the next passage. Cultures which have not been passed after 72 h will begin to differentiate and elongate or fuse before confluency can be reached. Plating these myoblasts at lower densities, in fact, can promote premature differentiation of the culture. Fusing myoblasts and myotubes will not proliferate after passage and do not plate well; as a consequence, even though quail myoblast cultures are relatively homogeneous, any cell existing in the culture that is more proliferative would be favored and selected. We also do not recommend using cultures after the tenth passage. Higher passage myoblasts no longer proliferate well and eventually lose the distinct differentiative phenotype. Higher passage cultures have markedly more irregular morphology and variable size, compared to the more uniform lower passage cultures. These cells will first appear to divide less briskly than lower passage myoblasts, and then become slightly wider and less defined. Cells plated for immediate experimentation should be seeded according to individual experimental protocols with respect to serum concentrations, cell density, substrate, etc.*

D. Expression Vectors

The human α5 cDNA in pRSVneo and the chicken α6 cDNA in pRSVneo have been described.[15] Vectors driven by the CMV promoter, such as pcDNA3.1 (Invitrogen) are also effective in quail myoblasts. The chicken α61044t truncation was constructed by first subcloning a 1.6 kb HindIII-SalI fragment of the chicken α6A DNA into M13, and then introducing an in-frame BclI site at amino acid position 1044.[16] Mutants were confirmed by restriction digestion of M13 clones with BclI and by single stranded DNA sequencing using the dideoxy-chain termination method according to the Sequenase™ protocol (USB, Cleveland, OH). An 800 bp BstXI-SalI fragment containing the mutation was subcloned into pRSVneoα6 partially digested with SalI and completely with BstXI. The human α6A and α6B cDNAs, in the expression plasmid pRc/CMV,[17] were a generous gift of A. Mercurio (Harvard Medical School, Boston, MA). The pRSVneo-CH8β1 plasmid was constructed by subcloning a 1kb HindIII fragment, containing the CH8 epitope tag, from the CH8β1 pBJ-1 construct received from Y. Takada (Scripps Research Institute, La Jolla, CA),[18] into pRSVneoβ1 expression vector.[19] pRSVIL2R-α5 and pRSVIL2R-β1A were constructed by cloning a Nhe1-Xba1 fragment from pCMVIL2R-α5cyto or pCMVIL2R-β1A plasmids received from Susan LaFlamme (Albany Medical College, Albany, NY),[20] into the Xba1 site of the pRSVneo vector. Clones were screened for orientation by restriction digests. Hemagglutinin (HA) tagged rat MEK1 and HA-tagged rat constitutively active (CA) MEK S218/220D in pCMVneo vector,[21] were received from M. Weber (University of Virginia, Charlottesville, VA). The chicken paxillin cDNA, the Y118F, and the S188/190A mutants in the pcDNA3.0neo vector,[22] were received from C. Turner. CD2FAK, CD2FAK(Y397F), and CD2FAK(K454R) in CDM8 vector,[23] were received from A. Aruffo.

III. Modulation of Mitogenic Signaling by Integrin Subunits

A. Overview

Previous studies demonstrate that integrins can decisively regulate terminal muscle differentiation. For instance, antibody ligation of the β1 integrin subunit in chick myoblasts maintains proliferation and prevents terminal differentiation.[24] Several other studies have also shown that particular integrin subunits display distinct and regulated patterns of expression during muscle development, further implicating integrins as regulators of myogenesis.[25,26] Some of these integrin subunits, such as α5 and α6, have also been implicated in transmembrane signaling,[27,28] which by now has been shown to affect molecules of well-known mitogenic pathways.[1] Hence, specific integrin subunits can regulate mitogenic signaling, which subsequently determine the onset of terminal muscle differentiation.

Although considerable research has been done to delineate the mechanisms underlying soluble peptide-regulation of mitogenic signaling pathways, much less is known about how anchorage-dependent signals, specifically through the integrins, modulate these same pathways. In order to investigate the molecular basis of this regulation, our lab examined the roles of specific integrin subunits in regulating the cell cycle transition from proliferation to differentiation in the primary quail myoblast system. Integrin α5 and α6 subunits were ectopically expressed in primary quail myoblasts and effects on myoblast proliferation and differentiation were measured. We found that ectopic expression of the α5 subunit promotes proliferation and delays cell cycle withdrawal and the onset of terminal differentiation, whereas the ectopic expression of α6 produces opposite effects. Further, anti-sense-α6 expression nullifies the differentiative α6 phenotype and results in increased cell proliferation. This indicates that the ratio of α5 vs. α6 (and perhaps other alpha subunits) expression can be decisive in terms of regulating the onset of terminal differentiation in myoblasts. Moreover, ectopic expression of chimeric alpha subunits in which the α5 and α6 subunit cytoplasmic domains were swapped with one another (α5/α6cyto, α6/α5cyto) suggested that the identity of the cytoplasmic domain determines, to a large extent, cell phenotype.

Hence, we then expressed various constructs of integrin cytoplasmic domains in order to elucidate the molecular mechanism through which unique integrin subunits differentially regulate signals that affect the decision to differentiate. Ectopic expression of an α6A truncation, α61044t, deletes the 11 C-terminal amino acid residues, restores proliferative signaling, and produces a phenotype similar to that of the hα5 subunit. In contrast, myoblasts transfected with α5 in which the cytoplasmic domain was truncated up to the conserved GFFKR region had a proliferative phenotype similar to that of wild-type α5 transfectants. We subsequently expressed the single-subunit cytoplasmic domain chimera, IL2R-α5,[29] in our system to determine whether or not the α5 cytoplasmic domain itself may produce proliferative signals. IL2R-α5 transfectants had a similar phenotype to that of mock-transfected control cells, indicating that the α5 cytoplasmic domain is not sufficient to promote cell proliferation.[49] Given that ectopic expression of wild-type α5 also conferred an increase in β1 subunit expression (unpublished data), and that the isolated α5 cytoplasmic domain (IL2R-α5) could not recapitulate the wild-type α5 phenotype, we suspected that the integrin signal which results in myoblast proliferation originates from the β1 cytoplasmic domain, presumably through its interaction with focal contact proteins.

To test this hypothesis, we assayed the effects of the single subunit chimera IL2R-β1, on myoblast proliferation and differentiation. Stable transfectants possessed a markedly proliferative phenotype, supporting the hypothesis. This is also consistent with our results that overexpression of wild-type β1 subunit also results in myoblasts favoring a proliferative phenotype. These results, taken together, support the view that the β1 cytoplasmic domain generates the proliferative signal responsible for the delay in differentiation. Further, that IL2R-β1 may generate proliferative signals in a monodimeric fashion suggests that differential interaction between the alpha subunits and β1 creates the contrasting α5 and α6 phenotypes, rather than differential signaling by the alpha subunits per se. Mechanistically, this is consistent with the "hinge model,"[30] in which the ligation of the αβ dimer releases the β cytoplasmic tail into an open conformation such that it can interact with enzymatic and adaptor proteins to generate signals; specific α subunits may promote or hinder this interaction, thereby modulating the outcome of β1 signaling.

B. Protocols

1. **Single stable transfections** — Proliferating myoblasts from passages 1 to 3 are seeded at 3 to 5 × 10^5 cells per 6 cm plate and grown for 16 h at 37°C. 8 μg of plasmid DNA and 25 μl Lipofectamine™ (Gibco-BRL) solution in 300 μl DMEM are incubated at room temperature for 45 min and added to each plate of cells, which are then incubated for another 12 h at 37°C in complete myoblast medium (2.5 mL per 6 cm plate). Depending on the level of confluency, cells are then either placed immediately into selection medium (myoblast medium + 0.50 mg/mL G418) or split 1:2 into maintenance medium (myoblast medium + 0.25 mg/mL G418 [Gibco-BRL]), which is then replaced with selection medium after cells have attached to substrate. Typically, after 3 days in selection medium, maintenance medium is then used to allow cells still living in the dish to expand into colonies. When cells within colonies start to align (i.e., before fusion occurs), the cells are replated into an appropriate sized dish (3 × 10^4 cells/cm^2) in maintenance medium and expanded until an appropriate number is available for cell sorting (at least 1 to 2 confluent 10 cm plates). This selection and expansion process typically takes 7 to 14 days.

Considerations: *To avoid clonal variations or artifacts, we routinely worked with mass populations of transfected cells that were enriched by cell sorting (see Step 3 below).*

2. **Transient transfections (ectopic expression of α6A, α6B, and α61044t)** — Cells are transfected with Lipofectamine (as in Step 1). Transiently transfected cells are sorted by flow cytometry 24 to 48 h after transfection (Step 3, below); the positive cells are sorted into complete myoblast medium and grown in an appropriate-size, gelatin-coated culture plate (3 × 10^4 cells/cm^2) overnight, and then are seeded onto 12 mm glass coverslips coated with the appropriate substrate for immunohistological assays. For our transfections of chicken or human α6A or α6B, and α61044t, positive, sorted cells were grown in a laminin coated plate, and then seeded onto laminin coated coverslips or made into extracts for immunochemical and biochemical assays.
3. **Flow cytometry** — Fluorescence-activated cell sorting (FACSing) is used to purify and enrich positive transfectant populations. Both transiently and stably transfected myoblasts are analyzed for surface expression by flow cytometry as described by Sastry et al., 1996.[15] Briefly, cells are washed twice with HEPES-Hanks solution, and are detached from substrate using 1 mL of 0.2 mM EDTA in HEPES-Hanks per confluent 6 or 10 cm plate (at 37°C for 15 to 20 min). Detached cells (in the EDTA-HEPES-Hanks solution) are pipetted into a 1.5 mL sterile microfuge tube and centrifuged at 2000 rpm for 2 min. The supernatant is carefully aspirated, and the pellet of cells is resuspended with 0.5 mL of the desired primary antibody in blocking buffer (HEPES-Hanks PBS-CMF with 2% BSA); cells are then incubated, with gentle agitation, for 30 min at 4°C. Cells are spun down in a microfuge for 2 min at 2000 rpm, washed twice with (i.e., resuspended in) block buffering (and spun down again), before the secondary antibody (in 500 μl blocking buffer) is added in a similar fashion. Cells are washed again in block buffer before fluorescence sorting or analysis takes place. Specifically, chicken α6A and α61044t transfected cells are stained with a chick α6 specific polyclonal antibody, α6ex, at 20 μg/ml in blocking buffer and FITC-labeled goat-anti-rabbit IgG (Cappell, Durham, NC). Human α6A or B transfected myoblasts are analyzed with the human α6 specific mAb, 2B7, at 10 μg/ml in blocking buffer. Human α5 transfected cells are stained with

VIF4 mAb (R. Isberg, Tufts University, Boston, MA). Cells transfected with CD2FAK or its mutants are stained with anti CD2 mAb TS2/18.1.1 (Developmental Studies Hybridoma Bank, Iowa City, IA). For our ectopic expression of β1, we used a plasmid encoding chick β1 attached to a human β1 epitope (CH8) tag; β1 transfected cells are stained with TS2/16 mAb (Developmental Studies Hybridoma Bank) against the human β1 epitope. IL2R-α5 or IL2R-β1A transfected cells are analyzed with an anti-IL2R antibody (Boehringer-Mannheim, Indianapolis, IN). *Note*: The FACS profiles of the IL2R-α5 and IL2-cyto-transfectants were not stable, and we were unable to obtain populations greater than 40% positive, which were used for analysis. Flow cytometry may be performed on an EPICS cell sorter (Coulter Electronics, Inc., Miami Lakes, FL) equipped with Cicero software for data analysis. Where indicated, cell surface expression levels are normalized among transfected cells by fluorescence activated cell sorting.

Considerations: *Unless otherwise indicated, reagents and cell preparations should all be kept on ice while preparing for FACS. Antibody-receptor complexes on cells may be internalized at warmer temperatures.*

4. **Cell differentiation assays** — We express cell differentiation using the fusion index, the percentage of total nuclei in myotubes (number of nuclei in myotube/total number of nuclei X 100%). Cells are also immunostained for muscle α-actinin as an early marker for terminal differentiation. For most assays with untransfected or transfected myoblasts, cells are grown on 12 mm glass coverslips to which FN or LM is crosslinked. This is to ensure that the substrate remains fixed to the coverslip as muscle cultures tend to be long term (72 or 96 h in some cases) and also contract, thereby removing the substrate from the coverslip. Briefly, the coverslips (Fisher) are washed in 20% H_2SO_4, followed by 0.1 M NaOH and distilled water. Dried coverslips are incubated in saline (Sigma), washed in distilled water and PBS, and then cross-linked with 0.25% glutaraldehyde for 30 min. After three washes with PBS, matrix proteins are diluted in PBS and added to coverslips for 1 h at room temperature. ECM-crosslinked coverslips are sterilized by dipping into 70% ethanol, placed in 24-well tissue culture clusters, washed, and stored in PBS at 4°C.[31] Cells are grown on FN or LM coated coverslips, seeded at a density of 1.6×10^4 cells/cm^2 in complete myoblast medium or 3.2×10^4 cells/cm^2 in serum-free medium (DMEM + 2% bovine serum albumin) and cultured for indicated times as before. Myoblasts are immunostained for the muscle-specific marker muscle α-actinin, to determine the extent of biochemical differentiation and myofibrillar organization in the various transfectants. At time points in which assays need to be done (e.g., 24 h, 48 h, 72 h), coverslips are washed with PBS and fixed with 3% formaldehyde in PBS for 15 min. Cells are then permeabilized with 0.2% Triton X-100 in PBS for 10 min at room temperature, washed, and blocked in 5% goat serum-PBS for 30 min. Next, cells are incubated with primary antibody 9A2B8 (a gift of D. Fishman, Cornell University) a mAb against muscle specific α-actinin (at a 1:5 dilution of hybridoma supernatant in 5% goat serum) for 30 min and then with FITC-sheep-anti-mouse IgG (Cappel), rhodamine phalloidin (Molecular Probes), and DAPI (1:2000 dilution; Sigma) to stain the total nuclei for an additional 30 min. Coverslips are washed and mounted in medium containing elvanol and p-phenylene diamine. We observed fluorescence on an Axioplan fluorescence microscope (Carl Zeiss, Inc., Thornwood, NY). The degree of morphological differentiation is expressed as the fusion index, the percentage of total nuclei in myotubes, and scored in three independent experiments for five random fields at the chosen timepoints.

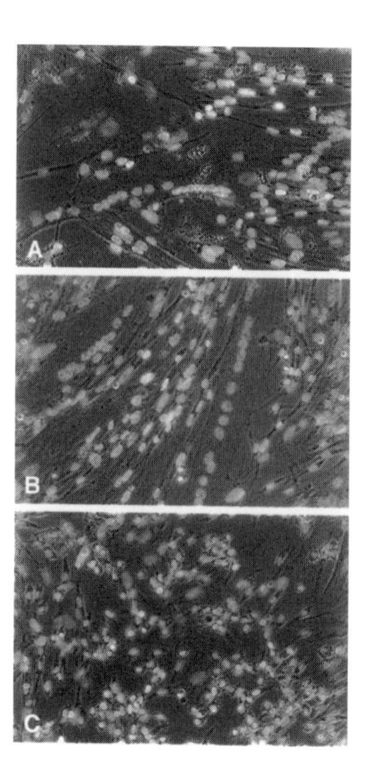

FIGURE 1
Scoring fusion indicews (4000X). Nuclei are stained with DAPI. Note that nuclei within myotubes are typically aligned, in contrast to the nuclei of individual myoblasts; myotubes also immunostain with muscle α-actinin (see Section III.B.4). (A) Mock transfectants; (B) II α5 transfectants; (C) ILβ1 transfectants. Cells transfected with monodimeric integrin β1 cytoplasmic domain delays terminal differentiation of myoblasts as shown by the decreased proportion of nuclei present in myotubes compared to controls (Section III).

Considerations: *Myotubes are α-actinin positive and contain more than 2 nuclei. Additionally, myotubes are wider and larger than elongated individual cells and may be distinguished from multiple aligned cells using a phase or DIC and fluorescence overlay on the microscope (Figure 1). The detection of the nuclear protein, myogenin, is also a popular marker for terminal differentiation. However, we prefer to detect α-actinin (Figure 2), which is indicative of the successful transactivation of myogenin, since under certain conditions myogenin may be present but functionally inhibited.*

5. **Cell proliferation and survival measurements** — The fraction of proliferating cells is determined by measuring incorporation of BrdU by myoblasts. Cells are seeded as described above (Step 4) in the appropriate medium and grown for 12, 24, 48, or 72 h. BrdU (50 mM in DMEM) is added to cultures for 12 h prior to fixation at indicated time points. Coverslips are fixed for 10 min in 95% ethanol, washed, denatured in 2N HCl for 30 min, and immunostained with an anti-BrdU mAb (Sigma) at a 1:750 dilution in 5% goat serum followed by FITC-sheep anti-mouse IgG and DAPI. The percentage of proliferating cells, the fraction of total nuclei immunopositive for BrdU, is scored for five random fields in three independent experiments. Cell survival for untransfected and α5 transfected myoblasts in serum-free medium or in serum-free medium including growth factors is expressed as a ratio of the average number of nuclei per field in the presence or absence of growth factor to the average number of nuclei per field in serum-free medium alone after 24 h. We calculated data for five random fields in three independent experiments.

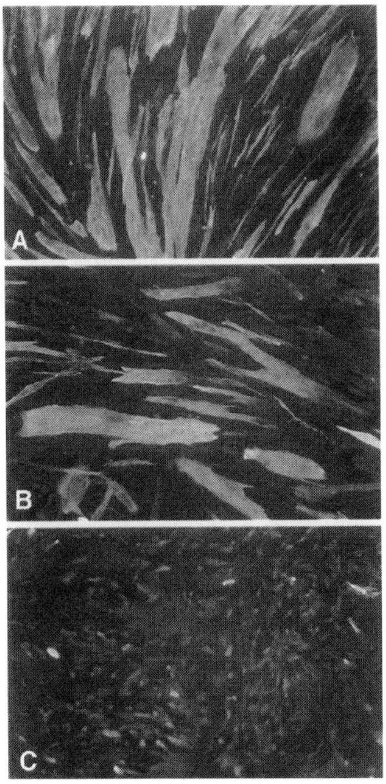

FIGURE 2
Muscle α-actinin as an early marker for terminal muscle differentiation (Section III.B.4). Myoblasts are plated for 60 h and stained with anti-muscle-α-actinin monoclonal antibody (4000X). (A) Mock tranfectants; (B) ILα5 transfectants; (C) ILβ1 transfectants. Cells expressing monodimeric integrin β1 cytoplasmic domain do not immunostain with anti-α-actinin mAb at 60 h in contrast to control cells, indicating a delay of the onset of terminal differentiation (Section III).

IV. Downstream Effectors of Integrin Signaling

A. Overview

Integrins possess no intrinsic enzymatic activity but instead interact, either directly or indirectly, with a number of intracellular proteins to initiate signals which might modulate cell cycle withdrawal. Among these proteins are focal adhesion kinase (pp125FAK) and paxillin. FAK, a protein tyrosine kinase, becomes tyrosine phosphorylated and activated upon integrin ligation, while paxillin possesses multiple phosphorylation and binding domains which may serve to link integrins with other effector molecules and propagate integrin-initiated signals.[32,33] Integrin-dependent activation of FAK also exposes phosphotyrosine binding sites for phosphatidylinositol 3-kinase (PI3K) and the src family of kinases.[34] In addition, FAK associates with Grb2,[35] and SH2-containing adaptor molecules such as Grb2 and SHC may provide a molecular basis through which integrins can cooperatively mediate the mitogenic ras/MAP kinase pathway with growth factors to affect the cell cycle.[36,37,44] These data suggest that FAK and paxillin, activated at adhesive sites, may regulate the level of activated MAP kinase and be proximal effectors for mitogenic integrin signaling in the mediation of myogenesis.

We determined whether the α5- and α6-transfectants, previously described to have contrasting phenotypes, also had contrasting signaling profiles with respect to FAK and paxillin activation. We found that ectopic α5 upregulated the expression and tyrosine phosphorylation of paxillin while not affecting FAK; in contrast, ectopic α6 downregulates the tyrosine phosphorylation of FAK while having a lesser effect on paxillin. Further, ectopically expressed paxillin recapitulates the α5 proliferative phenotype, while the transfection of CD2-FAK, which is targeted to the cell membrane, rescues the proliferation-inhibited α6 phenotype. These data suggest that the differential activation of FAK and paxillin by α5 and α6 are decisive downstream events that are sufficient to determine the characteristic cell phenotype associated with each subunit. Interestingly though, α5 and α6 localize to different regions of the cells; α5 localizes to focal adhesions but α6 disperses throughout the cell membrane.[49]

In addition, we also showed that the highly proliferative cells overexpressing wild-type α5 have increased ERK-1 activation and overall tyrosine phosphorylation. Conversely, cells expressing human α6 have decreased tyrosine phosphorylation and ERK-1 activation compared to untransfected cells. Hence, to test whether the differential regulation of the integrin alpha subunits on muscle differentiation acts through ERK-1, we expressed constitutively activated extracellular signal-related kinase kinase (MEK) in α6-transfected myoblasts, since the activation of ERK-1 has been shown to be downstream of MEK in the integrin pathway.[1] The results demonstrate that activated ERK-1 can prevent the typically differentiative phenotype of α6-overexpressing myoblasts. Moreover, MEK inhibitors attenuate the proliferative phenotype of α5 overexpressing myoblasts in a dose-dependent manner, further supporting that integrin-mediated signals act and may be regulated through the MAP kinases.

B. Protocols

1. **Stable double transfections of CD2-FAK+hα6A, MEK+hα6A** — Myoblasts are co-transfected with a pRSVneo or a pE-GFP-C1neo plasmid (Clontech, Palo Alto, CA) and CD2-FAKCDM8 plasmid at 1:7 ratio (neo resistance gene:CD2-FAK cDNA) and selected in G418 as described above (Section III, Step 1). For co-expression of hα6A integrin and CD2-FAK, cells are transfected with both pRc/CMVhα6A and CD2-FAK vectors at a ratio of 1:7, respectively. Cells are sorted for α6 expression 36 to 48 h after transfection. CDM8-CD2FAK vector does not carry a neo-resistance gene; therefore, only cells carrying both neo resistance (hα6A positive) and able to replicate (CD2FAK positive) will survive. Stable populations are analyzed both for CD2 expression as described above (Section III, Step 3). MAP kinase activity is manipulated in hα6A transfected myoblasts by co-expression of constitutively activate (CA) MEK1. Myoblasts are co-transfected with pRc/CMVhα6A and the HA-tagged pCMVneoMEK S218/220D vectors at a ratio of 1:7, respectively. Cells are selected in G418 and stable populations are sorted by flow cytometry for human α6A expression as described above (Section III, Step 2). Cell lysates are analyzed for hemagglutinin expression by Western blotting as described below (Section IV, Step 6).
2. **Cell extracts** — For Western blotting and immunoprecipitation experiments, untransfected and transfected myoblasts are plated on FN (UT and hα5 transfected cells), on LM (UT and chicken α6 and chicken α61044t or human α6A transfected cells) or on gelatin (UT, PAX, and CD2FAK transfected cells) for 24 h in complete myoblast medium. Cells are washed with ice-cold PBS containing 1 mM Na-orthovanadate and

lysed in ice cold modified RIPA extraction buffer (20 mM Tris pH 7.4, 150 mM NaCl, 0.5% NP-40, 1.0% Triton X-100, 0.25% NaDeoxycholate, 2 mM EDTA, 2 mM EGTA) with protease inhibitors (20 mg/ml leupeptin, 0.7 mg/ml pepstatin, 1 mM phenanthroline, 2 mM phenyl-methyl-sulfonyl-chloride, 0.05 units aprotinin) and phosphatase inhibitors (30 mM Napyrophosphate, 40 mM NaF, 1 mM Na-orthovanadate). Protein content of the clarified lysates is determined using the Pierce bicinchoninic acid (BCA) method with bovine serum albumin as the standard.

3. **Phosphotyrosine immunoblots** — For phosphotyrosine Western blots, 10 to 15 μg of lysates are separated on 10% SDS-PAGE gels,[38] under reducing conditions and transferred to nitrocellulose membranes.[39] Membranes are blocked in 1% heat denatured BSA in TST buffer (10 mM Tris pH 7.5, 100 mM NaCl, 0.1% Tween-20) overnight at 4°C. Phosphotyrosine containing proteins are detected by incubating the membranes with the anti-phosphotyrosine mAb PY20 and a secondary horseradish peroxidase (HRP) conjugated anti-mouse antibody (Jackson Labs, West Grove, PA) or with RC20H, a directly conjugated HRP anti-phosphotyrosine Ab. Blots are visualized by chemiluminescence (Pierce Super Signal, Pierce Chemical Co, Rockford, IL). Membranes are exposed to X-ray film (Kodak, X-OMAT AR) and developed in an automatic film processor. When indicated, membranes are stripped in stripping buffer (62.5 mM Tris-Cl pH-6.8, 2% SDS and 100 mM β-mercaptoethanol) for 30 min at 60°C and reprobed with a different antibody.
4. **Paxillin, FAK, CD2-FAK, and hemagglutinin immunoblot analysis** — 5 to 20 μg cell lysates are resolved on 7.5% SDS-PAGE gels under reducing conditions and proteins transferred to nitrocellulose membranes. Membranes are blocked in TST buffer containing 3% non-fat milk and the proteins are detected with 165 mAb (anti-paxillin) (C. Turner, SUNY, Syracuse, NY), BC3 pAb (anti-FAK) (T. Parsons, University of Virginia, Charlottesville, VA), TS2/18.1.1. mAb (anti-CD2) (Developmental Studies Hybridoma Bank) or 12CA5 mAb (antihemagglutinin) (Boehringer Mannheim). For FAK immunoprecipitations, 100 μg of RIPA lysate is mixed with 1 μl of anti-FAK mAb, 2A7, 50 μl of packed agarose-anti-mouse beads (blocked in 5% BSA, Sigma) in a final volume of 500 μl. The bead-antibody-antigen complex is incubated at 4°C for 2 h with continuous agitation. For paxillin immunoprecipitations, 100 μg of cell lysate and 1 μl of anti-paxillin mAb, 165 are incubated at 4°C with continuous agitation for one hour. In a separate tube, 50 μl of packed protein A-agarose beads and 30 μg/ml rabbit anti-mouse IgG are incubated in lysis buffer for 1 h. The antigen-antibody mixture is then added to rabbit-anti-mouse-protein A beads and incubated at 4°C an additional 2 h. The beads are pelleted gently and washed twice with lysis buffer. Bound protein is released from the beads by boiling in 100 μl Laemmli sample buffer containing 5% β-mercaptoethanol for 5 min. Equal aliquots of the precipitated protein for each antibody are loaded onto 7% SDS-PAGE gels. The FAK IP is blotted for FAK with C-20 mAb (Santa Cruz Biotechnology, Inc., Santa Cruz, CA) and for phosphoFAK withRC20H mAb (Transduction Laboratories, Lexington, KY). The paxillin IP is blotted for paxillin with the 165 mAb (C. Turner, SUNY) or for phosphopaxillin with RC20H. All immunoprecipitations and Western blots are detected by chemiluminescence.
5. **Bioassays with constitutively activated (CA)-MEK and MEK inhibitors** — Myoblasts stably cotransfected with CA-MEK and human α6A are plated on LM coated plates and observed for 96 h for muscle α-actinin expression, fusion into myotubes, and proliferation. To alter MAP kinase activity in hα5 transfected myoblasts, hα5 expressing cells are grown in the presence of the specific MEK inhibitor PD98059 (New England Biolabs, MA).[40] Transfected myoblasts are plated on FN coated cover-

slips and on FN coated TC plates. After 8 h in complete myoblast medium, the first dose of the inhibitor is added to the cells at 1, 10, 25, 50, or 100 μMfinal concentration. Cells are grown for an additional 24 h and a second dose of the inhibitor is added. After 24 and 48 h in presence of the inhibitor, coverslips are fixed and immunostained for DAPI and muscle α-actinin. At the same time, cells are extracted in RIPA buffer as described and lysates are analyzed by Western blotting for active MAPK and total erk1 expression as described below.

6. **MAPK immunoblots** — Activation of MAP kinase is determined by two methods. Active MAPK is detected biochemically by Western blotting and by its translocation to the cell nucleus by immunofluorescence staining described in the next section. For anti-MAPK Western blots, cells are trypsinized, washed once with soybean trypsin inhibitor (0.5 mg/ml), washed twice in Puck's Saline G (Gibco-BRL), and resuspended in serum-free medium containing 2% BSA. Cells are held in suspension for 1 h prior to plating on FN or LM in complete myoblast medium for 15, 30, or 60 min, 24 or 48 h. Cell extracts are prepared in RIPA buffer as described. 5 μg of cell lysates are separated on 12% SDS-PAGE gels under reducing conditions and the proteins transferred to nitrocellulose membranes. The membranes are blocked in 3% non-fat dry milk in TST overnight at 4°C. Active MAPK is detected by an anti-active MAPK polyclonal Ab (pAb) (Promega, Madison, WI). Membranes are then stripped with stripping buffer (62.5 mM Tris-Cl pH-6.8, 2% SDS and 100 mM β-mercaptoethanol) for 30 min at 60°C and reprobed for total MAPK with an anti-Erk1 mAb (Transduction Labs) or SC-94 anti-Erk1 pAb (Santa Cruz Biotechnologies).
7. **MAP kinase immunolocalization** — For immunofluorescence staining, UT, hα5 or hα6A transfected myoblasts are trypsinized, washed once with soybean trypsin inhibitor (0.5 mg/ml in PBS), washed twice in Puck's Saline G, and held in suspension in serum-free medium containing 2% BSA for 1 h. UT and hα5 transfected cells are plated on FN or poly-L-lysine coated coverslips in the presence or absence of serum. UT and hα6A transfected cells are plated on LM or poly-L-lysine coated coverslips in serum or in serum-free medium. Approximately 6×10^4 cells are plated per 12 mm coverslip. After 15, 30, 60, or 120 min, coverslips are removed and cells fixed in 3% formaldehyde in PBS. Cells are permeabilized in 0.4% Triton X-100 in PBS for 10 min, washed with PBS, and blocked for 30 min in 5% BSA in PBS. An anti-MAPK pAb in 5% BSA in PBS is incubated for 30 min at room temperature followed by 30 min incubation with FITC-conjugated goat-anti-rabbit antibody (Cappell, Durham, NC) and DAPI to stain the nuclei. The activation of MAPK is expressed as the percentage of total nuclei that contained MAPK staining in the nucleus as a function of time.

V. Integrins and Growth Factors Cooperatively Regulate the Onset of Terminal Differentiation in Myoblasts

A. Overview

Sustained inhibition of muscle-specific gene transcription by growth factors requires constitutive activation of the respective receptors,[41] suggesting that transient signaling changes mediate this suppression. This corresponds with early observations that myoblasts differentiated only upon serum withdrawal or depletion.[8,42] It has been well known that myogenic transcription factors, even when expressed at relatively high levels, fail to activate muscle-specific genes in the presence of soluble peptide mitogens such as fibroblast growth factor (FGF) and insulin.[10,43] Since integrins can mediate constituents of growth factor signaling pathways, it is likely that specific combinations of soluble peptide mitogens and integrin subunits can differentially determine myoblast phenotype in terms of cell cycle withdrawal and gene expression.

To begin demonstrating this, Sastry et al. (1996) assayed the effects of different serum conditions on the differentiation of untransfected, α5-, and α6-transfected myoblasts.[15,45] They discovered that untransfected and α6-transfected myoblasts exhibit substantial differentiation in a complete medium, containing a high serum concentration and embryo extract, as well as in low serum or serum-free conditions. In contrast, the α5 transfected myoblasts terminally differentiate only in a medium containing less than ~2% serum or serum-free conditions, although the organization of muscle α-actinin did not occur as in untransfected control cells. This indicates that a major effect of ectopic α5 expression is an enhanced susceptibility to the proliferation promoting effects of serum growth factors.

To examine the effects of specific growth factor and integrin signal integration on the cell cycle transition from proliferation to differentiation, α5-transfected myoblasts were exposed to various soluble peptide mitogens, and the extent to which the addition of these soluble mitogens affected the proliferative characteristics of these cells was assayed. In the presence of bFGF or TGF-β, fewer than 5% of α5 transfected myoblasts fuse into myotubes or express muscle α-actinin. Additionally, bFGF and TGF-β also produce opposite effects on proliferation of the hα5 transfected myoblasts. bFGF appears to mimic the mitogenic effects of a rich medium and stimulates the proliferation of α5 transfectants, while TGF-β inhibits proliferation of these myoblasts. Insulin, TGF-α, or PDGF-BB produces phenotypes distinct from those exhibited with bFGF or TGF-β. Insulin has no measurable effect on untransfected cells but promotes both proliferation and differentiation of α5-transfectants. However, the resultant myotubes are abnormally short and highly branched, while muscle α-actinin is not organized in striations. TGF-α, similar to insulin, promotes proliferation, although to a lesser extent, as well as differentiation of α5 transfected myoblasts, where the organization of muscle α-actinin into striations was present in the majority of hα5 transfected myotubes. PDGF-BB has no effect on proliferation or on the serum-free differentiation of α5 transfected myoblasts, but like TGF-α, promotes myofibril organization. Interestingly, among all growth factors tested, only TGF-α stimulates proliferation of the untransfected myoblasts growing under serum-free conditions. These data clearly indicate that the growth factor environment can modulate mitogenic signaling initiated by integrins, and that this interaction between adhesive and growth factor signaling pathways has implications on cell phenotype.

B. Protocols

1. **Growth factor supplementation**— Untransfected or transfected myoblasts are trypsinized, washed twice with Puck's Saline G (Gibco-BRL), and seeded onto fibronectin (α5 transfected cells; 20 μg/ml) or laminin (α6 transfected cells; 40 μg/ml) coated 12 mm glass coverslips at a density of 1.6×10^4 cells/cm^2 in complete myoblast medium or 3.2×10^4 cells/cm^2 in serum-free medium (DMEM + 2% bovine serum albumin) and cultured for indicated times in Sastry et al., 1996.[15] Myoblasts are grown in low serum medium (DMEM + 2% horse serum). Growth factors are added to untransfected and α5 transfected myoblasts in serum-free medium at the time of plating and twice daily at concentrations of: insulin (Sigma; 10 μg/ml), bFGF (from S. Hauschka and Sigma; 10 ng/ml), TGF-β (Sigma; 1 ng/ml), TGF-α (Calbiochem; 10 ng/ml), PDGF-BB (Sigma; 20 ng/ml).

VI. Discussion and Future Directions

One of the central themes produced by our research on primary myoblasts is that enhanced integrin signaling, either through the β1A cytoplasmic domain or proximal effectors such as (CD2)-FAK and paxillin, delays the onset of terminal differentiation by upregulating an adhesion-mediated mitogenic pathway that uses MAPK. Conversely, molecules that interfere with integrin signaling, such as inhibitors of extracellular signal-related kinase (MEK or MAPK kinase), functional mutants of paxillin, and the α6 integrin cytoplasmic domain, downregulate MAPK activity and promote cell cycle withdrawal and muscle differentiation. Enhanced integrin signaling may be activating a number of proto-oncogenes (see Reference 3 for review) either through its interaction with constituents of growth factor pathways and/or perhaps through an independent connection to more downstream effectors. Adaptor molecules at the focal adhesion have also been shown to link integrins to cell cycle regulatory mechanisms.[46] Thus, differential binding of the diverse number of adaptors and binding domains present on proteins at the focal contact site may also contribute to fine-tuning integrin regulation of cellular processes by modulating the total signaling strength contributed by the extracellular environment.

Further research using the primary quail cell culture model may focus on the molecular basis of integrin and growth factor synergy. There are several possible molecular mechanisms for the integrative response of myoblasts to stimuli from both matrix and soluble peptide mitogens. Although it seems increasingly unlikely, adhesive and growth factor signals may, through independent pathways, activate common downstream effectors. More likely, key components of both pathways are linked through physical interactions. The focal adhesion itself, for instance, may provide a site in which the recruitment of structural and signaling proteins could occur and assemble a signaling center where adhesive and mitogenic signals are cumulatively transduced.

This cell culture system provides a relatively straightforward and productive means to study the phenotypic effects of integrin signaling. Perturbing normal signaling processes required for normal myoblast differentiation, by ectopically expressing forms of integrins and putative intracellular signaling molecules, is a useful method in which to investigate integrin mediation of cell cycle withdrawal and gene activation. The results discussed previously are also consistent with results obtained from *in vivo* models of keratinocyte differentiation,[47] and largely concur with biochemical data obtained from other cell culture models used to investigate cell proliferation and survival.[48] Hence, data provided through this model may also give insight into integrin regulation of these processes in other tissue types.

References

1. Howe, A., Aplin, A. E., Alahari, S. K., and Juliano, R., Integrin signaling and cell growth control, *Curr. Opin. Cell Biol.*, 10, 220, 1998.
2. Roskelly, C., Srebrow., A., and Bissell, M., A hierarchy of ECM-mediated signaling regulates tissue-specific gene expression, *Curr. Opin. Cell Biol.*, 7, 736, 1995.
3. Alema, S. and Tato, F., Interaction of retroviral oncogenes with the differentiation program of myogenic cells. *Adv. Canc. Res.*, 49, 1987.
4. Konigsberg, I., Clonal analysis of myogenesis, *Science*, 140, 1273, 1963.
5. Okazaki, K. and Holtzer, H., Myogenesis: Fusion, myosin synthesis and the mitotic cycle, *Proc. Natl. Acad. Sci USA*, 56, 1484, 1966.
6. Lassar, A. B., Paterson, B. M., and Weintraub, H., Transfection of a DNA locus that mediates the conversion of 10T1/2 fibroblasts into myoblasts, *Cell*, 47, 649, 1986.
7. Pownell, M. E. and Emerson, C. E., Jr., Sequential activation of three myogenic regulatory genes during somite morphogenesis in quails, *Dev. Biol.*, 151, 67, 1992.
8. Bischoff, R. and Holtzer, H., Mitosis and the process of differentiation of myogenic cells *in vitro*, *J. Cell Biol.*, 36, 111, 1969.
9. Pownell, M. E. and Emerson, C. E., Jr., Molecular and embryological studies of avian myogenesis, *Semin. Dev. Biol.*, 3, 229, 1992.
10. Weintraub, H., Tapscott, S. J., Davis, R. L., Thayer, M. J., Adam, M. A., Lassar, A. B., and Miller, D., Activation of muscle-specific genes in pigment, nerve, fat, liver, and fiborblast cell lines by forced expression of MyoD, *PNAS*, 86, 5434, 1989.
11. Olson, E. N., Proto-oncogenes in the regulatory circuit for myogenesis, *Semin. Cell Biol.*, 2, 127, 1992.
12. Konigsberg, I., Skeletal myoblasts in culture, in *Methods in Enzymology*, Vol. 58, Academic Press, Inc., NY, 1979, 511.
13. Antin, P. and Ordahl, C., Isolation and characterization of an avian myogenic cell line, *Dev. Biol.*, 143, 111, 1991.
14. DiMario, J., Fernyak, S., and Stockdale, F., Myoblasts transferred to the limbs of embryos are committed to specific fibre fates, *Nature*, 362, 165, 1993.
15. Sastry, S., Lakonishok, M., Thomas, D., Muschler, J., and Horwitz, A. F., Integrin α subunit ratios, cytoplasmic domains, and growth factor synergy regulate muscle proliferation and differentiation, *J. Cell Biol.*, 133, 169, 1996.
16. DeCurtis, I., Quaranta, V., Tamura, R., and Reichardt, L., Laminin receptors in the retina: sequence analysis of the chick integrin $\alpha 6$ subunit, *J. Cell Biol.*, 113, 405, 1991.

17. Shaw, L., Lotz, M., and Mercurio, A., Inside-out integrin signaling in macrophages: analysis of the role of the $\alpha6A\beta1$ and $\alpha6B\beta1$ integrin variants in laminin adhesion by cDNA expression in an $\alpha6$ integrin-deficient macrophage cell line, *J. Biol. Chem.*, 268, 11401, 1993.
18. Takada, Y. and Puzon, W., Identification of a regulatory region of integrin $\beta1$ subunit using activating and inhibiting antibodies, *J. Biol. Chem.*, 268, 17597, 1993.
19. Reszka, A. A., Hayashi, Y., and Horwitz, A. F., Identification of amino acid sequences in the integrin $\beta1$ cytoplasmic domain implicated in cytoskeletal interactions, *J. Cell Biol.,* 117, 1321, 1993.
20. Tahiliani, P. D., Singh, L., Auer, K. L., and LaFlamme, S. E, The role of conserved amino acid motifs within the integrin beta3 cytoplasmic domain in triggering focal adhesion kinase phosphorylation, *J. Biol. Chem.*, 272, 7892, 1997.
21. Catling, A., Schaeffer, H., Reuter, C., Reddy, G., and Weber, M., A proline-rich sequence unique to MEK1 and MEK2 is required for raf binding and regulates MEK function, *Mol. Cell. Biol.*, 15, 5214, 1995.
22. Brown, M., Perrotta, J., and Turner, C., Identification of LIM3 as the principal determinant of paxillin focal adhesion localization and characterization of a novel motif on paxillin directing vinculin and focal adhesion kinase binding, *J. Cell Biol.*, 135, 1109, 1996.
23. Chan, P., Kanner, B., Whitney, G., and Aruffo, A., A transmembrane-anchored chimeric focal adhesion kinase is constitutively activated and phosphorylated at tyrosine residues identical to pp125FAK, *J. Biol. Chem.*, 269, 20567, 1994.
24. Menko, A. and Boettiger, D., Occupation of the extracellular matrix receptor, integrin, is a control point for myogenic differentiation, *Cell*, 51, 51, 1987.
25. Blaschuk, K. and Holland, P., The regulation of $\alpha5\beta1$ integrin expression in human muscle cells, *Dev. Biol.*, 164, 475, 1994.
26. Bronner-Fraser, M., Artinger, M., Muschler, J., and Horwitz, A., Developmentally regulated expression of the $\alpha6$ integrin in avian embryos, *Development*, 115, 197, 1992.
27. Juliano, R. and Haskill, S., Signal transduction from the extracellular matrix, *J. Cell Biol.*, 120, 577, 1993.
28. Jewell K., Kapron-Bras, C., Jeevaratnam, P., and Dedhar, S., Stimulation of tyrosine phosphorylation of distinct proteins in response to antibody-mediated ligation and clustering of $\alpha3$ and $\alpha6$ integrins, *J. Cell Sci.*, 108, 1165, 1995.
29. LaFlamme, S. E., Akiyama, S. K., and Yamada, K. M., Regulation of fibronectin receptor distribution, *J. Cell Biol.,* 117, 437, 1992.
30. Schwartz, M., Schaller, M., and Ginsberg, M., Integrins: emerging paradigms of signal transduction, *Annu. Rev. Cell Dev. Biol.,* 11, 549, 1995.
31. Crowley, E. and Horwitz, A.F., Tyrosine phosphorylation and cytoskeletal tension regulate the release of fibroblast adhesions, *J. Cell. Biol.*, 131, 525, 1995.
32. Burridge, K., Turner, C., and Romer, L., Tyrosine phosphorylation of paxillin and pp125(FAK) accompanies cell adhesion to extracellular matrix–a role in cytoskeletal assembly, *J. Cell Biol.*, 119, 893, 1992.
33. Turner, C., Glenney, J., and Burridge, K., Paxillin: a new vinculin-binding protein present in focal adhesions, *J. Cell Biol.*, 111, 1059, 1990.
34. Chen, Q., Kinch, M., Lin, T., Burridge, K., and Juliano, R., Integrin-mediated cell adhesion activates mitogen-activated protein kinases, *J. Biol. Chem.*, 269, 26602, 1994.

35. Schlaepfer, D., Hanks, S., Hunter, T., and van der Geer, P., Integrin-mediated signal transduction linked to Ras pathway by GRB2 binding to focal adhesion kinase, *Nature*, 372, 786, 1994.
36. Mainiero, F., Pepe, A., Wary, K., Spinardi, L., Mohamaddi, M., Schlessinger, J., and Giancotti, F., Signal transduction by the α6β4 integrin: distinct β4 subunit sites mediate the recruitment of shc/GRB2 and association with the cytoskeleton of hemidesmosomes, *EMBO J.*, 14, 4470, 1995.
37. Bennet, A. and Tonks, N., Regulation of distinct stages of skeletal muscle differentiation by mitogen-activated protein kinases, *Science*, 278, 1288, 1997.
38. Laemmli, U., Cleavage of the structural proteins during the assembly of the head of bacteriophage T4, *Nature*, 227, 680, 1970.
39. Towbin, H., Staehelin, T., and Gordon, J., Electrophoretic transfer of proteins from polyacrylamide gels to nitrocellulose sheets: procedure and some applications, *Proc. Natl. Acad. Sci. USA*, 76, 4350, 1979.
40. Alessi, D., Cuenda, A., Cohen, P., Dudley, D., and Saltiel, A., PD 098059 is a specific inhibitor of the activation of mitogen-activated protein kinase kinase *in vitro* and *in vivo*, *J. Biol. Chem.*, 270, 27489, 1995.
41. Florini, J. and Magri, K., Effects of growth factors on myogenic differentiation. *Am. J. Physiol.,* 256, 701, 1989.
42. Bonner, P. H. and Hauschka, S. D., Clonal analysis of vertebrate myogenesis. Early developmental events in the chick limb, *Dev. Biol.,* 37, 317, 1974.
43. Olson, E. N., MyoD family: a paradigm for development? *Genes Dev.*, 9, 1454, 1990.
44. Zhu, X. and Assoian, R., Integrin-dependent activation of MAP kinase: a link to shape-dependent cell proliferation, *Mol. Biol. Cell*, 6, 273, 1995.
45. Sastry, S. and Horwitz, A. F., Adhesion-growth factor interactions during differentiation: an integrated biological response, *Dev. Biol.*, 180, 455, 1996.
46. Wary, K., Mainiero, F., Isakoff, S., Marcantonio, E., and Giancotti, F., The adaptor protein shc couples a class of integrins to the control of cell cycle progression, *Cell*, 87, 733, 1996.
47. Carroll, J., Romero, M., and Watt, F., Suprabasal integrin expression in the epidermis of transgenic mice results in developmental defects and a phenotype resembling psoriasis, *Cell*, 83, 957, 1995.
48. Frisch, S., Vuori, K., Ruoslahti, E., and Chan-Hui, P., Control of adhesion-dependent cell survival by focal adhesion kinase, *J. Cell Biol.*, 134, 793, 1996.
49. Sastry, S., Lakonishok, M., Wu, S., Truong, T., Turner, C., Huttenlocher, A., and Horwitz, A., *J. Cell Biol.* (in press).

Chapter 11

Methods for Studying Anoikis

Steven M. Frisch

Contents

I. Introduction 161
II. Assaying the Effects of Genes on Anoikis in Stable Expression Experiments 162
A. Generating Stable Expressions 162
B. Assaying Stable Expressors for Anoikis 162
III. Transient Assays for the Effects of Genes on Anoikis 163
A. β-Galactosidase/DAPI Double-Staining Method 164
B. Keratin 18-Cleavage Transient Assay for Effects of Transgenes on Anoikis 164
IV. Materials 165
References 166

I. Introduction

Epithelial cells deprived of normal matrix contact through the appropriate integrin heterodimer undergo apoptosis; this phenomenon has been termed "anoikis," the ancient Greek word meaning "homelessness."[1] Anoikis prevents shed epithelial cells from colonizing elsewhere.[2] This process is important for normal development, and its loss can contribute significantly to tumor malignancy (reviewed in Reference 3).

Assays for the effects of compounds or genes on anoikis are currently under development, several of which are described here. Each has its advantages and disadvantages. The most time-consuming and conventional is stable expression of new genes in epithelial cells, and assaying the resulting cell lines for anoikis relative

0-8493-3385-7/99/$0.00+$.50

to the parental cells; this assay has been used successfully to show the role of certain signaling molecules such as FAK and MEKK-1.[4,5] This chapter will specify the stable gene transfer and apoptosis assay methods that work best in our experience with epithelial cells. In addition, we describe transient assays that are much faster but whose track record is less established.

II. Assaying the Effects of Genes on Anoikis in Stable Expression Experiments

A. Generating Stable Expressors

MDCK cells are a widely used epithelial cell model system. It has two drawbacks, though. First, being a canine cell line limits the choice of antibodies and nucleic acid probes, although many anti-human antibodies or probes work in this species. MDCK cells are also poorly transfectable, prompting the use of retroviral vectors.

We mainly use the vector pBABE.[6] This vector drives expression of the insert through the viral LTR enhancer and has an internal SV40 promoter to drive the expression of a puromycin-resistance gene for selection. Following subcloning of the gene of interest into pBABE (usually in a FLAG-, myc-, or HA-epitope-tagged form), the retrovirus vector is transfected by the standard calcium phosphate method into the amphotropic packaging cell line ϕNX,[7] which is much more efficient than previous packaging cell lines. Two to three days after transfection, the viral supernatant is removed, cleared by centrifugation ($3000 \times g$ for 10 min), polybrene is added to give 4 μg/ml final concentration, and it is applied to a subconfluent monolayer of MDCK cells. The MDCK cells can either be on tissue culture plastic, or, for higher efficiency, on permeable cell culture inserts of 25 mm diameter and 3.0 micron pore size (Falcon), which fit into 35 mm wells. For the latter, the viral supernatant is placed on the cells from which all media have been removed, and then allowed to flow through under gravity, which takes about 1 min per ml of supernatant. The infected cells are washed briefly with medium and incubated for one day, followed by being trypsinized and replated on tissue culture dishes at colony-forming densities. After cell attachment (6 to 8 h), puromycin is then added to give 1.5 μg/ml, and colonies are selected with refeeding every three days, for a total elapsed time of about two weeks. Colonies are then ring-isolated, transferred to 10 mm diameter wells, expanded, and lysates are tested for expression of the transgene by Western blotting using anti-epitope antibody. The total elapsed time for this process is about one month.

B. Assaying Stable Expressors for Anoikis

1. Plate out mdck cells on 2.4 cm cell culture inserts (Falcon 3090), grow to confluence. Continue growing one additional day (cells beyond confluence are preferred).
2. Trypsinize cells, spin down, resuspend in 2.0 ml of medium and count cells. Transfer 5×10^5 cells to a 2.0 ml microfuge tube and fill the tube to the top with medium.

3. Place on wheel in 37° incubator for 2.5 to 3.5 h.
4. Microfuge cells for 8 sec at 8000 rpm (in an Eppendorf model 5414 microfuge), suction off supernatant, wash with 600 μl of PBS by inverting, re-spin in microfuge.
5. Resuspend pellets in 600 μl of apoptosis lysis buffer (ALB) by pipetting up and down with a P1000. Transfer to a microfuge tube. (Note that the cells are extracted with Triton rather than SDS, thus leaving behind intact genomic DNA. Non-apoptotic cells will produce blank lanes on the gel as a result).
6. Vortex 20 sec. Spin out debris in cold microfuge 10 min at maximum speed.
7. Transfer supernatant to new microfuge tube. Phenol-chloroform extract three times with 550 μl of phenol-chloroform each time.
8. Transfer final aqueous phase (380 μl) to a microfuge tube, and add 40 μl of 2.5 M NaAc pH 7.3, 1.5 μl of 20 mg/ml glycogen and 1 ml cold ethanol. Precipitate overnight.
9. Spin in cold microfuge 14 min. Take off supernatant using P1000.
10. Wash with 300 μl cold 70% ethanol. Respin 4 min, take off supernatant with P200. Respin briefly, take off as much of remainder as possible with P20.
11. Redissolve pellets in 25 μl of 50 mM NaCl in TE containing 1 μl of RNase. (Boehringer, DNase-free) (make a master mix for all samples); incubate at 37° for 20 min.
12. Add 3.3 μl of Ficoll loading buffer, load onto a 1.5% agarose gel in TBE (can put 10 μl of ethidium bromide in 200 ml gel before pouring). Run blue dye halfway to end of gel, photograph gel.

III. Transient Assays for the Effects of Genes on Anoikis

We have employed two methods to detect anoikis of transiently transfected cells. In the first method, a vector containing the *E. coli* β-galactosidase gene under control of a mammalian promoter expresses the gene of interest. Following transfection, cells are placed in suspension for various periods of time, then doubly stained for nuclear morphology with the fluorescent nuclear stain DAPI and for β-galactosidase activity using the viable stain ImaGene Green. The β-galactosidase-positive cells are then scored for apoptotic nuclear morphology.

The second method requires no microscopy and is, therefore, more efficient. Epithelial cells undergoing apoptosis activate caspases that specifically cleave the keratin 18 protein to yield 2 discrete products.[8] We have taken advantage of this observation to develop a new assay for anoikis. We tagged the keratin 18 gene N-terminally with a myc epitope and substituted this into pcDNA3.1 in place of the neomycin-resistance gene. The gene of interest (i.e., the candidate anoikis-regulator) is subcloned into the polylinker downstream of the CMV promoter, and this resulting plasmid is transfected into epithelial cells. Following expression, the cells are placed in suspension, and total cell lysates are Western blotted, using myc epitope antibody. The ratio of cleavage product to intact K18 is a measure of the degree of anoikis. Note that the assay only scores transfected cells, thus providing a high degree of sensitivity and selectivity; it is also simple to perform. Although the transfection frequency may vary from one sample to the next, this does not affect the final results, as they are automatically normalized against intact, myc-K18.

A. β-Galactosidase/DAPI Double-Staining Method

We have inserted a β-galactosidase coding sequence into the Sfv-epitope transcription unit of pHook2 (Invitrogen), thus placing it under the control of the RSV promoter. This leaves most of the cloning sites in the polylinker downstream of the CMV available for inserting the candidate anoikis-regulating gene. The complete sequence of pHook2 can be found on Invitrogen's home page (www.invitrogen.com). The resulting plasmids can be transiently transfected to test the effect of any coding sequence on anoikis, using nuclear morphology of β-galactosidase-positive cells as an index.

1. Ethanol precipitate 75 μg of plasmid DNA and redissolve it in 0.5 ml of cytomix.
2. Trypsinize a 50% confluent 150 mm dish of MDCK cells and spin down.
3. Wash cells once with 5.0 ml of sterile PBS and re-spin.
4. Resuspend cells in the DNA solution and transfer to a 0.4 cm gap BioRad cuvette, keep on ice.
5. Electroporate at 0.3 kV, 960 μF (BioRad gene pulser), put cuvette back on ice and keep on ice for 10 min.
6. Resuspend cells and transfer to a tube containing 5.0 ml of complete medium, spin down cells.
7. Resuspend cells and plate onto one 25 mm cell culture insert (0.2 or 1.0 micron) for transient anoikis assay (see Steps 10 to 13 below).
8. Next day, wash dead cells off the top of the filter and re-feed.
9. Cells should be confluent enough the day after that (48 h post-electroporation) to perform anoikis assay (see Steps 10 to 13 below).
10. The cells are trypsinized and placed in suspension as described for the stable transfection protocol above.
11. During the final 15 min of suspension, both stains are added: ImaGene Green is added to give 33 μM final concentration, and DAPI to give 1 μM.
12. The cells are then centrifuged for 8 sec at 10,000 rpm in an Eppendorf 5414 microfuge, washed twice with PBS + 2 mg/ml BSA, resuspended in 30 uL of the latter, and transferred to a microscope slide.
13. The doubly stained cells are viewed on a fluorescence microscope, using an FITC filter to find transfected cells and a UV filter to assess normal vs. apoptotic nuclear morphology.

B. Keratin 18-Cleavage Transient Assay for Effects of Transgenes on Anoikis

We have created an anoikis reporter plasmid that contains the N-terminally tagged human K18 coding sequence with a myc epitope tag in place of the neomycin resistance gene (SmaI-BstBI site) of pcDNA3.1—(Invitrogen; the complete sequence of the parental plasmid is on the homepage www.invitrogen.com). Coding sequences candidate anoikis-regulators are blunt-end ligated into the PmeI site of this plasmid, resulting in their being transcribed from the CMV promoter.

Cells are electroporated and suspensed as described above. Following suspension for various times (generally, 0, 1 h, 2 h and 3 h provides a good time course for MDCK cells), cells are washed twice with PBS and lysed by boiling in 300 μL SDS-PAGE sample buffer per time point (4 time points can be obtained from one 25 mm diameter cell culture insert of cells). Western blots are run on 14% minigels (Novex), electroblotted onto PVDF filters and probed with anti-myc epitope (clone 9E10) monoclonal antibody (Oncogene Science). Following development using enzyme-linked chemiluminescence (Pierce) we scan the blot on a BioRad phosphorimager and quantitate the band intensities so as to reveal relative degrees of apoptosis, as judged by percentage of myc-K18 cleaved.

IV. Materials

MDCK cells: American Type Culture Collection

Tissue culture medium: DME-high glucose (Irvine Scientific) supplemented with 10% fetal bovine serum (Gibco/BRL) and 1X glutamine-penicillin-sreptomycin (from 100X stock, Gibco/BRL).

2.4 cm cell culture inserts: Falcon #3090

Trypsin (0.25%)-EDTA: Gibco-BRL

Apoptosis lysis buffer (ALB): 10 mM Tris pH 8, 10 mM EDTA, 0.5% Triton X-100.

Sodium acetate: 2.5 M NaOAc, Ph 7.3

Glycogen: nuclease-free, Boehringer-Mannheim

TE master mix: 50 mM NaCl in TE containing 10 uL RNase (Boehringer, Dnase-free) per ml.

Ficoll loading buffer: 10% Ficoll (MW 400,000; Sigma)/20 mM EDTA/0.05% bromphenol blue

1.5% agarose gel in 1X TBE (5X TBE = 54 g Tris, 27.5 g boric acid, 20 ml 0.5 ml of 0.5 M EDTA per liter).

Cytomix: 120 mM KCl, 0.15 mM $CaCl_2$, 10 mM potassium phosphate, pH 7.6, 25 mM HEPES, pH 7.6, 2 mM EGTA, 5 mM $MgCl_2$, Adjust to pH 7.6, before use, take out a 50 ml aliquot, and add ATP to give 2 mM and glutathione to give 5 mM, readjust the pH to 7.6 on pH meter and filter-sterilize

0.4 cm gap electroporation cuvettes: BioRad

Gene Pulser electroporator: BioRad

ImaGene Green reagent: Molecular Probes

DAPI: Sigma

PBS containing 2 mg/ml bovine serum albumin (Sigma)

SDS-PAGE sample buffer: 62.5 mM Tris, pH 6.8, 0.4% SDS, 10% glycerol, 0.05% bromphenol blue.

14% Tris-glycine minigels: Novex Corp.

Immobilon transfer membranes: Millipore Corp.
Anti-myc epitope antibody Ab-1: Oncogene Science/Calbiochem
Super-signal chemiluminescent detection kit: Pierce

References

1. Frisch, S. M. and Francis, H., Disruption of epithelial cell-matrix interactions induces apoptosis, *J. Cell Biol.*, 124, 619, 1994.
2. Bourdreau, N., Sympson, C., Werb, Z., and Bissell, M., Supression of ICE and apoptosis in mammary epithelial cells by extracellular matrix, *Science*, 267, 891, 1995.
3. Frisch, S. and Ruoslahti, E., Integrins and anoikis, *Curr. Opin. Cell Biol.*, 9, 701, 1997.
4. Frisch, S., Vuori, K., Ruoslahti, E., and Chan, P. Y., Control of adhesion-dependent cell survival by focal adhesion kinase, *J. Cell Biol.*, 134, 793, 1996.
5. Cardone, M., Salvesen, G., Widmann, C., Johnson, G., and Frisch, S., Regulation of anoikis: caspases activate MEKK-1, *Cell*, 90, 315, 1997.
6. Morgenstern, J. and Land, H., Advanced mammalian gene transfer: high titre retroviral vectors with multiple drug selection markers and a complementary helper-free packaging cell line, *Nucl. Acids Res.*, 18, 3587, 1990.
7. Kinsella, T. and Nolan, G., Episomal vectors rapidly and stably produce high-titer recombinant retrovirus, *Hum. Gene Ther.*, 7, 1405, 1996.
8. Caulin, C., Salvesen, G., and Oshima, R., Caspase cleavage of keratin 18 and reorganization of intermediate filaments during epithelial cell apoptosis, *J. Cell Biol.*, 138, 1379, 1997.

Chapter 12

Functional Analysis of FAK and Associated Molecules in Cell Migration

Leslie Cary and Jun-Lin Guan

Contents

I. Introduction 167
II. CHO Cell Stable Transfection 169
  A. General Considerations 169
  B. Protocol 169
  C. Material and Equipment 170
III. Cell Migration 171
  A. General Considerations 171
  B. Protocol 173
  C. Material and Equipment 174
IV. Biochemical Analysis of FAK and Associated Proteins 175
  A. General Considerations 175
  B. Protocol 176
  C. Material and Equipment 178
Acknowledgments 179
References 179

I. Introduction

Cell migration plays a critical role in many biological processes, including embryonic development, wound healing, and tumor metastasis.[1] Cell migration through extracellular matrix (ECM) proteins is generally mediated by cell surface integrin

0-8493-3385-7/99/$0.00+$.50

receptors.[2-5] In addition to serving as ECM adhesion receptors, integrins transduce a variety of biochemical signals across the plasma membrane.[6-8] Although many components of integrin signaling pathways have been identified, the molecular mechanisms by which integrins regulate cell migration processes are not well understood. Focal adhesion kinase (FAK),[9,10] a non-receptor protein tyrosine kinase which is localized to focal contacts in cultured fibroblasts, has been identified as an important mediator of integrin signal transduction pathways.[11] FAK is activated and tyrosine phosphorylated in response to integrin activation, such as by plating cells on the appropriate ECM.[9,12-14] FAK associates with a number of signaling molecules including Src family kinases,[15,16] Grb2,[17] phosphatidylinositol 3-kinase (PI 3-kinase),[18,19] and p130^{Cas}.[20] It also associates with the cytoskeletal proteins paxillin[21,22] and talin,[23] and is believed to associate with tensin.[24]

FAK is clearly an important mediator of integrin signaling, but its roles in specific cellular functions have only recently begun to be identified.[25-29] We have recently shown that stable overexpression of FAK in Chinese hamster ovary (CHO) cells increased cell migration on fibronectin (FN).[30] These results are consistent with other reports suggesting a role for FAK in cell migration using other approaches.[25,27] Our CHO cell transfection system has allowed us to dissect the molecular mechanisms of FAK-dependent cell motility by analyzing various FAK mutants. We have shown that FAK association with Src family members is crucial for its promotion of cell migration.[30] In addition, FAK association with p130^{Cas} plays a critical role for cell motility, while FAK binding to Grb2 is not important. We have also correlated p130^{Cas} tyrosine phosphorylation by the FAK/Src complex with increased cell motility.[31] Using this system, we are examining the roles of signaling molecules downstream of the FAK/Src/p130^{Cas} complex in integrin-mediated cell migration.

Our CHO cell transfection system for functional analysis of FAK provides many advantages over others. A stable overexpression system allows for repeated and reliable analysis both in cell biological and biochemical assays of a cell clone that has been screened for high levels of expression of the transfected gene. The use of a chemotaxis chamber assay for motility analysis is advantageous since results can be obtained within one day, and they are generally more reliable from one experiment to another than results from other motility assays. It also allows for more versatility in the types of questions asked about the motility, both in the type of attractant that is used and in the way it is presented to the cells. In this chapter, we describe methods for the stable expression of FAK in CHO cells, analysis of motility of transfected cell clones, and biochemical analysis of relevant proteins (FAK, Src, and p130^{Cas}). Although the methods described here are specific for FAK and associated proteins, after consideration of certain options, these methods may be adapted to other signaling molecules.

II. CHO Cell Stable Transfection

A. General Considerations

A system of functional analysis of a gene by its stable overexpression provides advantages over a transient transfection system, mainly that it allows for repeated and reliable analysis of a cell population, while a transiently overexpressing cell population will be more susceptible to changes based on varying expression levels or cell conditions. It also allows for selection of a clonal cell line in which a high percentage of cells are expressing the gene of interest, which is not possible in a transiently transfected population. The main disadvantage of the system is that a considerable amount of time is required for the establishment of stable cell clones.

If this method is chosen, several options need to be considered by the investigator. First, one needs to choose a cell line that is expected to both respond to the appropriate signals and be capable of overexpressing the gene of interest. Endogenous expression of the gene suggests that a particular cell line will meet the above criteria; however, the system will then be complicated by the presence of this endogenous protein. In this case, the use of an epitope tag should be considered in order to distinguish the transfected protein from the endogenous protein. Many epitope tags and monoclonal antibodies are available commercially. Varying degrees of success have been reported for these reagents, which may depend on the location of the epitope tag in the protein, among other factors. In addition, some epitope tag/antibody combinations work well for certain techniques but not for others (e.g., Western blotting, immunoprecipitation, or immunofluorescence). The expression vector to be used is another important consideration. Most important, one should choose an appropriate promoter for the cell line to be transfected. If expression of the gene is suspected to be toxic or growth inhibitory, an inducible expression system may be used. For example, vectors for tetracycline-regulated gene expression are available commercially, which have proven successful in a number of cell lines. Finally, whether the antibiotic resistance marker is located on the expression vector or on a different vector should be considered, since a cotransfection of two vectors may result in a lower percentage of positive clones (those expressing the gene) in the total population of antibiotic-resistant cells.

B. Protocol

1. Plate cells on a 35 mm tissue culture dish 1 day prior to transfection in growth medium so that they are 60 to 80% confluent on the day of transfection. Set up one plate for a mock transfection.
2. Follow the manufacturer's protocol for LipofectAMINE using 1 μg of pCDM8-FAK, 0.1 μg of pSV2neo, and 6 μl of LipofectAMINE. Incubate the DNA and LipofectAMINE for 45 min, and leave it on the cells for 5 h before adding optiMEM medium with 20% FBS. Incubate overnight.
3. Change the medium, replacing it with 2 ml of growth medium per plate. Incubate overnight.

4. Harvest the cells with trypsin and selection medium. Seed 10 cm plates with 1:20, 1:50, and 1:100 dilutions of the 35 mm culture. Dishes are seeded at different cell densities because transfection toxicity and efficiency vary from experiment to experiment. Seed one 10 cm plate of the mock transfection (1:20 dilution of the original culture) to determine the progress of selection. The remaining cells from the 35 mm dishes may be lysed and saved for future analysis.
5. Change the medium on each plate every 3 days, replacing it with 10 ml of selection medium.
6. When the antibiotic selection is complete (i.e., when any remaining mock transfection cells no longer appear healthy, generally between 10 to 14 days), clones may be selected from the pool. Harvest the cells with trypsin and selection medium, then count and dilute them in selection medium and seed a 96-well plate with 1 cell per well. After 4 to 5 days, observe all wells and note those which have only 1 colony. When these wells are approximately 70% confluent, harvest and transfer the cells to a 24-well plate.

   Alternative method: If the colonies on the 10 cm plate are well separated, then glass cloning rings may be used to isolate them. Mark each individual colony on the bottom of the dish. Wash the plate with PBS, and use sterile grease to attach a cloning ring around the colony. Harvest the cells in the cloning ring using trypsin and selection medium, and transfer the cells to a 24-well plate.
7. When the cells on the 24-well plate are nearly confluent, harvest them and transfer approximately 1/10 of the culture to a new 24-well plate. Screen the remaining cells for expression of the transfected gene by Western blotting of cell lysates. Pellet the remaining cells in an Eppendorf tube, wash them once with PBS, and lyse them with NP-40 lysis buffer (see Section IV.B, Steps 7 to 9). Analyze the lysates by SDS-PAGE and Western blotting with an appropriate antibody, such as that against the epitope tag (if present).

   Alternative method: If the antibody to be used for screening works well for immunofluorescence staining, then clones may be screened by this method. Harvest the cells and transfer 1/10 of the culture to a new 24-well plate. Plate the remaining cells on a sterile glass coverslip in a 24-well plate with selection medium, and allow the cells to attach and spread (several hours or overnight). Fix the cells, then permeabilize them (each step for 10 to 15 min at room temperature), and stain them with primary antibody followed by secondary antibody (each step for 30 min at 37°C). Mount the coverslips on slides using Slowfade and seal them with nail polish.
8. When positive clones have been identified and expanded, prepare and freeze multiple stock vials of them as soon as possible for future use. Maintain transfected cells in selection medium.

C. Material and Equipment

Growth medium: F-12 medium with 10% fetal bovine serum (FBS) (Gibco BRL)
Selection medium: growth medium with 0.5 mg/ml G418 (Gibco BRL)
Phosphate-buffered saline (PBS)
Trypsin (0.06% in PBS with 10 µg/ml phenol red and 0.2 mg/ml EDTA)
OptiMEM medium (Gibco BRL)
OptiMEM medium with 20% FBS

LipofectAMINE (Gibco BRL)
Plasmid DNA (pCDM8-FAK and pSV2neo), transfection quality
Glass cloning rings and grease (optional)
NP-40 lysis buffer (see Section IV.C.)
Reagents for SDS-PAGE and Western blotting

Immunofluorescence fixative solution (optional)

1. Dissolve 36 g of paraformaldehyde in 600 ml of distilled H_2O in a fume hood.
2. Add approximately 2 ml of 10N NaOH.
3. Stir for approximately 30 min while heating to approximately 45°C.
4. Add 90 ml of 10x PBS, then add distilled H_2O to a total volume of 900 ml.
5. Aliquot and either use immediately or store at –20°C.
6. Thaw aliquots in a heated water bath of at least 37°C.

Immunofluorescence permeabilization solution (0.5% Triton X-100 in PBS) (optional)
Primary and secondary antibodies for immunofluorescence staining, diluted in PBS with 10% goat serum (optional)
Slowfade mounting solution (optional) (Molecular Probes, Eugene, OR)
Nail polish (optional)

III. Cell Migration

A. General Considerations

The use of a chemotaxis chamber to analyze cell motility provides many advantages over other motility assays. Results can be obtained within one day, and they are generally very reliable from one experiment to another. Another popular method for analyzing cell motility is one in which a "wound" is created in a monolayer of confluent cells and migration of cells into the wound is analyzed. Results from this assay may be affected by the wound size or by proliferation of the cells upon release of cell-cell contacts, among other factors, and therefore results may vary significantly from one day to the next. In contrast, the chemotaxis chamber allows for cell motility analysis that is largely independent of cell-cell interactions. In addition, it allows for more versatility in the types of motility experiments carried out. One can determine the chemotactic response of cells to a soluble attractant such as growth factors. In the case of ECM proteins, one can distinguish between chemotaxis (soluble form) and haptotaxis (matrix form). By varying the relative concentration levels of the attractant in the upper and lower wells, one can distinguish between taxis (directed motility toward the attractant) and kinesis (random motility stimulated by the attractant). These questions are not possible with many other motility assays, such as the wound method.

If this method is chosen, several factors need to be considered. First, there are many types of chemotaxis chambers available. For example, some chambers are

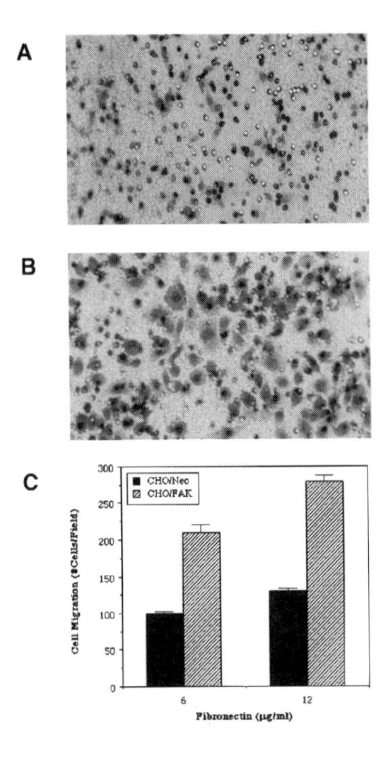

FIGURE 1
Transfected CHO cell migration on fibronectin. Cell migration of CHO cells stably transfected with FAK (CHO/FAK) or control cells (CHO/Neo) were analyzed using a chemotaxis chamber. CHO/Neo (A) or CHO/FAK (B) cells which migrated on 12 μg/ml fibronectin were fixed, stained, and photographed at 100× magnification, or were counted using a light microscope at 200× magnification (C). Mean cell counts from 5 experiments and at least 23 microscopic fields are shown. Error bars represent standard errors. No significant migration was observed in the absence of fibronectin (not shown).

compatible with 96-well microplates, allowing for a variety of quantitation methods using a microplate reader; others allow for smaller sample sizes and quantitation by simply counting the cells using a light microscope; still others require larger sample sizes, and allow for the collection and analysis of the migrated and/or unmigrated cell populations. Once the appropriate chamber has been chosen, the migration experiment will need to be optimized for any given cell type and attractant. Some of the most important factors to consider are the number of cells to load on the chamber, the attractant concentration and loading method (e.g., soluble ECM protein vs. ECM protein precoated in a matrix), and the incubation time. The following protocol is intended specifically for measuring the migration of transfected CHO cells using a 48-well chemotaxis chamber and soluble fibronectin as an attractant (results are shown in Figure 1), and should therefore be modified for other cell types or conditions. Note that if the ECM protein is loaded in its soluble form, during the course of the experiment it will likely coat the membrane to form a matrix. Also note that if a soluble attractant (such as a growth factor) is used, the membrane should be precoated (usually with collagen) to allow cells to attach.

Attention to certain steps in this migration assay will minimize problems. The chamber components were molded as one set and therefore fit together in a particular conformation. One should pay attention to the Neuro Probe (NP) trademark in the lower right corner of the lower and upper chamber components. One also needs to ensure that no air is trapped on either side of the membrane as this will prevent contact of the membrane with the reagents. It is recommended that a manual p200 pipettor is used to load both the lower and upper chambers, and the liquid should not be expelled completely from the pipette tip. For the lower chamber, one must be careful to load the correct volume of reagent so that a slight positive meniscus is formed over the well. For the upper chamber, a useful technique is to hold the pipette tip vertically against the side of the well and expel the liquid quickly from the pipette tip. These tips may prove useful for loading the chamber. Finally, the chamber components need to be cleaned and maintained properly to prevent the accumulation of debris. The components should be immersed in water immediately after disassembly of the chamber and washed thoroughly. After approximately five experiments, the components should be cleaned enzymatically and then washed thoroughly. Under no conditions should the chamber be autoclaved or heated to above 55°C.

B. Protocol

Loading the chamber

1. Plate the FAK-transfected or control CHO cell clones on 10 cm plates in selection medium and grow to 70 to 90% confluency.
2. Remove the medium and wash the cells once with PBS. Remove the PBS and add 1 ml of trypsin per plate until cells detach, then add 4 to 6 ml of trypsin inhibitor, pipette the cells off the plate, and transfer them to a centrifuge tube.
3. Pellet the cells by centrifugation. Remove the trypsin inhibitor, add 5 ml of F-12 medium, and centrifuge the cells again. Repeat this wash step two more times. After the final wash, resuspend the cells in F-12 medium, count and dilute them in F-12 medium to 2.5×10^5 cells per ml.
4. Serially dilute the fibronectin in F-12 medium to 12, 6, and 3 µg/ml and keep on ice until ready to use.
5. Set a pipettor to approximately 30 µl (the volume varies from one pipettor to another). Using a round gel-loading tip, pipette the fibronectin to the lower wells of the chamber. There should be a slight positive meniscus on the wells.
6. Using forceps to handle it, cut the polycarbonate membrane in the lower right corner (shiny side facing up) and gently place it over the wells of the lower chamber. Avoid extensive shifting of the membrane after it has been placed on the wells.
7. Place the silicone gasket over the membrane, with the cut corner on the lower right.
8. Place the top chamber over the gasket, with the NP trademark on the lower right. Hold the top chamber down with even (but not excessive) pressure on all sides while tightening the thumb nuts.

9. Resuspend the cells by pipetting briefly. Using round gel-loading pipette tips, load 50 µl of cells to each well, holding the pipette tip vertically against the side of the well and expelling the liquid quickly but not completely from the pipette tip.
10. Wrap the chamber in plastic, and incubate at 37°C and 5% CO_2 for 6 h.

Removing and staining the membrane

11. Remove the thumb nuts and disassemble the chamber.
12. Remove the membrane and immediately place it in methanol for 5 to 10 min to fix the cells.
13. Immediately and thoroughly wash the upper chamber, lower chamber, and silicone gasket in distilled H_2O, then allow the components to air-dry.
14. Remove the membrane from the methanol and allow it to air-dry. Incubate it in Giemsa stain for approximately 60 min, then rinse it briefly (10 sec) in distilled H_2O.
15. Drain the excess H_2O from the membrane, then cut it to separate each section of 12 wells, making sure to keep track of its orientation by cutting the lower right corner. Place each section on a glass coverslip, and use a damp cotton swab to wipe the unmigrated cells from the top of the membrane. Invert the coverslip and place it on a slide, attaching it with a small amount of nail polish.
16. Count the number of migrated cells per field under a light microscope at approximately 200× magnification.

Cleaning the chamber

17. After disassembling the chamber, immediately soak all components in distilled H_2O to prevent the accumulation of debris.
18. Clean the chamber enzymatically after approximately 5 experiments in order to remove debris. After rinsing briefly in distilled H_2O, immerse the upper chamber, lower chamber, and silicone gasket in Tergazyme solution and incubate at 55°C for 60 min.
19. Rinse the chamber components several times with distilled H_2O, soak them in distilled H_2O overnight, and then allow them to air-dry.

C. Material and Equipment

Selection medium (see Section II.C)
PBS
Trypsin
F-12 medium
Soybean trypsin inhibitor (0.5 mg/ml in F-12 medium), prepared fresh (Sigma)
Fibronectin (FN) (Gibco BRL or Sigma)

1. Dissolve human plasma fibronectin in room temperature PBS at 1 mg/ml.
2. Incubate at room temperature until FN is in solution.
3. Aliquot and store at –20°C or –80°C and avoid freeze/thaw cycles. Store the thawed aliquot at 4°C to prevent the FN from precipitating.

Neuro Probe 48-well chemotaxis chamber, including: (Neuro Probe Inc., Cabin John, MD)

- lower chamber
- silicone gasket
- upper chamber
- thumb nuts

Polycarbonate membranes (8 μm pore size) (Neuro Probe Inc.)
Methanol
Giemsa stain (diluted 1:10 in distilled H_2O) (Sigma)
Tergazyme solution (7.6 g/L in distilled H_2O) (Fisher)
Nail polish

IV. Biochemical Analysis of FAK and Associated Proteins

A. General Considerations

If an effect on cell motility is observed as a result of expression of a gene, the relevant biochemical pathways that are involved will likely be examined. The following protocols provide information about examination of FAK and associated proteins. These protocols may be modified to study other aspects of integrin signaling, although certain considerations must be made. For example, when looking for a biochemical response in cells plated on FN, the amount of time for which the cells are plated before lysis may vary depending on both the desired response (e.g., FAK phosphorylation vs. Erk activation) and the cell type. The lysis conditions can also be crucial; for example, the detection of FAK/p130^{Cas} association is greatly affected by both lysis buffer and cell type. Finally, the antibodies that are used in any experiment will greatly affect the results and should be chosen carefully. The use of an epitope tag on the expressed gene should be chosen carefully, since its effectiveness can vary for different techniques (see Section II).

A method outlined below describes the detection of *in vivo* FAK association with other molecules (Src or p130^{Cas}) by coimmunoprecipitation and Western blotting. In addition to this method, there are others which take advantage of the enzymatic activity of either FAK or the FAK-associated molecule, and for this reason they may be more sensitive than Western blotting. For example, *in vivo* FAK/Src association has been demonstrated by *in vitro* kinase assays on anti-FAK immunoprecipitates, followed by denaturation of the complex and immunoprecipitation of the phosphorylated Src with anti-Src.[32] Likewise, sensitive PI 3-kinase assays have been used on anti-FAK immunoprecipitates to demonstrate *in vivo* FAK/PI 3-kinase association.[18] It may therefore be advantageous to modify the protocols outlined below, particularly the detection methods within the protocol, to suit a particular experiment.

B. Protocol

Tyrosine Phosphorylation of FAK in Cells on Fibronectin

1. Grow cells to 70 to 90% confluency on 10 cm tissue culture plates in growth or selection medium.
2. Coat 10 cm tissue culture plates with 4 ml of either 10 µg/ml fibronectin or 50 µg/ml poly-L-lysine in PBS, and incubate them at room temperature for 2 h or at 4°C overnight. Wash the plates three times with PBS.
3. Add 4 ml of BSA solution to the coated plates. Incubate them at 37°C for 1 to 2 h. Wash the plates three times with PBS.
4. Harvest the cells with trypsin and soybean trypsin inhibitor (see migration protocol, Section III.C). Pellet them by centrifugation and wash them three times with F-12 medium. After the final wash, resuspend the cells in F-12 medium.
5. Remove the PBS from the coated plates and add the cells. Incubate the plates at 37°C and 5% CO_2 for 10 min.
6. Remove the medium and wash the cells with cold PBS. Remove the PBS and add approximately 500 µl of cold NP-40 lysis buffer. Keep the plate either on ice or at 4°C.
7. Use a cell scraper to collect cell lysates, then transfer them to Eppendorf tubes using a pipettor. Incubate either on ice or at 4°C for at least 20 min.
8. Centrifuge the lysates at 4°C for 10 min in a microfuge at maximum speed.
9. Transfer the supernatants to new Eppendorf tubes. Determine the total protein concentration of the lysates using the Protein Assay Dye. Store the lysates at –20°C (short term) or –80°C (long term) and minimize freeze/thaw cycles.
10. Immunoprecipitate FAK from equal amounts of cell lysate (approximately 500 µg) using an appropriate anti-FAK antibody (e.g., mAb KT3 against the epitope tag on transfected CHO cell FAK). Incubate the cell lysates with the anti-FAK antibody for 1 h or overnight in a tube rotator at 4°C. Add 40 to 50 µl of Protein A sepharose beads (if anti-FAK is from rabbit) or Protein A beads coupled to rabbit anti-mouse IgG (if anti-FAK is from mouse) and incubate for 1 h in a tube rotator at 4°C. Wash the beads four times with NP-40 lysis buffer and divide the samples in half.
11. Remove the lysis buffer and add concentrated SDS-PAGE sample buffer. Heat the samples at 95 to 100°C for 3 to 5 min, then pellet the beads by brief centrifugation. Load the supernatants onto an SDS-PAGE gel, separate the proteins by electrophoresis, and transfer them to a nitrocellulose membrane. Analyze each sample by Western blotting with either mAb PY20 or an anti-FAK antibody, followed by a HRP-conjugated secondary antibody and ECL detection.

In vitro FAK Kinase Assay

An example of *in vitro* FAK autophosphorylation and FAK phosphorylation of an exogenous substrate is shown in Figure 2.

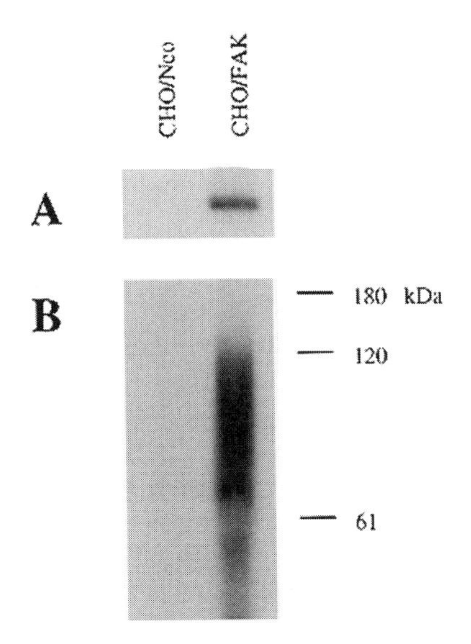

FIGURE 2
***In vitro* FAK kinase activity.** Lysates were prepared from CHO cells stably overexpressing FAK (CHO/FAK) or from control cells (CHO/Neo). Exogenous FAK was immunoprecipitated using the KT3 mAb and subjected to an *in vitro* kinase assay using [γ-^{32}P] ATP in the absence (A) or presence (B) of poly-Glu_4Tyr_1 (E_4Y_1) as an exogenous substrate. FAK autophosphorylation is shown in panel A, and FAK phosphorylation of E_4Y_1 is shown in panel B. Molecular mass positions (kDa) are shown on the right.

1. Prepare an SDS-PAGE gel appropriate for the experiment (e.g., 7.5% gel for detecting FAK autophosphorylation, 10% gel for detecting E_4Y_1 phosphorylation).
2. Immunoprecipitate FAK from approximately 500 µg of cell lysate, as described above (Section IV.B). Wash the beads three times with NP-40 lysis buffer and then once with kinase buffer. After the final wash, resuspend the beads in 40 µl of kinase buffer.
3. If E_4Y_1 will be used as a substrate, add 10 µg of it to the beads.
4. Add 10 µCi of [^{32}P]ATP to each sample. Incubate at room temperature for 20 min with occasional agitation. Stop the reaction by adding the appropriate volume of concentrated SDS-PAGE sample buffer.
5. Heat the samples at 95 to 100°C for 3 to 5 min. Pellet the beads by brief centrifugation in a microfuge. Load equal volumes (approximately one half of the sample) onto the SDS-PAGE gel and separate the proteins by electrophoresis. Store the remaining half of the samples at –20°C for future use (if necessary).
6. When the gel is finished, rinse it several times with distilled H_2O and dry it at 80°C for approximately 1 h.
7. Expose the gel to X-ray film with an intensifying screen at –80°C for approximately 1 to 16 h depending on the intensity of the signal.

In vivo FAK/Src or FAK/p130^{Cas} Association in CHO cells

These protocols describe the detection of exogenous FAK association with either endogenous Src or p130^{Cas} in CHO cells. Note that FAK/p130^{Cas} association can be difficult to detect, and the conditions for these coimmunoprecipitations will differ depending particularly on the cell type used.

1. Plate FAK-transfected or control CHO cells on 10 cm tissue culture plates and grow to 70 to 90% confluency in selection medium.
2. Remove the medium and wash the cells with cold PBS. Remove the PBS and add approximately 500 µl of cold lysis buffer (NP-40 or modified RIPA lysis buffer for FAK/Src or FAK/p130^{Cas} association, respectively). Prepare lysates as described above (Section IV.B, Steps 7 to 9).
3. Immunoprecipitate equal amounts of cell lysates (approximately 1 mg) with either anti-Src or anti-p130^{Cas}. Incubate the cell lysates with the primary antibody in a tube rotator at 4°C overnight. Add 40 to 50 µl of either Protein A sepharose beads (for anti-p130^{Cas}) or beads coupled to rabbit anti-mouse IgG (for anti-Src) to each sample. Incubate in a tube rotator at 4°C for 1 h. Wash the beads four times with lysis buffer and divide the samples into 3/4 and 1/4 fractions.
4. After the final wash, remove the lysis buffer and add concentrated SDS-PAGE sample buffer. Heat the samples at 95 to 100°C for 3 to 5 min, then pellet the beads by brief centrifugation. Load the supernatants onto an SDS-PAGE gel, separate the proteins by electrophoresis, and transfer them to a nitrocellulose membrane. Analyze the samples by Western blotting either the 3/4 fraction with anti-FAK (e.g., mAb KT3 against the epitope tag on transfected CHO cell FAK) or the 1/4 fraction with either anti-Src or anti-p130^{Cas}, followed by HRP-conjugated secondary antibodies and ECL detection.

C. Material and Equipment

Fibronectin (see Section III.C)
Poly-L-lysine (1 mg/ml in PBS, store at 4°C) (Sigma)
Bovine serum albumin (BSA) (2 mg/ml in PBS, heat inactivate at 55°C for 1 h and sterile filter, store at 4°C)
Growth or selection medium (see Section II.C)
PBS
Trypsin
F-12 medium
Soybean trypsin inhibitor (see Section III.C)
Phenylmethylsulfonylfluoride (PMSF) (200 mM in ethanol) (Sigma)
Aprotinin (Sigma)
Leupeptin (10 mg/ml) (Sigma)
NP-40 lysis buffer

1. Prepare lysis buffer (1% NP-40, 20 mM Tris-HCl pH 7.4, 137 mM NaCl, 10% glycerol, and 1 mM sodium vanadate) and store at 4°C.
2. To an aliquot of lysis buffer for 1 day's use, add protease inhibitors to final concentrations of 1 mM PMSF, 1% aprotinin, and 20 µg/ml leupeptin.

Modified RIPA lysis buffer

1. Prepare lysis buffer (50 mM Tris-HCl pH 7.4, 150 mM NaCl, 5 mM EDTA, 1% NP-40, 1 mM sodium vanadate, 0.5% sodium deoxycholate, and 0.05% SDS)

2. For 1 day's use, add protease inhibitors to an aliquot of lysis buffer, as above.

Protein Assay Dye (BioRad)

Antibodies:

anti-FAK (prepared in our lab[18])

KT3 mAb (gift of Dr. Gernot Walter,[33] and described previously[30,31])

anti-p130^{Cas} (Santa Cruz Biotechnology Inc.)

anti-Src mAb 2-17 (Quality Biotech Inc., Camden, NJ)

rabbit anti-mouse IgG (Sigma)

PY20 (Transduction Laboratories)

HRP-conjugated anti-mouse or anti-rabbit (Amersham Life Science Inc.)

Protein A beads in lysis buffer

1. Add lysis buffer to lyophilized Protein A sepharose beads (Sigma) and swell them in a tube rotator at 4°C for 1 h or overnight.
2. Wash the beads three times in lysis buffer.
3. After the final wash, add fresh lysis buffer so that the final packed bead volume is 50% of the total volume (i.e., a 50% slurry). Store at 4°C.

Reagents for SDS-PAGE and Western blots

ECL detection reagents (Amersham Life Science Inc.)

Kinase buffer (50 mM Tris pH 7.4, 10 mM $MnCl_2$), prepared fresh

Poly-Glu_4Tyr_1 (E_4Y_1) (optional) (10 mg/ml, store aliquots at –80°C) (Sigma)

[γ-^{32}P] ATP (3000 Ci/mmol, 10 mCi/ml) (NEN)

Acknowledgments

Work in authors' laboratory is supported by NIH and the American Heart Association.

References

1. Hynes, R. O. and Lander, A. D., Contact and adhesive specificities in the associations, migrations and targeting of cells and axons, *Cell*, 68, 303, 1992.
2. Hemler, M. E., VLA proteins in the integrin family: structures, functions, and their roles in leucocytes, *Annu. Rev. Immunol.*, 8, 365, 1990.
3. Ruoslahti, E., Integrins, *J. Clin. Invest.*, 87, 1, 1991.
4. Hynes, R. O., Integrins: versatility, modulation and signaling in cell adhesion, *Cell*, 69, 11, 1992.
5. Brown, K. E. and Yamada, K. M., The role of integrins during vertebrate development, *Semin. Dev. Biol.*, 6, 69, 1995.
6. Juliano, R. L. and Haskill, S., Signal transduction from the extracellular matrix, *J. Cell Biol.*, 120, 577, 1993.
7. Schaller, M. D. and Parsons, J. T., Focal adhesion kinase: an integrin-linked protein tyrosine kinase, *Trends Cell Biol.*, 3, 258, 1993.

8. Clark, E. A. and Brugge, J. S., Integrins and signal transduction pathways: the road taken, *Science*, 268, 233, 1995.
9. Hanks, S. K., Calalb, M. B., Harper, M. C., and Patel, S. K., Focal adhesion protein tyrosine kinase phosphorylated in response to cell spreading on fibronectin, *Proc. Natl. Acad. Sci. USA,* 89, 8487, 1992.
10. Schaller, M. D., Borgman, C. A., Cobb, B. C., Reynolds, A. B., and Parsons, J. T., pp125^{FAK}, a structurally distinctive protein tyrosine kinase associated with focal adhesions, *Proc. Natl. Acad. Sci. USA,* 89, 5192, 1992.
11. Schwartz, M. A., Schaller, M. D., and Ginsberg, M. H., Integrins: emerging paradigms of signal transduction, in *Annu. Rev. Cell Dev. Biol.*, 11, Spudich, J. A., Gerhart, J., McKnight, S. L., and Schekman, R., Eds., Annual Reviews Inc., Palo Alto, 1995, 549.
12. Guan, J.-L., Trevithick, J. E., and Hynes, R. O., Fibronectin/integrin interaction induces tyrosine phosphorylation of a 120-kDa protein, *Cell Regul.*, 2, 951, 1991.
13. Burridge, K., Turner, C. E., and Romer, L. H., Tyrosine phosphorylation of paxillin and pp125^{FAK} accompanies cell adhesion to extracellular matrix: a role in cytoskeletal assembly, *J. Cell Biol.*, 119, 893, 1992.
14. Kornberg, L., Earp, S. E., Turner, C. E., Procktop, C., and Juliano, R. L., Signal transduction by integrins: increased protein tyrosine phosphorylation caused by clustering of β1 integrins, *Proc. Natl. Acad. Sci. USA,* 88, 8392, 1991.
15. Cobb, B. S., Schaller, M. D., Leu, T.-H., and Parsons, J. T., Stable association of pp60^{src} and pp59^{fyn} with the focal adhesion-associated protein kinase, pp125^{FAK}, *Mol. Cell. Biol.*, 14, 147, 1994.
16. Xing, Z., Chen, H.-C., Nowlen, J. K., Taylor, S. J., Shalloway, D., and Guan, J.-L., Direct interaction of v-Src with the focal adhesion kinase mediated by the Src SH2 domain, *Mol. Biol. Cell*, 5, 413, 1994.
17. Schlaepfer, D. D., Hanks, S. K., Hunter, T., and van der Geer, P., Integrin-mediated signal transduction linked to ras pathway by Grb2 binding to focal adhesion kinase, *Nature*, 372, 786, 1994.
18. Chen, H.-C. and Guan, J.-L., Association of focal adhesion kinase with its potential substrate phosphatidylinositol 3-kinase, *Proc. Natl. Acad. Sci. USA,* 91, 10148, 1994.
19. Guinebault, C., Paytraste, B., Racaud-Sultan, C., Mazarguil, H., Breton, M., Mauco, G., Plantavid, M., and Chap, H., Integrin-dependent translocation of phosphoinositide 3-kinase to the cytoskeleton of thrombin-activated platelets involves specific interactions of p85α with actin filaments and focal adhesion kinase, *J. Cell Biol.*, 129, 831, 1995.
20. Polte, T. R. and Hanks, S. K., Interaction between focal adhesion kinase and Crk-associated tyrosine kinase substrate p130^{Cas}, *Proc. Natl. Acad. Sci. USA,* 92, 10678, 1995.
21. Turner, C. E. and Miller, J., Primary sequence of paxillin contains putative SH2 and SH3 domain binding motif and multiple LIM domains: identification of a vinculin and pp125^{FAK} binding region, *J. Cell Sci.*, 107, 1583, 1994.
22. Schaller, M. D. and Parsons, J. T., pp125^{FAK}-dependent tyrosine phosphorylation of paxillin creates a high-affinity binding site for Crk, *Mol. Cell. Biol.*, 15, 2635, 1995.
23. Chen, H.-C., Appeddu, P. A., Parsons, J. T., Hildebrand, J. D., Schaller, M. D., and Guan, J.-L., Interaction of focal adhesion kinase with cytoskeletal protein talin, *J. Biol. Chem.*, 270, 16995, 1995.

24. Miyamoto, S., Akiyama, S., and Yamada, K. M., Synergistic roles for receptor occupancy and aggregation in integrin transmembrane function, *Science*, 267, 883, 1995.
25. Ilic, D., Furuta, Y., Kanazawa, S., Takeda, N., Sobue, K., Nakatsuji, N., Nomura, S., Fujimoto, J., Okada, M., Yamamoto, T., and Aizawa, S., Reduced cell motility and enhanced focal contact formation in cells from FAK-deficient mice, *Nature*, 377, 539, 1995.
26. Frisch, S. M., Vuori, K., Ruoslahti, E., and Chan-Hui, P. Y., Control of adhesion-dependent cell survival by focal adhesion kinase, *J. Cell Biol.*, 134, 793, 1996.
27. Gilmore, A. P. and Romer, L. H., Inhibition of focal adhesion kinase (FAK) signaling in focal adhesions decreases cell motility and proliferation, *Mol. Biol. Cell*, 7, 1209, 1996.
28. Hungerford, J. E., Compton, M. T., Matter, M. L., Hoffstrom, B. G., and Otey, C. A., Inhibition of pp125^{FAK} in cultured fibroblasts results in apoptosis, *J. Cell Biol.*, 135, 1383, 1996.
29. Richardson, A. and Parsons, J. T., A mechanism for regulation of the adhesion-associated protein tyrosine kinase pp125^{FAK}, *Nature*, 380, 538, 1996.
30. Cary, L. A., Chang, J. F., and Guan, J.-L., Stimulation of cell migration by overexpression of focal adhesion kinase and its assocation with Src and Fyn, *J. Cell Sci.*, 109, 1787, 1996.
31. Cary, L. A., Han, D. C., Polte, T. R., Hanks, S. K., and Guan, J.-L., Identification of p130^{Cas} as a mediator of focal adhesion kinase-promoted cell migration, *J. Cell Biol.*, 140, 211, 1998.
32. Schlaepfer, D. D., Broome, M. A., and Hunter, T., Fibronectin-stimulated signaling from a focal adhesion kinase-c-Src complex: involvement of the Grb2, p130^{Cas}, and Nck adaptor proteins, *Mol. Cell. Biol.*, 17, 1702, 1997.
33. MacArthur, H. and Walter, G., Monoclonal antibodies specific for the carboxy terminus of simian virus 40 large T antigen, *J. Virol.*, 52, 483, 1984.

Chapter 13

Integrin Signaling in Pericellular Matrix Assembly

Chuanyue Wu

Contents

I. Introduction 184
A. Interactions Between Fibronectin and the Cell Surface Integrins are Required for the Initiation of Fibronecting Matrix Assembly 185
B. Specific Fibronectin-Binding Integrins Mediate Fibronectin Matrix Assembly 185
C. Integrin Cytoplasmic Domains Regulate Fibronectin Matrix Assembly 186
D. Regulation of Fibronectin Matrix Assembly by Intracellular Protein Kinases 186
E. Regulation of Fibronectin Matrix Assembly by Small GTPases 187
II. Analysis of Fibronectin Matrix Assembly by Fluorescence Microscopy .. 189
A. Overview 189
B. Protocols 189
C. Technical Comments 190
D. Materials 191
III. Biochemical Analyses of Fibronectin Matrix Assembly 192
A. Overview 192
B. Protocols 193
C. Technical Comments 194
D. Materials 195
IV. Conclusion Remarks 196
Acknowledgments 196
References 197

0-8493-3385-7/99/$0.00+$.50

I. Introduction

The interactions of cells with pericellular fibronectin matrix play critical roles in cell adhesion, migration, growth, survival, differentiation, and gene expression. At least two components contribute specificity to this important recognition system. Responsive cells must express suitable receptors recognizing pericellular fibronectin matrix. In addition, the assembly of pericellular fibronectin matrix must be properly controlled. Our understanding of the molecular mechanism by which cells control pericellular fibronectin matrix assembly, although still incomplete, has advanced significantly over the last several years. This was achieved through a combination of biochemical, cell biological, and genetic experimental approaches. In this chapter, I will first discuss, briefly, some of the recent findings on the signaling mechanisms involved in the cellular control of fibronectin matrix assembly, primarily focusing on the roles of integrin signaling. For detailed discussion of fibronectin matrix assembly, readers are referred to several excellent review articles that have been previously published.[1-5] Next, I will describe two basic experimental methods for analyzing fibronectin matrix assembly, and discuss some of the major technical considerations in designing experiments aimed at evaluating cellular activity of fibronectin matrix assembly. Due to space limitation, many other useful methods could not be included; readers are encouraged to consult the original publications listed in the reference sections of this chapter and other review articles.[1-5]

Fibronectins are modular, multidomain glycoproteins of approximately 500,000 daltons consisting of repeating homologies of three types.[2,6,7] During organogenesis, fibronectins are major constituents of the extracellular matrix. Inactivation of the mouse fibronectin gene causes early embryonic lethality,[8] demonstrating that fibronectin matrix is indispensable for vertebrate embryogenesis. In adult mammals, most fibronectin is synthesized by the liver and exists primarily as a soluble plasma protein present at high concentration (about 0.3 mg/ml in human). Only with wounding, or frequently during tumorigenesis, is fibronectin expressed locally and deposited into extracellular matrix. The importance of a tightly controlled fibronectin matrix assembly process is further demonstrated by numerous studies on oncogenically transformed cells. These studies have shown that fibronectin matrix assembly is altered in most types of oncogenically transformed cells. For example, fibronectin matrix assembly is dramatically stimulated in hairy cell leukemia cells,[9,10] which likely contributes to the bone marrow fibrosis observed in hairy cell leukemia patients. On the other hand, in several types of adenocarcinomas, loss of fibronectin matrix is a marker for invasive and metastatic properties.[11-14] Treatment of cells with a polymeric form of fibronectin or other reagents that correct the abnormality in fibronectin matrix assembly resulted in reduced malignant behavior of several types of tumors.[15] Thus, fibronectin matrix assembly is a fundamentally important process in embryogenesis, and control of fibronectin matrix assembly is a potentially promising therapeutic strategy in cancer treatment.

A. Interactions Between Fibronectin and the Cell Surface Integrins are Required for the Initiation of Fibronectin Matrix Assembly

Fibronectin matrix assembly is a dynamic cellular process in which the soluble dimeric fibronectin molecules are assembled into an insoluble, disulfide-bond stabilized, fibrillar polymeric matrix.[1,3,16,17] At least three fibronectin sites (the first five type I repeats, the first type III repeat, and the RGD-containing integrin binding site) are involved in fibronectin matrix assembly.[18-25] One of the earliest events in fibronectin matrix assembly is the binding of fibronectin to the cell surface, which is mediated primarily by interactions between the RGD-containing integrin binding domain and specific fibronectin binding integrins. The binding of cell surface integrins to fibronectin via the RGD-containing 10th type III repeat not only retains fibronectin on the cell surface but also could trigger a series of sequential fibronectin-fibronectin interactions which result in the polymerization of fibronectin.[25] Antibodies to the RGD-containing integrin binding site or fibronectin fragments containing the integrin binding site inhibited fibronectin matrix assembly by cultured cells, resulting in fewer fibrils that are of normal length.[19,26] Moreover, mutational studies showed that fibronectin mutants lacking the RGD-containing integrin binding site could not be assembled into a fibronectin matrix in the absence of fibronectin containing the RGD site.[27] These results suggest that the interactions between fibronectin and the cell surface integrins are required for the initiation of fibronectin matrix assembly. Recently, Hocking et al. have found that the amino-terminal type I repeats, which are indispensable for fibronectin matrix assembly, could also interact with the $\alpha5\beta1$ integrins.[28] Taken together, these results support a central role of integrins in fibronectin matrix assembly.

B. Specific Fibronectin-Binding Integrins Mediate Fibronectin Matrix Assembly

Studies perturbing the expression or functions of integrins provide direct evidence for an essential role of specific integrins in the initiation of fibronectin matrix assembly. Function-blocking anti-$\alpha5$ or anti-$\beta1$ integrin antibodies inhibited fibronectin matrix assembly by cultured fibroblasts and developing amphibian embryos.[26,29-31] Overexpression of $\alpha5\beta1$ integrin in CHO cells with endogenous $\alpha5\beta1$ integrin increased fibronectin deposition in extracellular matrix,[32] whereas CHO cells that are devoid of $\alpha5\beta1$ integrin could not assemble a fibronectin matrix.[33] Reconstituting $\alpha5\beta1$ integrin expression by transfecting the $\alpha5$ deficient cells with a full length cDNA encoding the human $\alpha5$ integrin completely restored fibronectin matrix assembly.[33] On the other hand, expression of the $\alpha4\beta1$ integrin,[34] which bind to the LDV sequence in the CS1 (or V25) site within the alternatively spliced IIICS (or V) region of fibronectin, or $\alpha v\beta1$ integrin,[35] in the $\alpha5$ deficient cells failed to restored fibronectin matrix assembly. Thus, $\alpha5\beta1$ integrin, but not $\alpha4\beta1$ or $\alpha v\beta1$ integrin, initiates and is indispensable for fibronectin matrix assembly by these cells.

Although compelling evidence links $\alpha5\beta1$ integrin with fibronectin matrix assembly, the $\alpha5\beta1$ integrin is not the only integrin capable of initiating fibronectin matrix assembly. Both members of the $\beta3$ integrin subfamily ($\alpha IIb\beta3$ and $\alpha v\beta3$), which recognize the RGD-containing integrin binding site of fibronectin, can initiate fibronectin matrix assembly.[36,37] Extensive gene knockout, gene transfer, and function blocking studies indicate that at least one of the matrix assembly competent integrins ($\alpha5$, αIIb, or αv integrins) is required for mediating fibronectin matrix assembly.[36-39] These studies further confirm an essential role of the integrins in fibronectin matrix assembly.

C. Integrin Cytoplasmic Domains Regulate Fibronectin Matrix Assembly

Mutational studies have demonstrated that integrin cytoplasmic domains regulate fibronectin matrix assembly.[36,37,40,41] There are at least two major roles that integrin cytoplasmic domains play in fibronectin matrix assembly.[17,36] First, fibronectin matrix assembly cannot be initiated unless the integrins are activated to a high affinity fibronectin binding state (inside-out signaling).[36,37] Second, the integrin β cytoplasmic domain must interact with an intact actin cytoskeleton. Integrin mutants that have high fibronectin binding affinity but impaired cytoskeleton binding activity are unable to mediate fibronectin matrix assembly.[36] Similarly, cells expressing high fibronectin binding affinity integrins but lacking an intact actin cytoskeleton cannot assemble a fibronectin matrix. The importance of integrin cytoplasmic domains in fibronectin matrix assembly is further supported by recent studies using cells or embryos expressing different integrin splice variants that differ only in their cytoplasmic domains. Fibronectin matrix assembly in cells that exclusively express the $\beta1D$ integrin, which is normally expressed in heart and skeletal muscles,[42-45] differs significantly from that of cells expressing the $\beta1A$ integrin.[46,47] Moreover, replacement of $\beta1A$ integrin with the $\beta1D$ integrin in mouse by "knockin" of the $\beta1$ exon D resulted in abnormality in the assembly of fibronectin matrix *in vivo*, which likely contributes to the lethality of the $\beta1D$ knockin embryos.[47]

D. Regulation of Fibronectin Matrix Assembly by Intracellular Protein Kinases

The importance of integrin cytoplasmic domains in cellular control of fibronectin matrix assembly suggest that the matrix assembly process is likely to be regulated by integrin cytoplasmic domain binding proteins.[17] Integrin-linked kinase (ILK) is a 59 kDa serine/threonine protein kinase that was initially identified from a yeast two-hybrid screen using a bait plasmid expressing the $\beta1$ cytoplasmic domain.[48] ILK is capable of phosphorylating the $\beta1$ integrins, and appears to be involved in both "inside-out" and "outside-in" integrin-mediated signal pathways.[48-51] A nearly identical molecule (99% identical to the human ILK at the amino acid level) has been identified in the mouse.[52] In mouse skins, ILK is abundantly expressed throughout the extracellular matrix-rich dermis.[53] In contrast, ILK expression is dramatically down-regulated in the suprabasal layers of keratinocytes that are undergoing terminal differentiation.[53] The high expression level of ILK in the matrix-rich dermis may reflect, among other things, a positive regulatory role of this integrin-linked kinase in promoting extracellular matrix deposition *in vivo*. Indeed, overexpression of ILK (but not a kinase-inactive ILK mutant) in cultured epithelial cells dramatically stimulated integrin-mediated fibronectin matrix assembly.[50] In addition, overexpression of ILK in the epithelial cells down-regulated E-cadherin, and induced tumor formation *in vivo*.[50] The stimulation of fibronectin matrix assembly induced by overexpression of ILK appears to play a role in promoting "anchorage independent" cell growth, as inhibition of fibronectin matrix assembly decreased the ILK induced cell growth in soft agar.[50] These results provide strong evidence indicating that ILK is an important regulator of pericellular fibronectin matrix assembly and cell-cell interaction, and suggest a critical role of this integrin-linked kinase in cell growth and tumorigenesis.

Another serine/threonine protein kinase that has been implicated in regulation of fibronectin matrix assembly is protein kinase C. Protein kinase C regulates cell adhesion and spreading on fibronectin,[54,55] and inhibition of protein kinase C reduces actin stress fiber and focal adhesion formation.[56] Moreover, inhibitors of protein kinase C rapidly decrease the binding of fibronectin and its amino-terminal fragments containing the "matrix assembly site" to cells.[57]

The third protein kinase that has been implicated in regulation of fibronectin matrix assembly is Raf-1, a cytoplasmic serine/threonine protein kinase that plays important roles in growth factor receptor tyrosine kinase and G protein coupled signal transduction pathways. Activation of Raf-1 suppressed integrin activation and inhibited fibronectin matrix assembly.[58] The inhibition of integrin activation and fibronectin matrix assembly induced by Raf-1 activation is independent of protein synthesis and probably involves activation of ERK MAP kinase.[58]

E. Regulation of Fibronectin Matrix Assembly by Small GTPases

Several small GTPases including R-Ras, H-Ras, and Rho, which function as molecular switches cycling between an inactive GDP-bound conformation and an active GTP-bound conformation,[59] have been implicated in cellular regulation of fibronectin matrix assembly. Overexpression of a constitutively active R-Ras in cells activated integrins and promoted fibronectin matrix assembly.[60] Interestingly, overexpression of H-Ras, which is closely related to R-Ras, suppressed integrin activation.[58] Thus, in contrast to R-Ras, H-Ras likely functions as a negative regulator of fibronectin matrix assembly. Indeed, as mentioned above, induced activation of Raf-1, which is a downstream effector of H-Ras, also inhibited fibronectin matrix assembly.[58]

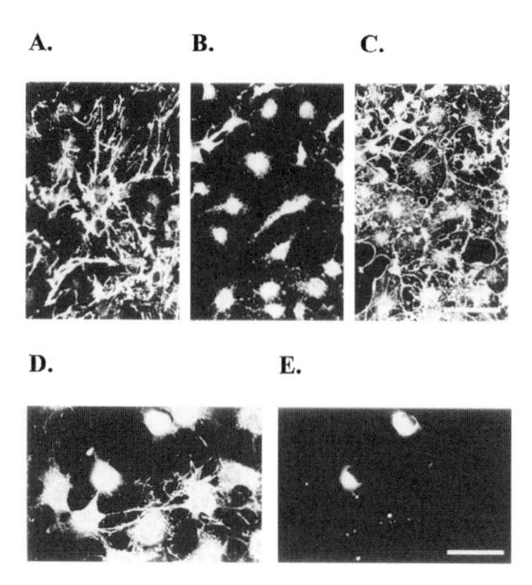

FIGURE 1

An active Rho is required for fibronectin matrix assembly. (A-C) Swiss 3T3 cells (1 ml of 5 × 10^4 cells/ml per 15 mm well) were cultured in DMEM containing 10% FBS for 8 days. The quiescent Swiss 3T3 cells were then cultured in DMEM media either in the presence (A) or in the absence (B and C) of serum for 16 hours, and fixed with 4% paraformaldehyde. Fibronectin matrix was visualized by staining the cells with a rabbit anti-fibronectin antibody (MC54) and a FITC-anti rabbit IgG secondary antibody (A and B). Actin filaments were visualized with TRITC-conjugated phalloidin (C). Note that serum-starved cells have very few fibronectin fibrils (B) and actin stress fibers. (D and F) Quiescent, serum-starved Swiss 3T3 cells were injected with 100 μg/ml C3 transferase and 0.5 mg/ml rat IgG, and then were grown in the presence of serum for 16 hours. The cells were fixed, and then stained with the rabbit anti-fibronectin antibody and an FITC-anti rabbit IgG secondary antibody (D) and a TRITC-conjugated anti-rat IgG secondary antibody (E). Panels D and F are images of an identical microscopic field observed with a FITC-filter set (D) or a TRITC-filter set (E). Note that two cells that were injected with the Rho inhibitor, as indicated by the presence of rat IgG (E), did not assemble fibronectin fibrils, while other cells did. The experiments described in this figure were designed by Dr. John A. McDonald, Dr. Alan Hall, and the author and performed by the author in Dr. Alan Hall's laboratory (MRC Laboratory For Molecular Cell Biology and Department of Biochemistry, University College London).

Several lines of evidence indicate that fibronectin matrix assembly requires Rho activity. First, lysophosphatidic acid, which activates Rho,[61] is a strong promoter of fibronectin matrix assembly.[62,63] Second, overnight-serum starvation of Swiss 3T3 cells, which suppresses Rho activity,[61] dramatically down-regulated fibronectin matrix assembly[64] (C. Wu, J. A. McDonald, and A. Hall, unpublished observations) (Figure 1). Third, direct inhibition of Rho activity with C3 transferase completely blocked fibronectin matrix assembly[63,64] (C. Wu, J.A. McDonald, and A. Hall, unpublished observations) (Figure 1). Finally, introducing a constitutively active Rho into quiescent Swiss 3T3 cells stimulated matrix assembly.[64]

Rho appears to regulate fibronectin matrix assembly via a mechanism that is distinctive from those of R-Ras and H-Ras. R-Ras and H-Ras regulate fibronectin matrix assembly through, at least in part, control of integrin activation.[58,60] Rho regulates fibronectin matrix assembly, on the other hand, primarily via its effect on actin cytoskeleton and cellular contractility.[63,64]

II. Analysis of Fibronectin Matrix Assembly by Fluorescence Microscopy

A. Overview

Fibronectin matrix is operationally defined as a fibronectin-rich, fibrillar polymeric extracellular matrix that is stabilized by disulfide-bonds and resistant to certain detergents (e.g., sodium deoxycholate) extraction. Two types of assays are frequently used in analysis of the assembly of a fibronectin matrix. The first one is based on the morphological criteria of a fibronectin matrix, which is typically carried out using an immunofluorescence staining procedure. The immunofluorescence staining procedure has been widely employed in investigating various aspects of fibronectin matrix assembly in cultured cells and embryos. A basic protocol is described below, which is based on what we have used for analysis of fibronectin matrix assembly in Chinese hamster ovary (CHO) cells. It may also be applied to other cell types, although modifications of culture media and other specific experimental conditions may be needed (see Section II.C., Technical Comments for discussion).

B. Protocols

All experimental procedures are performed at room temperature unless otherwise indicated.

1. Cell culture

CHO cells are cultured in the normal culture medium in 100 mm tissue culture plates placed in a 37°C incubator under 5% CO_2, 95% air atmosphere. For fluorescence microscopy analysis, cells are harvested with the trypsin-EDTA solution, washed once with serum containing medium and seeded in culture media (the fibronectin-depleted medium or the medium with defined concentration of fibronectin) in Lab-Tek 8-chamber slides (Nunc, Inc.; 200 to 400 µl/well) or 12-well HTC^R slides (Cel-Line Associates, Inc., Newfield, NJ; 30-50 µl/well) at a final density of 2×10^5 cells/ml. The cells are cultured at 37°C under 5% CO_2, 95% air atmosphere for 20 or more hours.

2. Cell fixation

At the end of culture time, the media are removed from the culture wells by aspiration, and the cell monolayers are fixed with 3.7% paraformaldehyde in PBS for 20 min.

3. Antibody staining

The fixed cells are rinsed three times with PBS, incubated with 1 mg/ml BSA in PBS for 10 min, and then with a PBS solution containing 1 mg/ml BSA and a primary anti-fibronectin antibody (e.g., polyclonal rabbit anti-fibronectin antibody, 170 μl/well of the Lab-Tek 8-chamber slides or 15 μl/well of the 12-well HTC[R] slides) for 60 min. After rinsing three times with PBS, the cells are incubated with a Cy3- or FITC-conjugated secondary antibody (e.g., Cy3-conjugated anti-rabbit IgG antibody, Jackson ImmunoResearch Laboratories, Inc.) for 60 min. At the end of incubation, the unbound antibodies are removed by rinsing with PBS three times, and the bound antibodies are detected by observing under a fluorescent light microscope equipped with proper filter units (for Cy3, peak absorption wave length = 550 nm and peak emission wave length = 570 nm). Representative microscopic fields are photographed using Kodak T-Max 400 or Ektachrome 1600 direct positive slide film. To obtain representative images, exposure times for different experimental conditions are fixed, using the positive, e.g., matrix forming cells, as the index exposure length.

The entire immunofluorescence staining procedure can be completed within three to four hours.

C. Technical Comments

Fibronectin matrix assembly is a cell-dependent process. There are at least three types of factors that could influence the matrix assembly process in cultured cells. The first is the activity of the cells to assemble a fibronectin matrix, which is determined by integrin expression level and activity, cytoskeleton organization, and other cellular signaling events. The second is the cell culture condition, which includes the seeding density and the culture time of the cells. The third one is the amount of soluble fibronectin that is available for assembly, which is somewhat related to the first and the second factors, but I single it out here due to its importance in influencing the matrix assembly process. For studies aimed at characterizing cellular signaling events that regulate fibronectin matrix assembly (the first type), it is important to perform experiments under conditions in which other factors (the second and the third type) do not vary significantly.

1. Cell density and culture time

Either seeding density or culture time could affect the amount of fibronectin matrix assembled by cultured cells. As a general rule, it normally takes a shorter period of time for cells seeded at a higher density to assemble a certain amount of fibronectin matrix. Similarly, increase of culture time often results in the assembly of a greater amount of fibronectin matrix. Thus, to compare fibronectin matrix assembly activities of different cells, it is critical to seed the cells at same density and to culture the cells for identical length of time. In addition, it is important to compare fibronectin staining pattern in areas where cell density is similar.

The minimal length of time required for fibronectin fibrillogenesis varies significantly between different cell lines. For human fiblablasts cells such as WI-38 or IMR-90, fibronectin fibrillar matrix can be detected within two hours after seeding.

CHO-K1 cells, on the other hand, take much longer to assemble a fibronectin fibrillar matrix under normal culture condition (we typically culture CHO cells for 20 hours or longer).

2. Soluble fibronectin

One of the major advantages of using CHO cells as a model system to investigate the cellular signaling pathways regulating fibronectin matrix assembly is that the amount of fibronectin synthesized by these cells is so low that the contribution of the endogenous fibronectin to the matrix assembly process is practically negligible. Because the concentration of soluble fibronectin in culture medium could vary between different serum preparations, it is recommended that the culture media with defined concentration of fibronectin (see Section D, Materials) are used. CHO cells cultured in the presence of exogenous fibronectin at a concentration as low as 5 μg/ml are capable of assembling a fibronectin matrix readily detectable by immunofluorescence staining. Either native (plasma or cellular) or recombinant fibronectin molecules may be used to supplement the culture media. Use of recombinant fibronectin mutants in the CHO cell culture system has revealed important information on fibronectin domains that are critically involved in the initiation of fibronectin matrix assembly.[21,24,27]

3. Anti-fibronectin antibodies

Most polyclonal anti-fibronectin antibodies are well suited for detection of fibronectin matrix in the immunofluorescent staining assay. Use of species-specific monoclonal anti-fibronectin antibodies allows the investigator to monitor the incorporation of fibronectin (or mutants of fibronectin) derived from a specific source into extracellular matrix. For example, the incorporation of endogenous fibronectin into extracellular matrix could be detected by staining human fibroblasts (e.g., WI-38 or IMR-90 cells) cultured in media containing fetal bovine serum with a monoclonal anti-fibronectin antibody that recognizes human but not bovine fibronectin. Recently, using a monoclonal antibody (L8) that recognizes a conformation-dependent epitope on fibronectin, Zhong et al. have shown that Rho-mediated contractility plays an important role in fibronectin matrix assembly.[64]

D. Materials

1. Normal culture medium

α-MEM (Life Technologies, Inc.)
10% (v/v) fetal bovine serum (HyClone Laboratories, Inc. or Atlanta Biologicals)
2 mM glutamine
1% (v/v) antibiotic-antimycotic mixture (Sigma Chemical Company) (optional)

2. Fibronectin-depleted culture medium

α-MEM (Life Technologies, Inc.)
10% (v/v) fibronectin-depleted fetal bovine serum*
2 mM glutamine
1% (v/v) antibiotic-antimycotic mixture (Sigma Chemical Company) (optional)

* *Fibronectin may be removed from fetal bovine serum by affinity chromatography using a gelatin-Sepharose column. The depletion of fibronectin in the flow-through fraction should be checked by, for example, immunoblotting with a polyclonal anti-fibronectin antibody that recognizes bovine fibronectin.*

3. Culture medium with defined concentration of fibronectin

α-MEM (Life Technologies, Inc.)
10% (v/v) fibronectin-depleted fetal bovine serum
Fibronectin (5-50 μg/ml)*
2 mM glutamine
1% (v/v) antibiotic-antimycotic mixture (Sigma Chemical Company) (optional)

* *see "Technical comments"*

4. PBS (1000 ml)

8 gram NaCl, 0.2 gram KCl, 1.44 gram Na_2HPO_4, 0.24 gram NaH_2PO_4, dissolved in H_2O (pH 7.4, final volume = 1000 ml)

5. Trypsin-EDTA solution

0.05% trypsin/0.53 mM EDTA (Mediatech/Cellgro®, Herndon, VA)

6. Fixative solution

3.7% (w/v) paraformaldehyde in PBS

III. Biochemical Analyses of Fibronectin Matrix Assembly

A. Overview

The second type of fibronectin matrix assembly assay is based on biochemical criteria of a fibronectin matrix, namely that the fibronectin matrix is stabilized by disulfide-bonds and resistant to certain detergents (e.g., sodium deoxycholate) extraction. In this type of assay, fibronectin matrix is typically isolated by sequential detergent extraction and the detergent-insoluble matrix fractions are then analyzed by ELISA, immunoblotting, and SDS gel electrophoresis under both reduced and unreduced conditions. Many of the technical considerations are similar to those discussed above. The following is a basic protocol that was used for the isolation and biochemical analyses of fibronectin matrix derived from CHO cells.

B. Protocols

1. Cell culture and harvest

CHO cells are cultured in the culture medium supplemented with human plasma fibronectin (e.g., 50 μg/ml) in 100 mm tissue culture plates placed in a 37°C incubator under 5% CO_2, 95% air atmosphere for 20 hours or longer. The cell monolayers are rinsed twice with PBS, and the cells are harvested with a cell scraper. The cells are suspended in PBS containing 1 mM [4-(2-aminoethyl)benzenesulfonylfluoride, HCl] and pelleted by centrifugation.

2. Isolation of detergent-insoluble fibronectin matrix

The fibronectin matrix is isolated by sequential extractions with detergent solutions based on a method described by Quade and McDonald.[20] The volumes of the solutions listed below are based on those used for isolation of fibronectin matrix from cells derived from one 100 mm plate.

a. The cell pellet is extracted with 1 ml of the 3% Triton X-100 solution, and the insoluble fraction is pelleted by centrifugation at 15,000× *g* for 20 min at 4°C.
b. The pellet is re-extracted with the 3% Triton X-100 solution (repeat Step 1).
c. The 3% Triton X-100 insoluble fraction is washed with 0.5 ml of the DNase I solution and pelleted by centrifugation at 15,000× *g* for 20 min at 4°C.
d. The pellet is extracted with 1 ml of the 2% deoxycholate solution, and the insoluble fraction is pelleted by centrifugation at 15,000× *g* for 30 min at 4°C.
e. Repeat Step d, and collect the pellet for analysis of fibronectin by ELISA or immunoblotting.

3. ELISA analysis of fibronectin

a. Coating of monoclonal anti-fibronectin antibody. Polystyrene ELISA plates (96-well; Corning, Inc.) are coated with mouse anti-human fibronectin monoclonal antibody N294[20,33] (5 μg/ml; 100 μl/well) in 100 mM $NaHCO_3$ buffer (pH 9.2) overnight at 4°C or three hours at room temperature. The remaining protein binding sites are blocked with 5 mg/ml BSA in 100 mM $NaHCO_3$ buffer (pH 9.2) containing 0.05% (v/v) Tween-20 (350 μl/well). The plates are rinsed twice with the 2 mg/ml BSA/Tween-20 solution, and then twice with the 0.2% SDS/Triton X-100 solution at room temperature.

b. Capture of fibronectin. The isolated 2% deoxycholate-insoluble fibronectin matrix is extracted with the SDS/DTT solution at 37°C for 30 min. After cooling to 20°C, the denatured and reduced fibronectin is carboxyamidomethylated by mixing with iodoacetamide (final concentration = 13 mM). The solubilized fibronectin sample is diluted 1:4 (v/v) with the 1.25% Triton X-100 solution and then added to the wells (100 µl/well) of the ELISA plates that have been pre-coated with the N294 anti-fibronectin antibody (Step 1). For standard curve, solutions containing known fibronectin concentrations are prepared by dissolving purified human plasma fibronectin in the SDS/DTT solution. The fibronectin standard samples are then carboxyamidomethylated with iodoacetamide, and diluted with the 1.25% Triton X-100 solution as described above. After incubation at room temperature for one hour, the wells are rinsed four times with the 2 mg/ml BSA/Tween-20 solution.

c. Detection of fibronectin. The captured fibronectin is detected with a rabbit polyclonal anti-fibronectin antibody. We have been using the IgG fraction (2 µg/ml, 100 µl/well) of MC54, a rabbit polyclonal antibody raised against human fibronectin developed in Dr. John A. McDonald's laboratory, in this assay, but other polyclonal anti-fibronectin antibodies could also be used. Between steps, the assay plates are washed four times with the 2 mg/ml BSA/Tween-20 solution. The bound rabbit anti-fibronectin antibody is detected by incubation with an alkaline phosphatase conjugated anti-rabbit IgG antibody that does not cross-react with mouse IgG (0.5 µg/ml, 100 µl/well; Kirkegaard & Perry Laboratories). After rinsing five times with the 2 mg/ml BSA/Tween-20 solution, bound alkaline phosphatase-conjugate is detected colorimetrically with p-nitrophenyl phosphate at 405 nm using a microplate reader. The amounts of fibronectin in the matrix fractions are determined by comparing the A_{405nm} values of the samples with those of the standards.

4. Immunoblotting

Another convenient way of analyzing the 2% deoxycholate-insoluble matrix fibronectin is by immunoblotting with anti-fibronectin antibodies. In this assay, the 2% deoxycholate-insoluble matrix fractions isolated in "2" are extracted with reducing SDS-PAGE sample buffer [62.5 mM Tris-HCl, pH 6.8, 2% (w/v) SDS, 100 mM dithiothreitol], separated by SDS-PAGE, and transferred onto a Immobilon®-P membrane (Millipore Corp.). The fibronectin is then detected with an anti-fibronectin antibody using a standard immunoblot protocol. The amount of fibronectin may be estimated by comparing the intensity of the fibronectin bands with those in the control lanes, onto which known amounts of fibronectin are loaded.

In parallel experiments, the 2% deoxycholate insoluble matrix fractions are extracted with non-reducing SDS-PAGE sample buffer [62.5 mM Tris-HCl, pH 6.8, 2% (w/v) SDS] and are then subject to immunoblotting with anti-fibronectin antibodies as described above.

C. Technical Comments

The biochemical assays provide quantitative information on the formation of a 2% deoxycholate-insoluble fibronectin polymer that is stabilized by disulfide bonds. They complement the morphologically based assays, and are often used in conjunction with the immunofluorescence staining assay in analyzing the fibronectin matrix assembly process. Many of the technical considerations discussed in Section A, which include cell density, culture time, exogenous fibronectin supplement, and anti-fibronectin antibody selection, are also applied to the biochemical assays.

The procedures described above could be modified to measure the incorporation of fibronectin that is derived from a specific source (or a specific form of fibronectin) into extracellular matrix. For example, to analyze the incorporation of exogenously supplied fibronectin into extracellular matrix, cells could be cultured in culture medium supplemented with radioactively labeled (e.g., ^{125}I-labeled) fibronectin. The 2% deoxycholate-insoluble matrix fractions are isolated as described above, and the amounts of the ^{125}I-labeled fibronectin incorporated into the extracellular matrix fraction are quantified by standard procedure of scintillation counting or autoradiography.[36,37]

D. Materials

1. 3% Triton X-100 solution (100 ml)

3 ml Triton X-100 mixed with PBS containing 1 mM [4-(2-aminoethyl)benzenesulfonylfluoride, HCl] (final volume = 100 ml).

2. DNase I solution (1 ml)

100 μg DNase I (Amersham Pharmacia Biotech) dissolved in 1 ml 50 mM Tris-HCl (pH 7.4) containing 10 mM $MnCl_2$, 1 M NaCl and 1 mM [4-(2-aminoethyl)benzenesulfonylfluoride, HCl].

3. 2% deoxycholate solution (100 ml)

2 gram sodium deoxycholate dissolved in 50 mM Tris-HCl (pH 8.8) containing 10 mM EDTA, 1 mM [4-(2-aminoethyl)benzenesulfonylfluoride, HCl] (final volume = 100 ml).

4. SDS/DTT solution (100 ml)

1 gram SDS dissolved in 50 mM Tris-HCl (pH 8.2) containing 5 mM DTT, 5 mM EDTA and 150 mM NaCl (final volume = 100 ml).

5. 1.25% Triton X-100 solution (100 ml)

1.25 ml Triton X-100 mixed with 50 mM Tris-HCl (pH 7.4) containing 5 mM EDTA, 190 mM NaCl (final volume = 100 ml).

6. 0.2% SDS/Triton X-100 solution (100 ml)

0.2 gram SDS dissolved in 50 mM Tris-HCl (pH 7.4) containing 1% (v/v) Triton X-100, 5 mM EDTA and 180 mM NaCl (final volume = 100 ml).

7. 5 mg/ml BSA/$NaHCO_3$ solution (100 ml)

0.5 gram BSA dissolved in 100 mM $NaHCO_3$ buffer (pH 9.2) containing 0.05% (v/v) Tween-20 (final volume = 100 ml).

8. 2 mg/ml BSA/Tween-20 solution (100 ml)

0.2 gram BSA dissolved in 50 mM Tris-HCl (pH 7.4) containing 0.2% (v/v) Tween-20, and 200 mM NaCl (final volume = 100 ml).

9. Culture medium

α-MEM (Life Technologies, Inc.)
10% (v/v) fibronectin-depleted fetal bovine serum
human fibronectin (5-50 μg/ml)*
2 mM Glutamine
1% (v/v) antibiotic-antimycotic mixture (Sigma Chemical Company) (optional)

* *see "Technical comments"*

IV. Conclusion Remarks

Proper control of fibronectin matrix assembly is essential for normal embryonic development and many physiological processes such as wound healing, and abnormality in fibronectin matrix deposition is closely associated with oncogenic transformation and other pathological conditions such as fibrosis. It becomes increasingly clear that fibronectin matrix assembly can be modulated by intracellular signaling events. However, while integrins and other signaling molecules that are critically involved in cellular control of fibronectin matrix assembly have been identified, the underlying signaling mechanism is still incompletely understood. Thus, an important area of future studies is to elucidate the signaling pathways by which cells regulate the pericellular fibronectin matrix assembly. Recent studies, through utilizing biochemical, immunological, cell biological, and genetic approaches, have significantly advanced our understanding of the integrin-mediated signaling pathways. It is expected that a combination of these methods (many of them are described in other parts of this book) with those assaying the matrix assembly process will be particularly fruitful in elucidating the intracellular signaling mechanisms controlling the pericellular matrix assembly.

Acknowledgments

I thank Dr. John A. McDonald for his support and encouragement, and for providing several protocols upon which the experimental methods described in this chapter are based. I also thank Drs. Rudy Juliano, Mark Ginsberg, Alan Hall, and Shoukat

Dedhar for their help, valuable discussions, and productive collaborations. The work in the author's laboratory was supported by research grants from the American Heart Association, American Lung Association, Francis Families Foundation, American Cancer Society, The V Foundation for Cancer Research, the Comprehensive Cancer Center, and the Cell Adhesion and Matrix Research Center of University of Alabama at Birmingham. The author is an Edward Livingston Trudeau Scholar of American Lung Association, Parker B. Francis Fellow in Pulmonary Research, the Francis Families Foundation, and a V Foundation Scholar.

References

1. McDonald, J. A., Extracellular matrix assembly, *Annu. Rev. Cell Biol.,* 4, 183, 1988.
2. Hynes, R. O., *Fibronectins,* Springer-Verlag, New York, 1990.
3. Mosher, D. F., Sottile, J., Wu, C., and McDonald, J. A., Assembly of Extracellular Matrix, *Curr. Opin. Cell Biol.,* 4, 810, 1992.
4. Schwartz, M. A., Schaller, M. D., and Ginsberg, M. H., Integrins: emerging paradigms of signal transduction, *Ann. Rev. Cell Dev. Biol.,* 11, 549, 1995.
5. Ruoslahti, E., Integrin signaling and matrix assembly, *Tumour Biol.,* 17, 117, 1996.
6. Mosher, D. F., *Fibronectin,* Academic Press, New York, 1989.
7. Ruoslahti, E., Fibronectin and its receptors, *Annu. Rev. Biochem.,* 57, 375, 1988.
8. George, E. L., Georges-Labouesse, E. N., Patel-King, R. S., Rayburn, H., and Hynes, R. O., Defects in mesoderm, neural tube and vascular development in mouse embryos lacking fibronectin, *Development,* 119, 1079, 1993.
9. Burthem, J. and Cawley, J. C., Specific tissue invasion, localisation and matrix modification in hairy-cell leukemia, *Leuk. Lymphoma,* 14, 19, 1994.
10. Burthem, J. and Cawley, J. C., The bone marrow fibrosis of hairy-cell leukemia is caused by the synthesis and assembly of a fibronectin matrix by the hairy cells, *Blood,* 83, 497, 1994.
11. Nishino, T., Ishida, T., Oka, T., Yasumoto, K., and Sugimachi, K., Distribution of fibronectin in adenocarcinoma of the lung: classification and prognostic significance, *J. Surg. Oncol.,* 43, 94, 1990.
12. Christensen, L., Nielsen, M., Andersen, J., and Clemmensen, I., Stromal fibronectin staining pattern and metastasizing ability of human breast carcinoma, *Cancer Res.,* 48, 6227, 1988.
13. Labat-Robert, J., Birembaut, P., Robert, L., and Adnet, J. J., Modification of fibronectin distribution pattern in solid human tumours, *Diagn. Histopathol.,* 4, 299, 1981.
14. Labat-Robert, J., Birembaut, P., Adnet, J. J., Mercantini, F., and Robert, L., Loss of fibronectin in human breast cancer, *Cell Biol. Int. Rep.,* 4, 609, 1980.
15. Pasqualini, R., Bourdoulous, S., Koivunen, E., Woods, V. L., and Ruoslahti, E., A polymeric form of fibronectin has antimetastatic effects against multiple tumor types, *Nature Medicine,* 2, 1197, 1996.
16. Wu, C., Integrin mediated fibronectin matrix assembly, *Trends Glycosci. Glycotech.,* 8, 315, 1996.
17. Wu, C., Roles of integrins in fibronectin matrix assembly, *Histol. Histopathol.,* 12, 233, 1997.

18. McKeown-Longo, P. J. and Mosher, D. F., Interaction of the 70,000-mol-wt aminoterminal fragment of fibronectin with the matrix-assembly receptor of fibroblasts, *J. Cell Biol.,* 100, 364, 1985.
19. McDonald, J. A., Quade, B. J., Broekelmann, T. J., LaChance, R., Forsman, K., Hasegawa, E., and Akiyama, S., Fibronectin's cell-adhesive domain and an amino-terminal matrix assembly domain participate in its assembly into fibroblast pericellular matrix, *J. Biol. Chem.,* 262, 2957, 1987.
20. Quade, B. J. and McDonald, J. A., Fibronectin's amino-terminal matrix assembly site is located within the 29-kDa amino-terminal domain containing five type I repeats, *J. Biol. Chem.,* 263, 19602, 1988.
21. Schwarzbauer, J. E., Identification of the fibronectin sequences required for assembly of a fibrillar matrix, *J. Cell Biol.,* 113, 1463, 1991.
22. Chernousov, M. A., Fogerty, F. J., Koteliansky, V. E., and Mosher, D. F., Role of the I-9 and III-1 modules of fibronectin in formation of an extracellular fibronectin matrix, *J. Biol. Chem.,* 266, 10851, 1991.
23. Morla, A. and Ruoslahti, E., A fibronectin self-assembly site involved in fibronectin matrix assembly: Reconstruction in a synthetic peptide., *J. Cell Biol.,* 118, 421, 1992.
24. Sottile, J. and Wiley, S., Assembly of amino-terminal fibronectin dimers into the extracellular matrix, *J. Biol. Chem.,* 269, 17192, 1994.
25. Hocking, D. C., Smith, R. K., and McKeown-Longo, P. J., A novel role for the integrin-binding III-10 module in fibronectin matrix assembly, *J. Cell Biol.,* 133, 431, 1996.
26. Roman, J., LaChance, R. M., Broekelmann, T. J., Kennedy, C. J., Wayner, E. A., Carter, W. G., and McDonald, J. A., The fibronectin receptor is organized by extracellular matrix fibronectin: implications for oncogenic transformation and for cell recognition of fibronectin matrices, *J. Cell Biol.,* 108, 2529, 1989.
27. Sechler, J. L., Takada, Y., and Schwarzbauer, J. E., Altered rate of fibronectin matrix assembly by deletion of the first type III repeats, *J. Cell Biol.,* 134, 573, 1996.
28. Hocking, D. C., Sottile, J., and McKeownlongo, P. J., Activation of distinct alpha(5)beta(1)-mediated signaling pathways by fibronectins cell adhesion and matrix assembly domains, *J. Cell Biol.,* 141, 241, 1998.
29. Akiyama, S. K., Yamada, S. S., Chen, W. T., and Yamada, K. M., Analysis of fibronectin receptor function with monoclonal antibodies: roles in cell adhesion, migration, matrix assembly, and cytoskeletal organization, *J. Cell Biol.,* 109, 863, 1989.
30. Fogerty, F. J., Akiyama, S. K., Yamada, K. M., and Mosher, D. F., Inhibition of binding of fibronectin to matrix assembly sites by anti-integrin (alpha 5 beta 1) antibodies, *J. Cell Biol.,* 111, 699, 1990.
31. Darribere, T., Guida, K., Larjava, H., Johnson, K. E., Yamada, K. M., Thiery, J. P., and Boucaut, J. C., *In vivo* analyses of integrin beta 1 subunit function in fibronectin matrix assembly, *J. Cell Biol.,* 110, 1813, 1990.
32. Giancotti, F. G. and Ruoslahti, E., Elevated levels of the alpha 5 beta 1 fibronectin receptor suppress the transformed phenotype of Chinese hamster ovary cells, *Cell,* 60, 849, 1990.
33. Wu, C., Bauer, J. S., Juliano, R. L., and McDonald, J. A., The alpha 5 beta 1 integrin fibronectin receptor, but not the alpha 5 cytoplasmic domain, functions in an early and essential step in fibronectin matrix assembly, *J. Biol. Chem.,* 268, 21883, 1993.

34. Wu, C., Fields, A. J., Kapteijn, B. A., and McDonald, J. A., The role of alpha 4 beta 1 integrin in cell motility and fibronectin matrix assembly, *J. Cell Sci.,* 108, 821, 1995b.
35. Zhang, Z., Morla, A. O., Vuori, K., Bauer, J. S., Juliano, R. L., and Ruoslahti, E., The alpha v beta 1 integrin functions as a fibronectin receptor but does not support fibronectin matrix assembly and cell migration on fibronectin, *J. Cell Biol.,* 122, 235, 1993.
36. Wu, C., Keivens, V. M., O' Toole, T. E., McDonald, J. A., and Ginsberg, M. H., Integrin activation and cytoskeletal interaction are essential for the assembly of a fibronectin matrix, *Cell,* 83, 715, 1995a.
37. Wu, C., Hughes, P. E., Ginsberg, M. H., and McDonald, J. A., Identification of a new biological function for the integrin avb3: Initiation of fibronectin matrix assembly., *Cell Adhes. & Commun.,* 4, 149, 1996.
38. Yang, J. T. and Hynes, R. O., Fibronectin receptor functions in embryonic cells deficient in alpha 5 beta 1 integrin can be replaced by alpha v integrins, *Mol. Biol. Cell,* 7, 1737, 1996.
39. Wennerberg, K., Lohikangas, L., Gullberg, D., Pfaff, M., Johansson, S., and Fassler, R., Beta 1 integrin-dependent and -independent polymerization of fibronectin, *J. Cell Biol.,* 132, 227, 1996.
40. Hughes, P. E., Diaz-Gonzalez, F., Leong, L., Wu, C., McDonald, J. A., Shattil, S. J., and Ginsberg, M. H., Breaking the integrin hinge. A defined structural constraint regulates integrin signaling, *J. Biol. Chem.,* 271, 6571, 1996.
41. Sakai, T., Zhang, Q. H., Fassler, R., and Mosher, D. F., Modulation of beta-1a integrin functions by tyrosine residues in the beta-1 cytoplasmic domain, *J. Cell Biol.,* 141, 527, 1998.
42. Zhidkova, N. I., Belkin, A. M., and Mayne, R., Novel isoform of beta 1 integrin expressed in skeletal and cardiac muscle, *Biochem. & Biophy. Res. Comm.,* 214, 279, 1995.
43. van der Flier, A., Kuikman, I., Baudoin, C., van der Neut, R., and Sonnenberg, A., A novel beta 1 integrin isoform produced by alternative splicing: unique expression in cardiac and skeletal muscle, *FEBS Letters,* 369, 340, 1995.
44. Belkin, A. M., Zhidkova, N. I., Balzac, F., Altruda, F., Tomatis, D., Maier, A., Tarone, G., Koteliansky, V. E., and Burridge, K., Beta 1D integrin displaces the beta 1A isoform in striated muscles: localization at junctional structures and signaling potential in nonmuscle cells, *J. Cell Biol.,* 132, 211, 1996.
45. van der Flier, A., Gaspar, A. C., Thorsteinsdottir, S., Baudoin, C., Groeneveld, E., Mummery, C. L., and Sonnenberg, A., Spatial and temporal expression of the beta1D integrin during mouse development, *Dev. Dyn.,* 210, 472, 1997.
46. Belkin, A. M., Retta, S. F., Pletjushkina, O. Y., Balzac, F., Silengo, L., Fassler, R., Koteliansky, V. E., Burridge, K., and Tarone, G., Muscle beta1D integrin reinforces the cytoskeleton-matrix link: modulation of integrin adhesive function by alternative splicing, *J. Cell Biol.,* 139, 1583, 1997.
47. Baudoin, C., Goumans, M.-J., Mummery, C., and Sonnenberg, A., Knockout and knockin of the beta1 exon D define distinct roles for integrin splice variants in heart function and embryonic development, *Gene & Dev.,* 12, 1202, 1998.
48. Hannigan, G. E., Leung-Hagesteijn, C., Fitz-Gibbon, L., Coppolino, M. G., Radeva, G., Filmus, J., Bell, J. C., and Dedhar, S., Regulation of cell adhesion and anchorage-dependent growth by a new beta 1-integrin-linked protein kinase, *Nature,* 379, 91, 1996.

49. Radeva, G., Petrocelli, T., Behrend, E., Leunghagesteijn, C., Filmus, J., Slingerland, J., and Dedhar, S., Overexpression of the integrin-linked kinase promotes anchorage-independent cell cycle progression, *J. Biol. Chem.*, 272, 13937, 1997.
50. Wu, C., Keightley, S. Y., Leung-Hagesteijn, C., Radeva, G., Coppolino, M., Goicoechea, S., McDonald, J. A., and Dedhar, S., Integrin-linked protein kinase (ILK) regulates fibronectin matrix assembly, E-cadherin expression and tumorigenicity, *J. Biol. Chem.*, 273(1), 1998.
51. Novak, A., Hsu, S. C., Leunghagesteijn, C., Radeva, G., Papkoff, J., Montesano, R., Roskelley, C., Grosschedl, R., and Dedhar, S., Cell adhesion and the integrin-linked kinase regulate the Lef-1 and beta-catenin signaling pathways, *Proc. Natl. Acad. Sci. USA*, 95, 4374, 1998.
52. Li, F., Liu, J., Mayne, R., and Wu, C., Identification and characterization of a mouse protein kinase that is highly homologous to human integrin-linked kinase., *Biochim. Biophys. Acta*, 1358, 215, 1997.
53. Xie, W., Li, F., Kudlow, J. E., and Wu, C., Expression of the Integrin-linked kinase (ILK) in mouse skins: loss of expression in suprabasal layers of the epidermis and up-regulation by erbB-2, *Am. J. Pathol.*, in press.
54. Danilov, Y. N. and Juliano, R. L., Phorbol ester modulation of integrin-mediated cell adhesion: a postreceptor event, *J. Cell Biol.*, 108, 1925, 1989.
55. Vuori, K. and Ruoslahti, E., Activation of protein kinase C precedes alpha 5 beta 1 integrin-mediated cell spreading on fibronectin, *J. Biol. Chem.*, 268, 21459, 1993.
56. Woods, A. and Couchman, J. R., Protein kinase C involvement in focal adhesion formation, *J. Cell Sci.*, 101, 277, 1992.
57. Somers, C. E. and Mosher, D. F., Protein kinase C modulation of fibronectin matrix assembly, *J. Biol. Chem.*, 268, 22277, 1993.
58. Hughes, P. E., Renshaw, M. W., Pfaff, M., Forsyth, J., Keivens, V. M., Schwartz, M. A., and Ginsberg, M. H., Suppression of integrin activation: a novel function of a Ras/Raf-initiated MAP kinase pathway, *Cell*, 88, 521, 1997.
59. Hall, A., Small GTP-binding proteins and the regulation of the actin cytoskeleton, *Annu. Rev. Cell Biol.*, 10, 31, 1994.
60. Zhang, Z., Vuori, K., Wang, H., Reed, J. C., and Ruoslahti, E., Integrin activation by R-ras, *Cell*, 85, 61, 1996.
61. Ridley, A. J. and Hall, A., The small GTP-binding protein rho regulates the assembly of focal adhesions and actin stress fibers in response to growth factors, *Cell*, 70, 389, 1992.
62. Zhang, Q., Checovich, W. J., Peters, D. M., Albrecht, R. M., and Mosher, D. F., Modulation of cell surface fibronectin assembly sites by lysophosphatidic acid, *J. Cell Biol.*, 127, 1447, 1994.
63. Zhang, Q. H., Magnusson, M. K., and Mosher, D. F., Lysophosphatidic acid and microtubule-destabilizing agents stimulate fibronectin matrix assembly through Rho-dependent actin stress fiber formation and cell contraction, *Mol. Biol. Cell*, 8, 1415, 1997.
64. Zhong, C. L., Chrzanowskawodnicka, M., Brown, J., Shaub, A., Belkin, A. M., and Burridge, K., Rho-mediated contractility exposes a cryptic site in fibronectin and induces fibronectin matrix assembly, *J. Cell Biol.*, 141, 539, 1998.

SECTION III

Inside-Out Signaling by Integrins

Chapter 14

Studies of Integrin Signaling Through Platelet $\alpha_{IIb}\beta_3$

Takaaki Hato, Petra Maschberger, and Sanford J. Shattil

Contents

I. Introduction 203
II. Inside-Out Signaling 204
A. Affinity Modulation 204
B. Avidity Modulation of $\alpha_{IIb}\beta_3$ in CHO Cells 207
III. Outside-in Signaling 209
A. CHO Cell Adhesion to Immobilized Ligands 209
B. Analysis of Protein Tyrosine Phosphorylation/Kinase Activity 210
C. Integrin-Dependent Morphological Changes 212
References 213

I. Introduction

Integrin $\alpha_{IIb}\beta_3$, which is expressed only in megakaryocytes and platelets, is essential for normal platelet adhesion and aggregation during hemostasis and for the development of occlusive, platelet-rich thrombi in vascular diseases. Studies in platelets over the past two decades have demonstrated that $\alpha_{IIb}\beta_3$ is engaged in bidirectional signaling. "Inside-out" signals convert the receptor from a low to a high affinity/avidity state. This enables platelets to bind soluble, multivalent adhesive ligands, such as fibrinogen and von Willebrand factor; in turn, these bridge the integrins on adjacent platelets, culminating in cell-cell aggregation. Inside-out signals also enable

0-8493-3385-7/99/$0.00+$.50

the receptor to recognize or to bind more tightly to Arg-Gly-Asp ligands immobilized in the vessel wall, thus contributing to firm platelet adhesion during hemostasis and thrombosis.[1] "Outside-in" signals are triggered by the binding of soluble and immobilized ligands to $\alpha_{IIb}\beta_3$, and they are responsible for mediating or facilitating a number of important post-ligand binding events, including platelet spreading, cytoskeletal reorganization, exocytosis, and development of platelet membrane procoagulant activity.[2]

Platelets are anucleate cells and consequently are not amenable to study by the usual recombinant methods. Accordingly, in the past several years, investigators have turned to model cell systems to study the adhesive and signaling functions of $\alpha_{IIb}\beta_3$.[3-6] In such systems, $\alpha_{IIb}\beta_3$ and/or other recombinant molecules are expressed transiently or in a stable manner in tissue culture cells, and various aspects of bidirectional signaling are studied. These model systems have been quite useful in characterizing the signaling pathways that can interact with $\alpha_{IIb}\beta_3$, and they have established the critical importance of the relatively short integrin cytoplasmic tails in bidirectional signaling. We have focused our own studies in this area on a CHO cell model system that was first popularized for $\alpha_{IIb}\beta_3$ by O'Toole and colleagues.[7,8] With this system, we have been able to reconstitute many of the signaling properties of $\alpha_{IIb}\beta_3$ normally found in platelets.[9-12] Here, we describe the principal methods we have used to characterize $\alpha_{IIb}\beta_3$ signaling in CHO cells. The reader is also referred to review articles that summarize methods for studying $\alpha_{IIb}\beta_3$ signaling in platelets.[13,14]

II. Inside-Out Signaling

A. Affinity Modulation

1. Overview

Affinity modulation refers to that aspect of inside-out signaling that facilitates ligand binding by causing conformational changes within individual $\alpha_{IIb}\beta_3$ heterodimers, resulting in the formation and/or exposure of ligand binding sites in the extracellular domain of the receptor.[15] This is in contrast to "avidity modulation," a process presumably mediated by lateral clustering of heterodimers into oligomers within the plane of the plasma membrane, leading to increased ligand binding as the result of chelate or rebinding effects.[16] The usual readout for inside-out signaling through $\alpha_{IIb}\beta_3$ is increased binding of a soluble ligand, typically fibrinogen, von Willebrand factor, or a ligand-mimetic monoclonal antibody; strictly speaking, however, this does not discriminate between affinity and avidity modulation. However, we have developed multivalent (IgMκ) and monovalent (Fab) forms of a ligand-mimetic monoclonal antibody specific for $\alpha_{IIb}\beta_3$ (PAC1) that has been useful in differentiating between these two causes of ligand binding. As might be predicted, the multivalent form of PAC1 is sensitive to both affinity and avidity modulation, while the monovalent form is sensitive only to affinity modulation.[17] In reality, inside-out signaling is a combination of affinity and avidity modulation, with the relative contributions of each probably varying with the integrin and the cell.[16,17]

2. Specific Protocols

a. Transient and stable transfection of CHO cells. CHO-K1 cells are plated onto 100 mm dishes so that they are 40 to 60% confluent on the day of transfection (day 1). On day 1, cells are washed with Dulbecco's Modified Eagle Medium (DMEM) to remove fetal bovine serum from culture medium. Plasmid DNA is purified by Maxi-prep (Qiagen, Santa Clarita, CA). A total of 4 μg plasmid DNA (0.1 to 2 μg each of α_{IIb} and β_3 subunit, and, if necessary, empty vector pcDNA3) are added to 200 μl of DMEM, and incubated with 20 μl of Lipofectamine solution into a polystyrene tubes* for 15 min. The incubation mixture is diluted with 3.8 ml of DMEM and overlaid onto the cells. After the cells are incubated for 6 h in a 37°C, 5% CO_2 incubator, the medium is changed to DMEM containing 10% fetal bovine serum, 1% nonessential amino acids, 2 mM L-glutamine, 100 U/ml penicillin and 100 μg/ml streptomycin (complete growth medium). On day 2, the medium is changed to fresh complete growth medium. On day 3, the cells are subjected to flow cytometry and biochemical experiments.**

To make stable CHO transfectants, 0.4 μg of pcDNA3*** is co-transfected with α_{IIb} and β_3 DNA as described above. On day 3, 800 μg/ml of G418 is added to the medium and the cells are cultured for 2 weeks to select G418-resistant cells. The growing cells are stained with a complex-dependent anti-$\alpha_{IIb}\beta_3$ monoclonal antibody, (e.g., A2A9 or D57),[3,18] followed by FITC-labeled goat or rabbit anti-mouse heavy and light chain IgG, and $\alpha_{IIb}\beta_3$-expressing cells are individually sorted into microtitre wells by FACS. The cells are cultured in the presence of 400 μg/ml G418 and after suitable growth, the clones are eventually screened with A2A9 or D57 by FACS and positive clones are established as stable cell lines.

b. FACS analysis of soluble ligand binding to CHO cells. CHO cells transiently or stably expressing $\alpha_{IIb}\beta_3$ are harvested with 0.05% trypsin/0.53 mM EDTA solution and resuspended in Tyrode's buffer containing 2 mM $CaCl_2$ and $MgCl_2$ at 1×10^7 cells/ml. We usually use the fibrinogen-mimetic monoclonal antibody PAC1 as a ligand specific for $\alpha_{IIb}\beta_3$.[19] 4×10^5 cells are added to tubes containing a final concentration of 0.4% PAC1 ascites or 40 nM purified PAC1 in a final Vol. of 50 μl and incubated for 30 min at room temperature. When desired, parallel cell incubations are also carried out in the presence of 150 μg/ml anti-LIBS6 Fab, a β_3 "activating antibody" that converts $\alpha_{IIb}\beta_3$ into a high affinity form through conformational changes,[20] and/or with 10 μM Integrilin, an $\alpha_{IIb}\beta_3$ antagonist to specifically block PAC1 binding.[21] Cells are then washed and incubated for 30 min on ice in the presence of 5 μg/ml biotin-labeled D57. After washing, the cells are incubated for 30 min on ice with phycoerythrin-streptavidin (1:25 dilution) and FITC-labeled goat anti-mouse μ heavy chain antibody (1:20 dilution). After washing, the cells are resuspended in 0.5 ml Tyrode's buffer containing 2 μg/ml propidium iodide and analyzed by FACS. After electronic compensation, PAC1 binding (FL1 channel) is analyzed on the gated subset of live cells (propidium iodide-negative, FL3) that are

---

* Lipofectamine is believed to attach to polypropylene.

** We usually achieve 20 to 40% of transfection efficiency.

*** This vector has neomycin resistance gene for selection with G418.

positive for $\alpha_{IIb}\beta_3$ expression (FL2). Specific PAC1 binding is defined as binding inhibitable by Integrilin (or by 1 mM RGDS peptide). To control for variations in integrin expression from transfection to transfection, PAC1 binding, measured as mean fluorescence intensity in arbitrary units, is expressed relative to the levels of $\alpha_{IIb}\beta_3$, measured simultaneously with biotin-D57.

c. FACS analysis of integrin-dependent CHO cell aggregation. CHO cells stably expressing $\alpha_{IIb}\beta_3$ are labeled with either a red fluorescence tracer, hydroethidine (HE) or green fluorescent protein (GFP), either unfused or fused to a protein under study.[22] To label with HE, cells suspended in Tyrode's buffer at 1×10^7/ml are incubated with an equal volume of 20 μg/ml HE* for 45 min at room temperature under gentle agitation. The labeled cells are washed three times in cation-free Tyrode's buffer and resuspended in Tyrode's buffer containing 2 mM $CaCl_2$ and $MgCl_2$ at 4×10^6 cells/ml. To label with GFP or GFP fusion proteins, CHO cells are transfected with mammalian expression vector, 0.02 μg pEGFP-C1, according to the transfection protocol described above and cultured for 48 h. Then 250 μl of HE-labeled cells are added to siliconized glass cuvettes containing 250 μl of GFP-labeled cells (2×10^6/ml). After addition of 300 μg/ml fibrinogen, the cells are stirred with a magnetic stir bar at 1000 rpm for 20 min at room temperature. Incubations are stopped by addition of 0.25% formaldehyde, and the samples are kept on ice for 30 min. Mixed red-green cellular aggregates are detected by FACS as originally described elsewhere.[22,23]

3. Materials

a. Lipofectamine; GIBCO BRL (Gaithersburg, MD).

b. pcDNA3; Invitrogen (Carlsbad, CA).

c. G418; GIBCO BRL, Geneticin®.

d. anti-$\alpha_{IIb}\beta_3$ monoclonal antibodies.

We use PAC1, D57, and anti-LIBS6 antibodies. PAC1 (Becton-Dickinson) is an anti-$\alpha_{IIb}\beta_3$ complex-specific IgM antibody, and binds only to the active form of $\alpha_{IIb}\beta_3$. D57 is an anti-$\alpha_{IIb}\beta_3$ complex-specific, non-function-blocking IgG antibody. Anti-LIBS6 recognizes the β_3 subunit and converts $\alpha_{IIb}\beta_3$ to a high affinity state. Antibodies D57 and anti-LIBS6 were generously provided by Dr. Mark Ginsberg (Scripps Research Institute).

e. Integrilin.

This peptide was generously provided by Dr. David Phillips (Cor Therapeutics, Inc., South San Francisco, CA). RGDS peptide was obtained from Sigma (St, Louis, MO).

f. Tyrode's buffer.

137 mM NaCl, 12 mM $NaHCO_3$, 2.6 mM KCl, 5 mM HEPES, 5.5 mM glucose, and 1 mg/ml bovine serum albumin (BSA), pH 7.4.

g. Fluorescent probes.

---

* Stock solution of HE is prepared by dissolving in DMSO at 80 mg/ml. This solution is stored at –20°C for a maximum of 4 weeks and thawed just before use. Working solution is prepared by diluting stock with Tyrode's buffer to 20 μg/ml and filtering through a 0.22-μm filter.

FITC-labeled goat anti-mouse heavy and light chain IgG and FITC-goat anti-mouse μ chain IgG were from Biosource International (Camarillo, CA). Phycoerythrin-streptavidin was from Molecular Probes (Eugene, OR), propidium iodide from Sigma, hydroethidine from Polysciences (Warrington, PA), and pEGFP-C1 vector from Clontech (Palo Alto, CA).

h. FACS: A Beckton-Dickinson (Mountain View, CA) FACStar® was used for single cell sorting and a FACSCalibur® for ligand binding analysis.

B. Avidity Modulation of $\alpha_{IIb}\beta_3$ in CHO Cells

1. Overview

As mentioned previously, avidity modulation is thought to occur as the result of clustering of $\alpha_{IIb}\beta_3$ into oligomers within the plane of the plasma membrane, mediated either by bound multivalent ligands or by reorganization of cytoskeletal components that are in linkage with the cytoplasmic tails of the integrin. Experimentally, receptor clustering is typically initiated from outside the cell by antibody crosslinking. Recently, we developed a method to cluster $\alpha_{IIb}\beta_3$ from inside the cell using a "chemical-inducer of dimerization approach," originally conceived by Crabtree, Schrieber and colleagues.[24]

2. Specific protocols

a. Construction of integrin subunits with dimerization domains. Full-length α_{IIb} cDNA is isolated from pCDM8 expression vector encoding wild-type α_{IIb}[25] using the polymerase chain reaction (PCR) with *Pfu* polymerase. The PCR primers are 5′-GGAGCTCTAGAATGGCCAGAGCTTTGTGT-3′ and 5′-GGTGGTACTAGTCTCCCCCTCTTCATCATC-3′. XbaI and SpeI sites are introduced at the 5′ and 3′ ends of the PCR fragments for cloning purposes.* The PCR product is cut with XbaI and SpeI, and ligated into an XbaI-cut, and dephosphorylated CMV-based mammalian expression vector, pCF1E. This vector encodes a single FK506-binding protein, FKBP12 (FKBP) at the downstream of the XbaI site.[26] Plasmids with inserts in the correct orientation are determined by colony PCR using primers located at α_{IIb} and pCF1E, and then amplified and purified. The resulting αIIb(FKBP)/pCF1E plasmid encodes α_{IIb} fused in-frame to FKBP, which in turn is fused in-frame to a hemagglutinin epitope tag (Figure 1).[17]

To construct α_{IIb} fused to two tandem FKBP repeats [α_{IIb}(FKBP)$_2$], a single FKBP is removed from pCF1E with XbaI and SpeI and ligated into SpeI-cut, and dephosphorylated α_{IIb}(FKBP)/pCF1E. The orientation of inserts are determined by colony PCR.

b. FACS analysis: the effects of $\alpha_{IIb}\beta_3$ clustering on ligand binding to the integrin. CHO cells are transiently transfected with 2 μg of α_{IIb}(FKBP) or α_{IIb}(FKBP)$_2$ and 2 μg of β_3 DNA.** The protocol for transfection and FACS analysis

* XbaI and SpeI sites are compatible cohesive ends.

** We find that α_{IIb}(FKBP)β_3 and α_{IIb}(FKBP)2β_3 are expressed to the same extent as wild-type $\alpha_{IIb}\beta_3$ after transient or stable transfection.

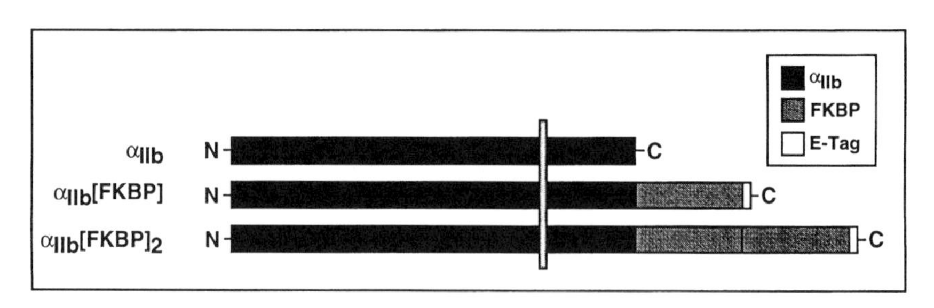

FIGURE 1

α_{IIb} constructs engineered to contain either one or two FKBP dimerization repeats. The vertical bar represents the cell membrane. Wild-type α_{IIb} is depicted on top. Extracellular domains of α_{IIb} are to the left of the bar and intracellular domains to the right. The relative sizes of the various domains are not drawn to scale. For example, the cytoplasmic tail of α_{IIb} contains 20 amino acid residues and a single FKBP repeat contains approximately 100 residues. Each of these various α_{IIb} subunits were cotransfected into CHO cells along with the β_3 integrin subunit to enable studies of integrin clustering using a bivalent chemical FKBP dimerizer.[17]

of ligand binding to the transfected CHO cells are essentially the same as described above. To induce clustering of α_{IIb}(FKBP)β_3 or α_{IIb}(FKBP)$_2\beta_3$, 10 to 5000 nM AP1510,* a membrane-permeable, bivalent FKBP ligand, is added to the incubation of the transfected cells during the PAC1 binding assay.[26]

c. Confocal microscopy to detect integrin clustering. CHO cells stably expressing α_{IIb}(FKBP)$_2\beta_3$ are harvested with trypsin/EDTA and resuspended in DMEM at 1×10^7/ml. The cell suspension is incubated with 1000 nM AP1510 or 0.5% ethanol as a vehicle control for 30 min at room temperature to induce clustering of α_{IIb}(FKBP)$_2\beta_3$. The cells are then incubated with 10% goat serum in phosphate-buffered saline (PBS), pH 7.4, for 30 min at room temperature to block non-specific binding, and further incubated with 10 μM FITC-labeled D57 antibody or unlabeled D57 for 30 min on ice. After the cells are washed twice in cold PBS, the sample containing unlabeled D57 is incubated for 30 min with FITC-labeled goat anti-mouse heavy and light chains (1:500 dilution in PBS containing 10% goat serum) to deliberately cluster the integrin as a positive control. All samples are fixed in 4% paraformaldehyde in PBS for 20 min at room temperature. After washing, fixed cells are resuspended in mounting media (Fluorsave™ reagent**), mounted onto glass slides, and analyzed with a confocal microscopy (Figure 2).

3. Materials

a. *pfu* DNA polymerase; Stratagene (La Jolla, CA).

b. FKBP expression plasmid, pCF1E, and a synthetic dimerizer, AP1510 are provided by Material Transfer Agreement from ARIAD Pharmaceuticals (Cambridge, MA).

c. Goat serum; GIBCO BRL.

---

* 1000 nM AP1510 is sufficient to achieve its maximum effect.

** Fluorsave, unlike other mounting media, does not contain glycerol. The cell suspension in this medium is easily prepared because of the low viscosity of the medium.

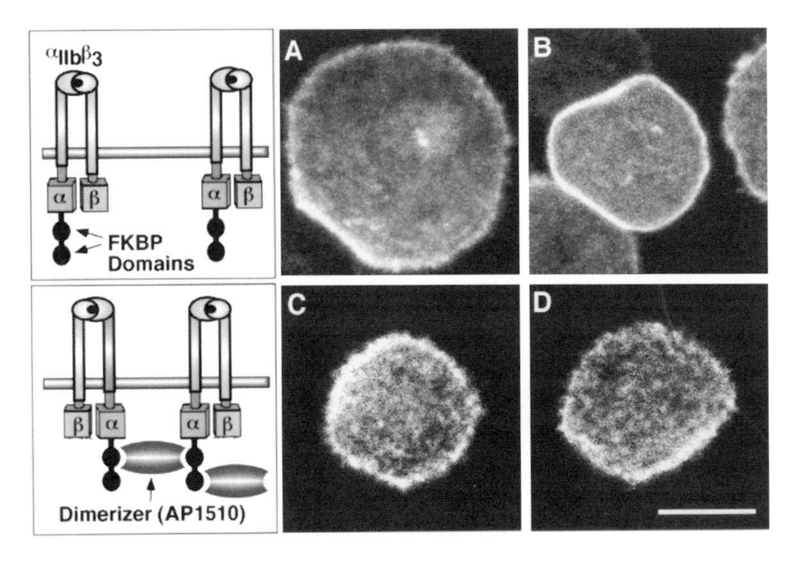

FIGURE 2
Conditional clustering of $\alpha_{IIb}(FKBP)_2\beta_3$ from within CHO cells using a chemical dimerizer of FKBP domains. The chemical dimerization strategy is depicted in the cartoons on the left. CHO cells stably-expressing $\alpha_{IIb}(FKBP)_2\beta_3$ were incubated in suspension in the absence (panels A and B) or presence (panels C and D) of 1000 nM AP1510. They were then incubated with FITC-D57 to stain the integrin, fixed, and deposited onto glass slides for confocal microscopy. Note that integrin staining is in a fine distribution in the absence of AP1510 but in a coarse, patchy distribution after treatment with AP1510. Bar = 10 μm. Adapted from Reference 17, with permission.

d. Fluorsave™; Calbiochem-Novabiochem Corporation (La Jolla, CA).

e. Confocal microscopy.

   We use MRC 1024 Bio-Rad laser scanning confocal imaging system Bio-Rad Laboratories (Hercules, CA) attached to a Leitz Diaplon microscope.

III. Outside-in Signaling

A. CHO Cell Adhesion to Immobilized Ligands

1. Overview

Experimentally, outside-in signaling through $\alpha_{IIb}\beta_3$ can be triggered in either of two ways: 1) induction of soluble fibrinogen binding to CHO cells maintained in suspension by addition of an "activating antibody,"[27] or 2) induction of CHO cell adhesion to an immobilized $\alpha_{IIb}\beta_3$ ligand, e.g., fibrinogen, von Willebrand factor or antibody D57. The former method is particularly useful for studying early events after ligand binding; the latter is useful for studying events that require integrin-triggered morphological changes and cytoskeletal reorganization. In the case of immobilized ligands, the extent of cell adhesion can itself be a useful readout.

2. Specific protocol

This is an example of an assay used to assess the effects of integrin affinity and avidity modulation on the extent of cell adhesion to an immobilized $\alpha_{IIb}\beta_3$ ligand. A 96-well microtiter plate is coated with purified fibrinogen, von Willebrand factor or D57 overnight at 4°C at coating concentrations ranging from 0.01 to 2 μg/well (100 μl of 0.1 to 20 μg/ml protein in coating buffer/well). The wells are incubated with 20 mg/ml BSA in PBS for 2 h at room temperature to block nonspecific binding. CHO cells stably expressing $\alpha_{IIb}(FKBP)_2\beta_3$ are harvested with trypsin/EDTA and resuspended in DMEM at 1.5×10^6/ml. The cells are labeled for 30 min at 37°C with 2 μM BCECF-AM.* After washing the cells three times with DMEM, they are resuspended in Tyrode's buffer containing 2 mM $CaCl_2$ and $MgCl_2$ at 1×10^6/ml and incubated for 30 min at room temperature in the presence or absence of 1000 nM AP1510 (to induce receptor clustering). Then, 100 μl aliquots are added to the coated microtiter wells in the presence or absence of 150 μg/ml anti-LIBS6 Fab (to induce affinity modulation) and incubated for 90 min at 37°C. The wells are washed three times with 150 μl of PBS, and 100 μl PBS is added to each well. Then, fluorescence is determined by cytofluorimetry at 485/530 nm. The number of adherent cells is related to the fluorescence intensity in a linear manner and adhesion is expressed as a percentage of fluorescence of total cells added.

3. Materials

a. Coating buffer: 150 mM NaCl, 5 mM Na_2HPO_4, 5 mM NaH_2PO_4, pH 8.0.

b. Microtiter plate; Immulon-2 HB microtiter plate (Dynex Laboratories, Chantilly, VA).

c. Fibrinogen, von Willebrand factor: Plasminogen-depleted human fibrinogen is purchased from Enzyme Research Laboratories (South Bend, IN). Purified human von Willebrand factor is kindly provided by Dr. Zaverio Ruggeri (Scripps Research Institute).

d. BCECF-AM; Molecular Probes. This fluorescent indicator is widely used for estimating intracellular pH and we used it to label CHO cells.

e. Cytofluorimeter; Cytofluor II from Biosearch (Bedford, MA).

B. Analysis of Protein Tyrosine Phosphorylation/Kinase Activity

1. Overview

One prominent early response in platelets to the ligation of $\alpha_{IIb}\beta_3$ is tyrosine phosphorylation of numerous substrates and activation of several protein tyrosine kinases.[2,28,29] The same is true in the CHO cell system.[9,10] Detailed protocols for assessing integrin-dependent tyrosine phosphorylation and tyrosine kinase activation in platelets have been provided by Clark and Brugge,[13] and these are applicable to CHO cells as well. Tyrosine dephosphorylation and activation of protein tyrosine phosphatases are equally important facets of integrin signaling.[29,30] However, we have not yet studied these in the CHO cell system.

---

* The stock solution is prepared by dissolving BCECF-AM in DMSO at 2 mM and stored at -20°C.

2. Specific protocol

As an example, we describe a protocol for the detection of tyrosine phosphorylation of recombinant Syk and endogenous FAK in $\alpha_{IIb}\beta_3$-expressing CHO cells in response to fibrinogen binding to cells maintained in suspension. These cells are transiently transfected with 0.5 μg of a Syk expression plasmid. Six hours later, the cells are placed into complete growth medium. 30 h after transfection, the amount of bovine serum in the culture medium is reduced to 0.5%; 48 h after transfection, the cells are harvested with trypsin/EDTA and resuspended in DMEM to 3×10^6/ml. The suspension is slowly rotated for 45 min at 37°C in the presence of 20 μM cycloheximide. Then cells are placed onto dishes which have been coated with 10 mg/ml of heat-denatured BSA* to block adhesive sites on the plastic, and then incubated for 10 min at 37°C with one or more of the following: 1000 nM AP1510 to stimulate receptor clustering, 150 μg/ml anti-LIBS6 Fab to increase $\alpha_{IIb}\beta_3$ affinity, or 250 μg/ml fibrinogen to achieve ligand binding. The cells, which remain in suspension, are then diluted with PBS, spun down, and the cell pellets are lysed in RIPA buffer for 30 min at 4°C. After clarification, the protein concentrations of the lysates are determined by BCA assay. Then, 200 μg of protein are immunoprecipitated with rabbit antiserum specific for Syk or FAK, and protein A-Sepharose CL4B beads. Immunoprecipitates are subjected to sodium dodecyl sulfate polyacrylamide gel electrophoresis and then transferred from the gel to nitrocellulose membranes. The membranes are blocked with 6% non-fat dry milk and then incubated with anti-phosphotyrosine monoclonal antibodies, 4G10 and PY20, followed by incubation with horseradish peroxidase-conjugated anti-mouse IgG. Immunoreactive bands are detected by enhanced chemiluminescence. To monitor loading of gel lanes, blots are stripped of antibody and reprobed with anti-Syk monoclonal antibody, 4D10, or rabbit anti-FAK serum.[10] *In vitro* kinase activity in Syk immunopreciptates is determined as previously described in detail.[13]

3. Materials

a. EMCV-Syk vector

This plasmid is kindly provided by Dr. Joan Brugge (Harvard Medical School, Boston, MA).

b. RIPA buffer

158 mM NaCl, 10 mM Tris-HCl, pH 7.4, containing 1% Triton X-100, 1% sodium deoxycholate, 0.1% SDS, 1 mM Na_2EGTA, 0.5 mM leupeptin, 0.25 mg/ml Pefabloc® (Boehringer Mannheim, Indianapolis, IN), 5 μg/ml aprotinin, 20 mM NaF, 3 mM β-glycerophosphate, 10 mM sodium pyrophosphate, and 5 mM sodium vanadate.

c. BCA protein assay; Pierce (Rockford, IL).

d. Antibodies

4G10 and PY20 are from Upstate Biotechnology (Lake Placid, NY) and Transduction Laboratories (Lexington, KY), respectively. 4D10 is from Santa Cruz Biotechnology (Santa Cruz, CA). Rabbit anti-Syk serum was raised against a synthetic peptide

---

* The BSA solution is denatured by the incubation for 20 min at 90°C.

corresponding to a linear sequence in the interdomain B region of human Syk (residues 324-339; EPELAPWAADKGPQRE). Rabbit anti-FAK serum is a gift from Dr. Thomas Parsons (University of Virginia).

e. Enhanced chemiluminescence (ECL); Amersham (Arlington Heights, IL).

C. Integrin-Dependent Morphological Changes

1. Overview

One major post-ligand binding response involves actin polymerization and reorganization mediated, in part, by Rho family GTPases.[31] In CHO cells stably expressing $\alpha_{IIb}\beta_3$, these cytoskeletal events are instrumental in promoting focal integrin complexes and morphological changes, including the formation of filopodia and lamellipodia, and the development of focal adhesions.[12,32] These changes are readily analyzed by confocal microscopy.

2. Specific protocol

Glass coverslips (18 mm in diameter) are placed in a 12-well plate and coated with 100 μg/ml fibrinogen in PBS overnight at 4°C. The coverslips are then blocked with 10 mg/ml BSA in PBS for 1 h at 37°C. CHO cells expressing $\alpha_{IIb}\beta_3$ are harvested with trypsin/EDTA and resuspended in DMEM at 1 to 5 × 10^5/ml.* The coated coverslips can be incubated with 1 ml cell suspensions for periods of time up to 2 h in a CO_2 incubator, during which time there is progressive cell spreading. The coverslips are then washed twice with PBS and fixed with 3.7% formaldehyde in PBS for 10 min at room temperature. After washing three times with PBS, the coverslips are incubated with 0.2% Triton X-100 in PBS for 5 min at room temperature to permeabilize the fixed cells. The coverslips are then washed three times with PBS, and blocked with 10% goat serum in PBS for 30 min at room temperature. To stain $\alpha_{IIb}\beta_3$, we typically use rabbit anti-α_{IIb} serum and FITC-labeled goat anti-rabbit IgG as a secondary antibody. To stain F-actin, we use rhodamine-labeled phalloidin. A stock solution of Rhodamine-phalloidin is prepared by dissolving in methanol at a concentration of 200 units/ml (6.6 μM). To make a working solution, 50 μl stock solution is evaporated and dissolved in 1 ml PBS containing 10% goat serum. The coverslips are incubated with anti-α_{IIb} serum for 45 min at 37°C. After washing with PBS, the coverslips are incubated with FITC-labeled secondary antibody and Rhodamine-phalloidin for 45 min at 37°C. After washing, the coverslips are rinsed with distilled water, and mounted on glass slides with mounting medium. Localization of $\alpha_{IIb}\beta_3$ and F-actin is observed with confocal microscopy. When desired, the cells can be stained with antibodies to other focal complex or focal adhesion-containing proteins; for example, vinculin or protein tyrosine kinases, and co-staining with rhodamine-phalloidin or antibodies to α_{IIb}.

---

* When stable cell lines are used, the cell concentration should be less than 2 × 10^5/ml to avoid overlapping of adherent cells.

3. Materials

a. Formaldehyde; Methanol-free 10% formaldehyde solution from Polysciences.

b. Rabbit anti-α_{IIb} serum

This serum is kindly provided by Dr. Mark Ginsberg (Scripps Research Institute) or can be obtained from Pharmingen (San Diego, CA).

c. FITC-labeled goat anti-rabbit IgG; Biosource International.

d. Rhodamine-labeled phalloidin; Molecular Probes.

e. Mounting medium; Immuno Floure mounting medium from ICN Biomedicals (Aurora, OH).

References

1. Savage, B., Almus-Jacobs, F., and Ruggeri, Z. M., Specific synergy of multiple substrate-receptor interactions in platelet thrombus formation and flow, *Cell*, 94, 657, 1998.
2. Shattil, S. J., Kashiwagi, H., and Pampori, N., Integrin signaling: the platelet paradigm, *Blood*, 91, 2645, 1998.
3. O'Toole, T. E., Katagiri, Y., Faull, R. J., Peter, K., Tamura, R., Quaranta, V., Loftus, J. C., Shattil, S. J., and Ginsberg, M. H., Integrin cytoplasmic domains mediate inside-out signaling, *J. Cell Biol.*, 124, 1047, 1994.
4. Loh, E., Qi, W. W., Vilaire, G., and Bennett, J. S., Effect of cytoplasmic domain mutations on the agonist-stimulated ligand binding activity of the platelet integrin $\alpha_{IIb}\beta_3$, *J. Biol. Chem.*, 271, 30233, 1996.
5. Lyman, S., Gilmore, A., Burridge, K., Gidwitz, S., and White, G. C., II, Integrin-mediated activation of focal adhesion kinase is independent of focal adhesion formation or integrin activation — Studies with activated and inhibitory β_3 cytoplasmic domain mutants, *J. Biol. Chem.*, 272, 22538, 1997.
6. Schaffner-Reckinger, E., Gouon, V., Melchior, C., Plancon, S., and Kieffer, N., Distinct involvement of β_3 integrin cytoplasmic domain tyrosine residues 747 and 759 in integrin-mediated cytoskeletal assembly and phosphotyrosine signaling, *J. Biol. Chem.*, 273, 12623, 1998.
7. O'Toole, T. E., Loftus, J. C., Du, X., Glass, A. A., Ruggeri, Z. M., Shattil, S. J., Plow, E. F., and Ginsberg, M. H., Affinity modulation of the $\alpha_{IIb}\beta_3$ integrin (platelet GPIIb-IIIa) is an intrinsic property of the receptor, *Cell Regulation*, 1, 883, 1990.
8. O'Toole, T. E., Mandelman, D., Forsyth, J., Shattil, S. J., Plow, E. F., and Ginsberg, M. H., Modulation of the affinity of integrin $\alpha_{IIb}\beta_3$ (GPIIb-IIIa) by the cytoplasmic domain of α_{IIb}, *Science*, 254, 845, 1991.
9. Leong, L., Hughes, P. E., Schwartz, M. A., Ginsberg, M. H., and Shattil, S. J., Integrin signaling: Roles for the cytoplasmic tails of $\alpha_{IIb}\beta_3$ in the tyrosine phosphorylation of $pp125^{FAK}$, *J. Cell Sci.*, 108, 3817, 1995.
10. Gao, J., Zoller, K., Ginsberg, M. H., Brugge, J. S., and Shattil, S. J., Regulation of the $pp72^{Syk}$ protein tyrosine kinase by platelet integrin $\alpha_{IIb}\beta_3$, *EMBO J.*, 16, 6414, 1997.
11. Leng, L., Kashiwagi, H., Ren, X.-D., and Shattil, S. J., RhoA and the function of platelet integrin $\alpha_{IIb}\beta_3$, *Blood*, 91, 4206, 1998.

12. Miranti, C., Leng, L., Maschberger, P., Brugge, J. S., and Shattil, S. J., Identification of a novel integrin signaling pathway involving the kinase Syk and the gnanine nucleotide exchange factor VAV1. *Curr. Biol.*, 8: 1289, 1998.
13. Clark, E. A. and Brugge, J. S., Tyrosine phosphorylation, in *Platelet Function and Signal Transduction: A Practical Approach*, Watson, S. and Authi, K. Eds., Oxford, Oxford Press, 1996.
14. Michelson, A. D. and Shattil, S. J., The use of flow cytometry to study platelet activation, in *Platelets: A Practical Approach,* Watson, S. P. and Authi, K. S., Eds., Oxford, Oxford University Press, 1996, p 111.
15. Loftus, J. C. and Liddington, R. C., Cell adhesion in vascular biology. New insights into integrin-ligand interaction, *J. Clin. Invest.*, 99, 2302, 1997.
16. Bazzoni, G. and Hemler, M. E., Are changes in integrin affinity and conformation overemphasized?, *Trends Biochem. Sci.*, 23, 30, 1998.
17. Hato, T., Pampori, N., and Shattil, S. J., Complementary roles for receptor clustering and conformational change in the adhesive and signaling functions of integrin $\alpha_{IIb}\beta_3$, *J. Cell Biol.*, 141, 1685, 1998.
18. Bennett, J. S., Hoxie, J. A., Leitman, S. S., Vilaire, G., and Cines, D. B., Inhibition of fibrinogen binding to stimulated platelets by a monoclonal antibody, *Proc. Natl. Acad. Sci. USA,* 80, 2417, 1983.
19. Shattil, S. J., Hoxie, J. A., Cunningham, M., and Brass, L. F., Changes in the platelet membrane glycoprotein IIb-IIIa complex during platelet activation, *J. Biol. Chem.*, 260, 11107, 1985.
20. Du, X., Gu, M., Weisel, J., Nagaswami, C., Bennett, J. S., Bowditch, R., and Ginsberg, M. H., Long range propagation of conformational changes in integrin $\alpha_{IIb}\beta_3$, *J. Biol. Chem.*, 268, 23087, 1993.
21. Scarborough, R. M., Naughton, M. A., Teng, W., Rose, J. W., Phillips, D. R., Nannizzi, L., Arfsten, A., Campbell, A. M., and Charo, I. F., Design of potent and specific integrin antagonists. Peptide antagonists with high specificity for glycoprotein IIb-IIIa, *J. Biol. Chem.*, 268, 1066, 1993.
22. Kashiwagi, H., Schwartz, M. A., Eigenthaler, M. A., Davis, K. A., Ginsberg, M. H., and Shattil, S. J., Affinity modulation of platelet integrin $\alpha_{IIb}\beta_3$ by β_3-endonexin, a selective binding partner of the β_3 integrin cytoplasmic tail, *J. Cell Biol.*, 137, 1433, 1997.
23. Gawaz, M. P., Loftus, J. C., Bajt, M. L., Frojmovic, M. M., Plow, E. F., and Ginsberg, M. H., Ligand bridging mediates integrin $\alpha_{IIb}\beta_3$ (platelet GPIIB-IIIA) dependent homotypic and heterotypic cell-cell interactions, *J. Clin. Invest.*, 88, 1128, 1991.
24. Klemm, J. D., Schreiber, S. L., and Crabtree, G. R., Dimerization as a regulatory mechanism in signal transduction, *Annu. Rev. Immunol.*, 16, 569, 1998.
25. O'Toole, T. E., Loftus, J. C., Plow, E. F., Glass, A., Harper, J. R., and Ginsberg, M. H., Efficient surface expression of platelet GPIIb-IIIa reguires both subunits, *Blood*, 74, 14, 1989.
26. Amara, J. F., Clackson, T., Rivera, V. M., Guo, T., Keenan, T., Natesan, S., Pollock, R., Yang, W., Courage, N. L., Holt, D. A., and Gilman, M., A versatile synthetic dimerizer for the regulation of protein-protein interactions, *Proc. Natl. Acad. Sci. USA,* 94, 10618, 1997.

27. Huang, M.-M., Lipfert, L., Cunningham, M., Brugge, J. S., Ginsberg, M. H., and Shattil, S. J., Adhesive ligand binding to integrin $\alpha_{IIb}\beta_3$ stimulates tyrosine phosphorylation of novel protein substrates before phosphorylation of $pp125^{FAK}$, *J. Cell Biol.*, 122, 473, 1993.
28. Clark, E. A., Shattil, S. J., and Brugge, J. S., Regulation of protein tyrosine kinases in platelets, *Trends Biochem. Sci.*, 19, 464, 1994.
29. Jackson, S. P., Schoenwaelder, S. M., Yuan, Y. P., Salem, H. H., and Cooray, P., Non-receptor protein tyrosine kinases and phosphatases in human platelets, *Thromb. Haemost.*, 76, 640, 1996.
30. Liu, F., Sells, M. A., and Chernoff, J., Protein tyrosine phosphatase 1B negatively regulates integrin signaling, *Curr. Biol.*, 8, 173, 1998.
31. Hall, A., Rho GTPases and the actin cytoskeleton, *Science*, 279, 509, 1998.
32. Ylanne, J., Chen, Y., O'Toole, T. E., Loftus, J. C., Takada, Y., and Ginsberg, M. H., Distinct functions of integrin-alpha and integrin-beta subunit cytoplasmic domains in cell spreading and formation of focal adhesions, *J. Cell Biol.*, 122, 223, 1993.

Chapter 15

Tracking Integrin-Mediated Adhesion Using Green Fluorescent Protein and Flow Cytometry

Wendy J. Kivens and Yoji Shimizu

Contents

I. Overview 217
II. Protocols 219
III. Materials 226
IV. Buffers 229
V. Summary and Future Directions 229
Acknowledgments 231
References 231

I. Overview

It is well established that the functional activity of integrin receptors can be regulated by activation of intracellular signaling pathways. This is particularly evident in hematopoietic cells, which undergo cycles of adhesive and non-adhesive states that are dependent on the activation state of the cells. For example, T lymphocytes express β1 and β2 integrins on the cell surface, but adhere poorly to relevant cell surface counter-receptors and extracellular matrix ligands. However, T cells can exhibit increases in integrin-mediated adhesion within seconds to minutes following activation.[1-4] This acti-

0-8493-3385-7/99/$0.00+$.50

vation-dependent regulation of integrin-mediated adhesion is critical to both T cell migration and T cell recognition of foreign antigen.[5] A number of activation stimuli have been identified that can regulate integrin adhesiveness on T cells, including: (1) pharmacological agents, such as the phorbol ester PMA and Ca^{2+} ionophore;[1,3,6] (2) ligand engagement of the CD3/T cell receptor and other immunoglobulin superfamily member co-receptors;[1,3,6] (3) ligand engagement of G protein-coupled chemokine receptors;[4] (4) activation by integrin-specific antibodies;[7] and (5) certain divalent cations, particularly Mn^{2+}.[8]

A number of technical approaches can be used to analyze the intracellular signaling pathways that regulate integrin activity. One approach is the use of inhibitors that block the activity of specific signaling pathways. Inhibitors have been used to implicate a number of signaling molecules in activation of integrin function, notably protein kinase C (PKC) and phosphoinositide 3-OH kinase (PI 3-K).[1,9,10] The advantage of inhibitors is their ease of use and applicability to the analysis of integrin activation in many different cell types. However, the investigator assumes a high degree of specificity in the molecular target of the inhibitor, an assumption that is often not warranted as the inhibitory mechanisms are investigated in more detail. One example of this inverse relationship between specificity and degree of use in the field is wortmannin, a fungal metabolite that blocks PI 3-K activity, but has recently been shown to block the activity of other enzymes, particularly when used at concentrations well above the half-maximal inhibitory concentration of wortmannin for PI 3-K.[11,12]

Another approach for disrupting signaling pathways is to overexpress wild-type or mutant forms of kinases and adapter proteins in cells, and determine the effects of the expression of these proteins on the functional response of interest. This approach has been particularly illuminating in many signaling studies, due to the identification of mutations that either block the function of an endogenous molecule ("dominant-negative") or endow the transfected molecule with activity that is not regulated normally ("constitutively active"). Although the precise mechanism of action of dominant-negative and constitutively active mutations is beyond the scope of this chapter, this approach has nevertheless been widely used to dissect the role of specific molecules in signal transduction, as well as to define the relationship between two molecules in a specific signaling pathway.[13-15] Several studies have recently implicated proteins such as PI 3-K, Rho GTPases, and adapter proteins such as $p130^{Cas}$ in integrin-mediated adhesion, migration, and signaling.[16-19] Dominant-negative and constitutively active molecules have also been useful in the analysis of developmental processes *in vivo* with the tissue-specific expression of these constructs in transgenic mice.[20]

There are a number of technical issues that must be resolved in any experimental system to effectively utilize dominant-negative and constitutively active constructs. First, the effects of these constructs on signaling pathways often requires high levels of overexpression in the cell type under study. Second, many of these constructs may be toxic to cells when constitutively expressed, making it difficult to generate stable cell lines expressing these molecules. Thus, the use of transient transfection is often preferred and/or required in this type of experimental analysis. The ability to disrupt signaling pathways with dominant-negative and constitutively active constructs has been well documented in certain cell types, such as fibroblasts, where transient transfection results in a high percentage of cells expressing the transfected cDNA clone. However, many other cell types, such as hematopoietic cells, are difficult to transfect, resulting in only a low percentage of cells expressing the molecule of interest.

We have developed an assay that overcomes many of these technical problems so that the signaling pathways regulating integrin activity in hematopoietic cells can be analyzed. This approach exploits the quantitative capabilities of the flow cytometer and the recent development of green fluorescent protein (GFP) as a reporter for detecting transfected cells by flow cytometry.[21-24] The general strategy is to transiently transfect a plasmid containing a cDNA that encodes a GFP-tagged fusion protein into cells, and then to perform an adhesion assay with this heterogeneous population of transiently transfected cells. This variegated population will contain a small percentage of cells expressing the GFP fusion protein and a much larger percentage of cells that do not express the fusion protein. Following completion of the adhesion assay, the cells that remain adherent to the integrin ligand are collected and resuspended in a known volume of flow cytometry buffer. A known number of microbeads that can be distinguished by scatter or fluorescence properties from the cells is then added to the sample before acquisition on the flow cytometer. With a known sample volume and a known number of microbeads, the total number of adherent cells in the sample tube can be calculated on the basis of the ratio of the microbeads to cells. Thus, this approach can be used to quantitatively assess the adhesive behavior of a small percentage of transiently transfected cells expressing the construct of interest. We have recently used this system to implicate PI 3-K and the Cbl adaptor protein in the regulation of β1 integrin activity in myeloid and lymphoid cells.[25-27] This chapter provides a detailed description of the assay and the subsequent data analysis, using CD2$^+$ HL-60 cells, a GFP fusion protein expressing a dominant-negative form of the p85 subunit of PI 3-K, and adhesion to fibronectin (FN) as an example.

II. Protocols

1. **Cell maintenance** — The HL-60 monocytic cell line is maintained in RPMI 1640 medium supplemented with 10% fetal calf serum (FCS), 2 mM L-glutamine, 100 U/ml penicillin, and 100 µg/ml streptomycin. All tissue culture cells are maintained in a humidified 37°C, 5% CO_2 incubator. For higher electroporation efficiencies, cell lines should be kept healthy and growing in 150 cm^2 tissue culture flasks at a logarithmic rate in RPMI supplemented with 10% FCS and kept under selection, if required. All subcloned HL-60 cell lines expressing CD2 must be grown in the presence of 1 mg/ml G418 (GIBCO/BRL) to maintain cell surface expression.[9] Cell lines should not be allowed to grow beyond a density of 0.8×10^6 cells/ml.
2. **DNA quality and quantity** — The pEGFP-C1, pEGFP-C2, and pEGFP-C3 vectors were purchased from Clontech. GFP-fusions were created by fusing the coding region of a particular gene in frame within the multiple coding site located at the carboxy terminus of the GFP reading frame, as previously described.[25,27] We find that plasmid

DNA purified on a CsCl gradient, rather than with a matrix purification system, gives the highest transfection efficiency. We do not notice any increase in transfection efficiency by banding the DNA through more than one CsCl gradient. The banded DNA should be dissolved in sterile glass-distilled water rather than TE in order to reduce salt concentrations and improve transfection efficiency. The final concentration of the DNA preparation should be at least 2 mg/ml for optimum transfection and expression efficiency.

3. **Optimal cell to DNA ratio** — We have successfully used a DNA:cell ratio of 30 to 60 μg of uncut GFP fusion vector mixed with 10×10^6 cells. However, the optimal DNA:cell ratio for each cell type must be determined empirically. For transient transfection adhesion assays, where 3 to 6×10^6 live cells are required, we routinely transfect 40×10^6 cells with 240 μg of CsCl banded GFP-fusion DNA. We observe that the electroporation causes 75 to 80% cell death, resulting in a 20 to 25% cell recovery. For HL-60 cells, approximately 5 to 15% of the living cells that remain after electroporation normally express detectable levels of the GFP-fusion protein. With this technique, a basal GFP-fusion protein expression level of 1 to 2% is more than sufficient to reveal any potential effect of the GFP-fusion protein upon the modulation of integrin-mediated adhesion.

4. **Transfection** — Sterile technique should be used when handling the cells and the GFP-fusion vectors throughout the transient transfection protocol. Between 30×10^6 cells and 40×10^6 cells per transient transfection are pelleted (1500 rpm, 5 min, room temperature) and washed once with Opti-MEM medium. The cells are pelleted a second time and the Opti-MEM supernatant is aspirated off without disturbing the cell pellet. The washed cells are resuspended in 0.8 ml (50×10^6 cells/ml) of Opti-MEM containing 120 to 240 μg of the GFP vector or the GFP-fusion vector. The cell/DNA mixture is allowed to incubate for 10 min at room temperature in a sterile 4 mm gapped electroporation cuvette. For electroporation, the voltage and capacitance that gives the highest transfection efficiency must be determined empirically for each cell type. HL-60 cells can be electroporated either with a BTX Square Wave electroporator set to LV mode, at 230V with one 15 ms pulse, or with a Bio-Rad Gene Pulser set to 280V, 960 μF, resulting in a time constant of approximately 20 ms. After electroporation, cells are incubated for 30 min at room temperature before resuspension at 1×10^6 cells/ml in a 75 cm^2 tissue culture flask containing RPMI 1640 medium supplemented with 10% FCS, 2 mM L-glutamine, 100 U/ml penicillin, and 100 μg/ml streptomycin. Cells are harvested after 16 to 18 h of incubation in a 37°C, 5% CO_2 tissue culture incubator, as described in Section 6, and used in the static adhesion assay described below.

5. **Preparation of FN-coated 96-well plate** — The adhesion of GFP^+ transiently transfected cells to FN (1 μg/well) is set up as previously described.[25,27] On the same day that the cells are electroporated, the appropriate number of wells of a flat-bottomed 96-well microtiter plate(s) is coated with 50 μl of a 20 μg/ml solution of human FN. We employ a repeating pipet to aliquot the FN in each well in order to minimize well-to-well differences in FN concentration. The plate is left overnight at 4°C. The next morning, the plate is washed twice with 200 μl of PBS + Ca^{2+}/Mg^{2+}. As previously described, we have found that a hand-washing plate apparatus works well for this experimental set-up.[28] Briefly, unbound ligand is washed away from wells with PBS + Ca^{2+}/Mg^{2+} using a syringe pipet attached to an 8-tip bent manifold. Direct the wash buffer onto the side of each well. Remove the wash buffer from the wells using an 8-tip straight manifold attached to an aspiration device. Again, minimize contact

between manifold tip and the bottom of the well. The microtiter dish is rotated 180° between washes so that both sides of the wells are washed down.

Any remaining non-specific binding sites within the well are blocked with 50 μl of 2.5% BSA (prepared in PBS + Ca^{2+}/Mg^{2+}). The BSA blocking solution should be added using a repeating pipet to minimize well-to-well volume variation. The BSA blocking solution is left in the wells for 1 to 2 h and incubated in a 37°C incubator. The plate is then washed twice with PBS + Ca^{2+}/Mg^{2+}. The plate is now ready for the cells and various stimulating reagents required for the experiment to be added to each well.

6. **Ficoll purification of transfected cells** — After overnight incubation, 30 to 40 ml of each transfected cell culture is transferred to a 50 ml conical tube. If needed, the volume is brought up to 40 ml with PBS lacking Ca^{2+}/Mg^{2+}. Add 10 ml of room temperature Ficoll to the bottom of the 50 ml conical tube using a blunt ended, 6-in., 14-gauge sterile needle attached to a 12 cc sterile syringe. Live cells are banded between the media-Ficoll interface by spinning the transfected cells at 2000 rpm for 20 min at room temperature, without brake engagement. It is imperative that the centrifuge brake is not activated in order to retain the live-cell interface between the media and Ficoll. After the spin is completed, 20 to 25 ml of the media above the live cell interface is removed by aspiration. The live cells are then carefully removed from the interface with a pipet and placed in a fresh 50 ml conical tube.

   The live cells are washed twice in PBS lacking Ca^{2+}/Mg^{2+}. The cells are then counted and washed once in ice-cold PBS/0.5% human serum albumin (PBS/HSA). After the final wash, the cells are resuspended at a density of 1.6×10^6 cells/ml in ice-cold PBS/HSA. The cells are now ready to be added to the appropriate FN-coated well, unless the stimulation conditions require cross-linking a cell-surface receptor. A description of how to treat cells to be cross-linked is described in Section 7. Otherwise, add 50 μl of cells at a density of 1.6×10^6 cells/ml (an equivalent to 80,000 transfected, live cells) to each well.

   Duplicate aliquots of 200 μl of each cell sample are set aside on ice for flow cytometric analysis in order to verify the cell numbers added per well, and to determine the level of GFP-fusion expression within the pre-adherent cell population. If possible, we recommend that prior to adding the cells in the appropriate wells and proceeding with the adhesion assay, an aliquot of pre-adherent GFP-fusion transfected cells should be analyzed and compared to the analysis of untransfected cells to ensure that the GFP-fusion protein is being expressed at a detectable level.

7. **Cross-linking** — This is a generalized protocol for integrin activation conditions that require cross-linking of a cell-surface receptor, such as the antigen-specific CD3/T cell receptor complex.[1,3] Cells are suspended at a density of 1.6×10^6 cells/ml in PBS/HSA. If necessary, cells are incubated on ice for 15 min with 1 μg human IgG per 1×10^6 cells to block non-specific activation through Fc receptors. The monoclonal antibody directed against the receptor to be cross-linked is then added at a concentration of 0.5 to 2.0 μg/1×10^6 cells. The amount of antibody added must be empirically determined for each mAb. The cell suspension is incubated on ice for an additional 30 min. The excess unbound Ab is washed away with two cell washes in ice-cold PBS/HSA. Cells are brought up to a density of 1.6×10^6 cells/ml in ice-cold PBS/HSA and 50 μl aliquots are placed in the appropriate FN-coated microtiter wells.

8. **Stimulation conditions** — There should be twelve replicate wells for each stimulation condition (this is equivalent to one of the eight horizontal rows of a 96-well microtiter plate). Using the repeating pipet, 50 μl of ice-cold PBS/HSA is added to wells where the cells will not be stimulated. If cells will be stimulated with phorbol ester, 50 μl of

a 20 ng/ml solution of PMA (diluted in PBS/HSA) is added to wells. For CD2 stimulation, add an activating combination of CD2 mAb pairs at a final dilution of 1:10 for the CD2-specific mAb 95-5-49 and 1:6000 for the CD2-specific mAb 9-1. If cells require cross-linking of an activating antibody, 50 µl of a 2 µg/ml solution of goat anti-mouse IgG (diluted in PBS/HSA) is added to wells. After addition of the stimuli to the appropriate wells, the microtiter plate is incubated for 30 to 60 min on ice to allow the cells to settle to the bottom of the FN-coated well. The plate is then carefully floated in a 37°C water bath for the required activation time. Our standard activation time for maximal stimulation of β1 integrin-mediated adhesion is 10 min. After incubation at 37°C, the plate is removed from the water bath, and the wells are washed four times with PBS + Ca^{2+}/Mg^{2+}. For washing with a manual syringe-manifold and aspirator, we rotate the plate orientation 180° between each wash, so that the wells are washed evenly. It is important that neither the plate-washer manifold tips nor the aspiration tips touch the bottom of the wells, as this contact could unintentionally dislodge the adherent cells present in each well. We have found that washing with an automated plate-washer is also effective with certain cell types.

9. **Preparation of pre-adherent and post-adherent cells** — Adhesion is quantitated by the collection of adherent cells and analysis by flow cytometry. 200 µl of versene is added to each well with an 8-channel pipettor. The plate is then incubated at 37°C for 5 to 10 min. The adherent cells are dislodged from the bottom of the plate by vigorously pipetting the versene solution up and down within the well using an 8-channel pipettor. Throughout the cell dislodging process, it is essential to only use the same pipet tip in well rows that contain the identical cells and stimulation conditions, as any cross-contamination between well rows (stimulation conditions) will alter the adhesion results. Cells from six replicate wells are pooled and resuspended in 1 ml Hank's Balanced Salt Solution (HBSS) supplemented with 10% bovine calf serum (BCS) and 0.2% sodium azide. Since there are twelve wells for each stimulus, pooling six wells will give two replicate samples per activation condition.

   The cells are spun down and resuspended in 200 µl HBSS supplemented with 1% BCS and 0.2% sodium azide (FACS Media). The 200 µl aliquots containing the pre-adherent cell samples are also spun down and resuspended in 200 µl of FACS media. A 50 µl aliquot (10,082 beads) of PKH26 reference microbeads and 25 µl of 40 µg/ml propidium iodide (PI) are added to each tube, for a total sample volume of 275 µl.

10. **Acquisition of data on the flow cytometer** — Each sample is analyzed on a FACScan or a FACSCalibur flow cytometer. For sample acquisition, forward scatter vs. side scatter dot plots should be established so that the cells and reference beads can be clearly discriminated (Figures 1 and 2). Gates should be drawn around the beads and cells. Since the 7.5 µ sized beads that we typically use are impregnated with PKH26 dye, which can be detected in the FL2 channel, we also establish a FL2 histogram profile based on the gate containing the beads (Figure 2). Since GFP fluorescence is detected on the FL1 channel and PI is detected on the FL2 channel, we also establish a FL1 vs. FL2 dot or density plot based on the gate containing the cells (Figure 2). This plot distinguishes live, PI-negative cells from any dead, PI-positive cells as well as GFP fluorescence of the live cells. It is critical at this step to have the appropriate compensation settings between the FL1 and FL2 channels. Once the appropriate parameters and settings have been set up on the flow cytometer (Figures 1 and 2), event acquisition can vary from 30 sec to 4 min per sample. We typically acquire a minimum of 30,000 total events.

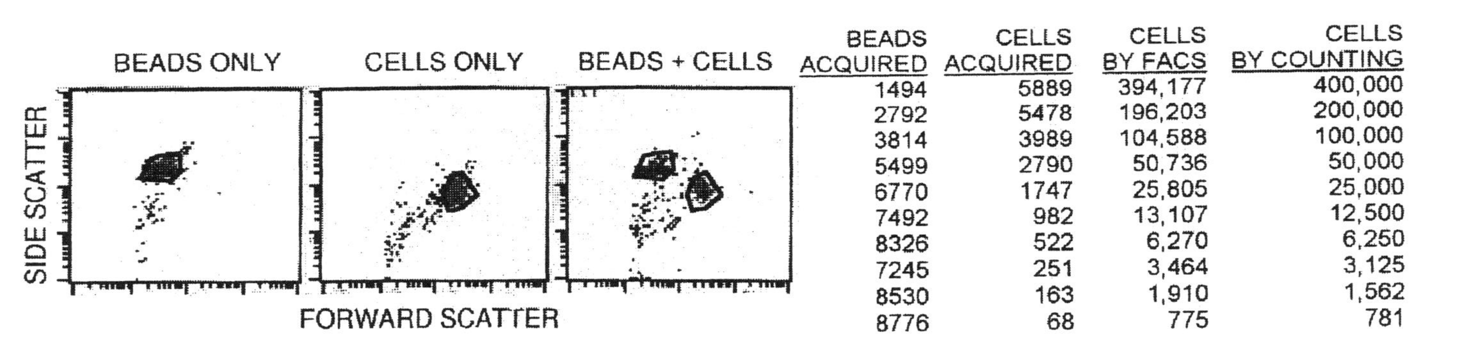

BEADS ACQUIRED	CELLS ACQUIRED	CELLS BY FACS	CELLS BY COUNTING
1494	5889	394,177	400,000
2792	5478	196,203	200,000
3814	3989	104,588	100,000
5499	2790	50,736	50,000
6770	1747	25,805	25,000
7492	982	13,107	12,500
8326	522	6,270	6,250
7245	251	3,464	3,125
8530	163	1,910	1,562
8776	68	775	781

FIGURE 1
Quantitation of cell number by flow cytometry. HL-60 cells were suspended in a 1 ml volume of FACS media containing 100,000 5 micron latex microspheres ("beads") and analyzed on a Becton-Dickinson FACScan. Tubes containing beads only (left panel) and HL-60 cells only (middle panel) were run for the purposes of setting gates. In the right panel is a representative dot plot containing beads and HL-60 cells. To assess the accuracy of the technique over a range of HL-60 cell amounts, 1:1 dilutions of HL-60 cells (beginning with 400,000 cells/ml) were made and analyzed by FACS as described in the text. The equations described in the text were used to calculate the total number of HL-60 cells in each dilution; these values are compared to the number of expected cells (based on the initial cell number of 400,000 cells, as assessed by counting with a hemacytometer) and are shown on the right.

11. **Data analysis** — Analysis of adherent transfected cells is performed in the following manner. Since the volume of the sample is known (275 µl) and there is a known number of microbeads, the total number of transfected cells in the sample can be determined as follows. For each sample, the forward scatter and side scatter profiles are used to discriminate and gate the fluorescent reference microbeads and the cells. An example is shown in Figure 1, which also demonstrates the ability of the flow cytometer to quantitate the number of cells in a sample using a known number of microbeads as a reference. FL2 events in the bead gate (Beads Gate in Figure 2) are used to calculate the total number of reference microbeads in each acquired sample. Using the cell gate (Cells Gate in Figure 2), FL1/FL2 dot plots are established to determine the number of live, PI-negative transfected cells that are acquired. Gating on the FL1 fluorescence, which reflects expression of GFP in either the pre-adherent or the post-adherent cells (Figure 2), is used to quantitate the total number of cell events in each sample (Figure 3). For each sample, the total number of reference microbeads acquired (Figure 2) is divided by the bead density in the sample (36,662 beads/ml) to obtain the total volume of sample acquired (Figure 3). The total number of cells in each sample is then determined by the following equation:

$$\text{\# of Cells per Sample} = [(\text{Cells acquired})/(\text{volume of acquired sample})](0.275 \text{ ml})$$

The pre-adherent cell samples are used to determine the number of $CD2^+$ HL-60 cells that were added to each well at the start of the adhesion assay. This number is then multiplied by six to give the number of pre-adherent cells added to six identical wells at the start of the assay (Figure 4A). The post-adherent cell samples are used to determine the number of $CD2^+$ HL-60 cells that adhered to the FN from six pooled wells (Figure 4B). The percent of GFP^+ and GFP–cell adhesion in each sample is determined by the total number of adherent cells divided by the total number of pre-adherent cells (Figure 4C).

Using the data presented in Figure 2, the calculations depicted in Figures 3 and 4 provide detailed examples of the data analysis for CD2-stimulated $CD2^+$ HL-60 cells transiently expressing a GFP protein fused to a dominant-negative form of the p85 subunit of PI 3-K (GFP-Δp85).[27] In this specific example, three GFP gates are drawn:

FIGURE 2 (opposite)
Flow chart of FACS data acquisition and data analysis. A depiction of the pre-adherent and adherent FACS samples is shown on the top left of the figure, where cells (either 200 µl of preadherent cells or an unknown number of adherent cells from 6 replicate wells) are mixed together with 50 µl of PKH26 reference microbeads and 25 µl of PI. The top-right panel is a representative density plot containing beads and $CD2^+$ HL-60 cells. The beads and cells are separately gated (dark circles surrounding each population) and each is analyzed. As shown in the bottom left histogram panels, an FL2 histogram profile can be established from the gated bead population, and the number of beads acquired from the sample determined. As seen in the bottom right density plots, GFP fluorescence is detected on the FL1 channel and PI is detected on the FL2 channel. An FL1 vs. FL2 density plot based on the gate containing the cells can be established. These plots distinguish live, PI-negative cells from any dead, PI-positive cells as well as GFP fluorescence of the live cells. It is critical here to have the appropriate compensation settings between the FL1 and FL2 channels in order to be able to gate for GFP fluorescence of live cells vs. dead cells. Once these settings are established and all the data has been acquired, the GFP gates can be drawn. To ensure reproducible adhesion results and to lower sample-to-sample variation, we recommend that at least 100 cells should be present in any given GFP gate.

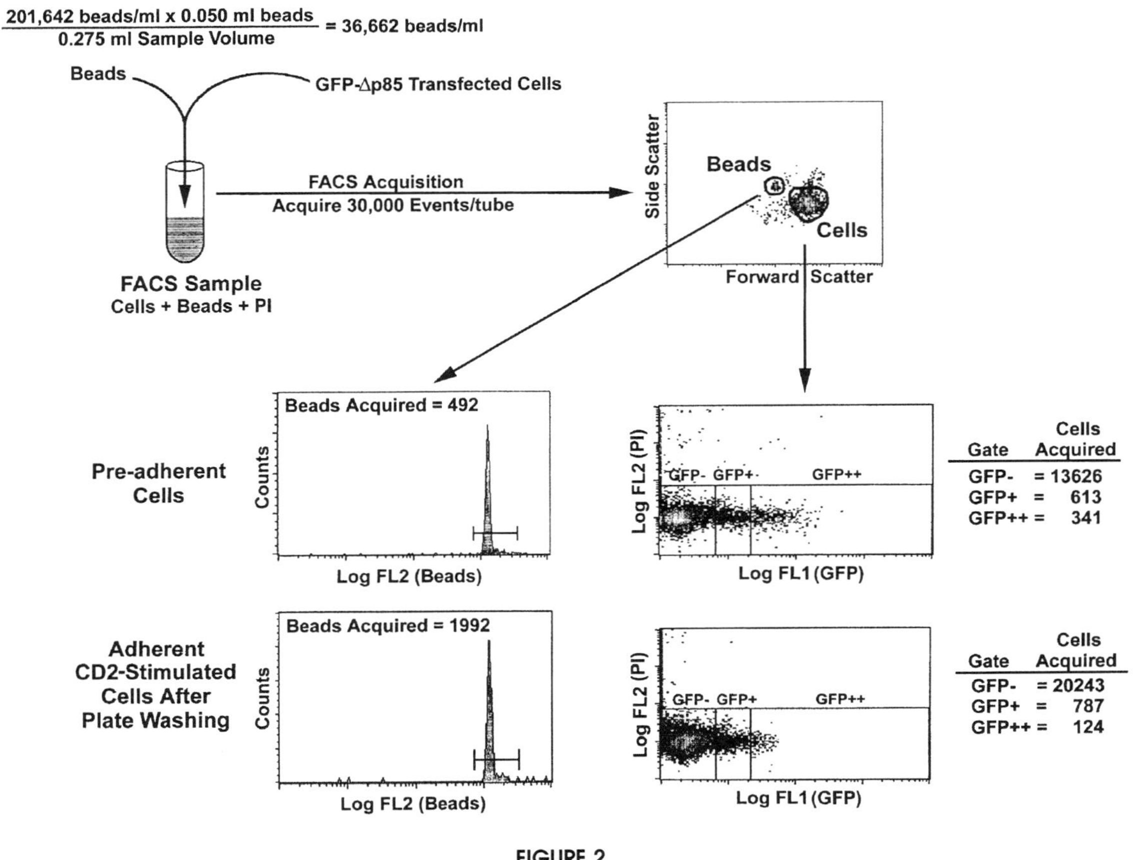
201,642 beads/ml x 0.050 ml beads / 0.275 ml Sample Volume = 36,662 beads/ml
Beads
GFP-Δp85 Transfected Cells
FACS Acquisition
Acquire 30,000 Events/tube
FACS Sample
Cells + Beads + PI
Side Scatter
Forward Scatter
Beads
Cells
Pre-adherent Cells
Adherent CD2-Stimulated Cells After Plate Washing
Counts
Beads Acquired = 492
Beads Acquired = 1992
Log FL2 (Beads)
Log FL2 (PI)
Log FL1 (GFP)
GFP- GFP+ GFP++
Gate Cells Acquired
GFP- = 13626
GFP+ = 613
GFP++ = 341
Gate Cells Acquired
GFP- = 20243
GFP+ = 787
GFP++ = 124

FIGURE 2

Calculation of Cell Number per FACS Sample:

FACS Sample Volume = 0.025 ml PI + 0.050 ml Beads + 0.200 ml Pre-adherent Cells = 0.275 ml

$$\text{Beads/ml in FACS Sample} = \frac{\text{(201,642 beads/ml [stock concentration])(0.050 ml beads)}}{\text{0.275 ml FACS Sample Volume}} = \text{36,662 beads/ml}$$

$$\text{Volume (in ml) of Acquired Sample} = \frac{\text{Beads Acquired}}{\text{Beads/ml in FACS Sample}}$$

$$\text{Number of Cells Present in FACS Sample Tube} = \frac{\text{(Number of Cells Acquired by FACS)(0.275 ml Sample Volume)}}{\text{Volume (in ml) of Acquired Sample}}$$

FIGURE 3
Calculation of the cell number present in each FACS sample. The calculations used to determine the number of cells per FACS sample are outlined here. The following information is required in order to quantify the number of cells present per FACS sample: the total FACS sample volume, stock concentration of beads, the number of bead events acquired from the FACS sample, and the number of live cell events acquired from the FACS sample.

one for the large population of GFP-negative cells in each sample, and two distinct gates for the GFP-positive cells that allow us to compare the adhesion of cells expressing low (GFP^{+}) and high (GFP^{++}) levels of the GFP-Δp85 construct.

III. Materials

1. **Cell culture** — Humidified 37°C, 5% CO_2 incubator; aspirator; 75 cm^2 and 150 cm^2 tissue culture flasks; sterile 50 ml conical tubes; sterile 1 ml, 2 ml, 5 ml, 10 ml, and 25 ml pipets; cell counting microscope; hemacytometer; trypan blue (Mediatech, Inc., Cat. #MT 25-900); and cell counter.
2. **Repeating pipet** — We routinely use a repeating pipet from Oxford Labware (product #8885-881006) outfitted with 1.5 ml syringes (product #8885-881022).
3. **8-channel pipettor** — Any continuously adjustable 8-channel pipet should suffice. We use a multi-channel pipettor from Oxford Labware (product #500077).

FIGURE 4 (opposite)
Detailed data analysis for CD2-stimulated $CD2^{+}$ HL-60 cells transiently expressing a GFP fusion protein expressing a dominant-negative form of the p85 subunit of PI 3-K. The data depicted in Figure 2 is analyzed here to determine the level of CD2-stimulated adhesion within the GFP^{-}, GFP^{+}, and GFP^{++} gated cell populations. **A.** The total number of pre-adherent GFP-gated, GFP^{+}-gated, and GFP^{++}-gated cells added to six wells of a microtiter plate are determined using the number of beads and cells acquired from the pre-adherent FACS sample. **B.** The total number of CD2-stimulated adherent GFP-gated, GFP^{+}-gated, and GFP^{++}-gated cells present in six wells after plate washing are determined using the number of beads and cells acquired from the post-adherent FACS sample. **C.** The percent of CD2-stimulated adhesion to FN in the presence of increasing levels of GFP- Δp85 expression. The percent adhesion is determined by dividing the number of post-adherent cells pooled from six wells (the cell numbers determined in Section B) by the number of pre-adherent cells originally added to six wells prior to the adhesion assay (cell numbers determined in Section A) and multiplying by 100%. The results suggest a dose-dependent inhibition of CD2-mediated upregulation of HL-60 adhesion to FN by the GFP-Δp85 fusion protein (compare the 5.98% adhesion in the GFP^{++} gate to the 21.12% adhesion in the GFP^{+} gate, and the 24.44% adhesion in the GFP^{-} gate).

For Pre-adherent Cell Samples:

$$\frac{\text{(Number of Cells in Sample)(0.050 ml Cells/well)}}{\text{0.200 ml Pre-adherent Cells Loaded into FACS Sample Tube}} \times \text{(6 pooled wells)} = \text{Total Number of Cells Added to Six Wells Prior to Adhesion Assay}$$

Gate	GFP-	GFP+	GFP++
Beads Acquired	492 beads	492 beads	492 beads
Volume of Sample Acquired	$\frac{\text{492 beads}}{\text{36,662 beads/ml}} = 0.0134$ ml	$\frac{\text{492 beads}}{\text{36,662 beads/ml}} = 0.0134$ ml	$\frac{\text{492 beads}}{\text{36,662 beads/ml}} = 0.0134$ ml
Total Number of Cells in Sample	$\frac{\text{(13626 Cells)(0.275 ml)}}{\text{0.0134 ml of Sample}} = 279{,}638$ Cells	$\frac{\text{(613 Cells)(0.275 ml)}}{\text{0.0134 ml of Sample}} = 12{,}580$ Cells	$\frac{\text{(341 Cells)(0.275 ml)}}{\text{0.0134 ml of Sample}} = 6{,}998$ Cells
Total Number of Pre-adherent Cells Added to Six Wells	$\frac{\text{(279,638 Cells)(0.050 ml/well)(6 wells)}}{\text{0.200 ml Pre-adherent cells}} = 419{,}457$ Cells	$\frac{\text{(12,580 Cells)(0.050 ml/well)(6 wells)}}{\text{0.200 ml Pre-adherent cells}} = 18{,}870$ Cells	$\frac{\text{(6,998 Cells)(0.050 ml/well)(6 wells)}}{\text{0.200 ml Pre-adherent cells}} = 10{,}497$ Cells

For Post-adherent Samples:

Number of Cells in the Sample = Total Number of Adherent Cells from Six Pooled Wells after Washing

Gate	GFP-	GFP+	GFP++
Beads Acquired	1992 beads	1992 beads	1992 beads
Volume of Sample Acquired	$\frac{\text{1992 beads}}{\text{36,662 beads/ml}} = 0.0543$ ml	$\frac{\text{1992 beads}}{\text{36,662 beads/ml}} = 0.0543$ ml	$\frac{\text{1992 beads}}{\text{36,662 beads/ml}} = 0.0543$ ml
Total Number of Cells in Sample	$\frac{\text{(20,243 Cells)(0.275 ml)}}{\text{0.0543 ml of Sample}} = 102{,}519$ Adherent Cells	$\frac{\text{(787 Cells)(0.275 ml)}}{\text{0.0543 ml of Sample}} = 3{,}986$ Adherent Cells	$\frac{\text{(124 Cells)(0.275 ml)}}{\text{0.0543 ml of Sample}} = 628$ Adherent Cells

To Determine the Percent Adhesion:

$$\text{\% Adhesion} = \frac{\text{Total Number of Post-adherent Cells Pooled from Six Wells}}{\text{Total Number of Pre-adherent Cells Added to Six Wells}} \times 100\%$$

Gate	GFP-	GFP+	GFP++
% Adhesion	$\frac{\text{102,519 GFP- Post-adherent Cells}}{\text{419,457 GFP- Pre-adherent Cells}} \times 100\%$ = 24.44% Adhesion	$\frac{\text{3,986 GFP+ Post-adherent Cells}}{\text{18,870 GFP+ Pre-adherent Cells}} \times 100\%$ = 21.12% Adhesion	$\frac{\text{628 GFP++ Post-adherent Cells}}{\text{10,497 GFP++ Pre-adherent Cells}} \times 100\%$ = 5.98% Adhesion

FIGURE 4

4. **Monoclonal antibodies** — We obtained the CD2-specific mAb 95-5-49 as culture supernatant from Dr. R. Gress (National Institutes of Health, Bethesda, MD), and the CD2-specific mAb 9-1 as a dilution of ascites fluid from Dr. S.Y. Yang (Memorial Sloan-Kettering Cancer Center, New York, NY).
5. **Goat anti-mouse IgG** — Goat anti-mouse IgG can be purchased from ICN-Cappel (Cat. #55479).
6. **Human IgG** — We purchased Human IgG from ICN-Cappel (Cat. #55908).
7. **Human fibronectin** — Human plasma fibronectin can be purchased from GIBCO/BRL (Cat. #33016-015).
8. **Ficoll** — Ficoll is purchased from Mediatech, Inc. (Cat. #MT 25-25-072-LV).
9. **Ficoll gradient needles** — We use 14-gauge, 6-in. Popper Laboratory pipetting needles purchased from Fisher Scientific (Cat. #14-825-16N). We sterilize the needles by autoclaving prior to setting up the Ficoll gradient.
10. **Bovine Serum Albumin** — We purchase BSA, Fraction V, from Calbiochem (Cat. #12659).
11. **Tissue culture media** — Opti-MEM (Cat. #31985-070) was purchased from GIBCO/BRL. RPMI-1640 (Cat. #MT10-040), Penicillin-Streptomycin mix (Cat. #MT30-002), and L-glutamine (Cat. #MT25-005) were purchased from Mediatech, Inc. FCS was purchased from Atlanta Biologicals (Cat. #S11150). BCS can be purchased from Hyclone (Cat. #A-2151-D).
12. **PMA** — Phorbol 12-myristate 13-acetate (PMA) can be purchased from Alexis Corporation (Cat. #445-004-M001) in 1 mg quantities. We dilute the PMA to a stock concentration of 100 μg/ml with DMSO and store aliquots of this stock concentration at –70°C.
13. **Human Serum Albumin (HSA)** — HSA can be purchased from GIBCO/BRL (Cat. #10697-027).
14. **HL-60 cells** — HL-60 cells can be obtained from ATCC (#CCL-240). As previously described, we isolated a stable, transfected subclone of HL-60 that expresses wild-type human CD2 on the cell surface.[9,27]
15. **Electroporation device and cuvettes** — We recommend using either a BTX Square Wave Electroporator or a Bio-Rad Gene Pulser. Electroporation cuvettes (0.4 cm) can be purchased from BTX, Bio-Rad, or Invitrogen.
16. **GFP fusion vectors** — pEGFP-C1, pEGFP-C2, pEGFP-C3 vectors are available individually or as a set from Clontech (Cat. #K6000-1 for the set).
17. **FACS tubes** — 12 × 75 mm polystyrene Falcon tubes (Cat. #2052).
18. **FACScan or FACSCalibur** — Either flow cytometer is adequate for this experimental technique.
19. **Low speed table-top centrifuge** — We use either the IEC Centra-7R refrigerated centrifuge with a #216 rotor, or the Beckman GS-6R table-top refrigerated centrifuge with a #GH 3.8 rotor.
20. **Plate washing apparatus** — We use a 5 ml Micromatic Syringe Pipet (Thomas Scientific, Cat. #7726-M20), attached to an 8-tip bent Manifold (Thomas Scientific, Cat. #7726-Q10) to wash wells. For aspiration, use a Drummond 8-place, straight manifold (Thomas Scientific, Cat. #7691-R45) attached to an aspirator. Alternatively, we use an automatic plate-washer (EL404, Bio-Tek Instruments).

21. **96-well microtiter plate** — 96-well, flat-bottom, tissue culture treated Costar plate (Cat. #3596).
22. **PKH26 reference microbeads** — We find that these 7.5 μ sized beads from Sigma (Cat. #P-7458) can be easily distinguished from blood cells or blood cell lines when looking at a Forward vs. Side scatter density plot (see Figure 1 or Figure 2). Other reference microbeads can be substituted, depending on the cell type being analyzed.
23. **Propidium Iodide (PI)** — PI is purchased from Sigma (Cat. #P-4170). Dilute the powder in sterile H_2O to make a stock solution of 2 mg/ml. As PI is light sensitive, it should be kept away from excess light exposure. Prior to flow cytometric analysis, make a 1:50 dilution using FACS media and use 25 μl of this dilution for each tube.
24. **Computer spreadsheet program** — We find the Microsoft Excel program to work well for analyzing this data.

IV. Buffers

1. **Phosphate-buffered saline (PBS)** — 137 mM NaCl, 2.7 mM KCl, 4.3 mM $Na_2HPO_4 \cdot 7H_2O$, 1.4 mM KH_2PO_4. The pH of the PBS should be adjusted to between 7.2 and 7.4. PBS can be sterilized by autoclaving.
2. **100X Ca^{2+}/Mg^{2+} solution to add to PBS** — 90 mM $CaCl_2$ and 49 mM $MgCl_2 \cdot 6H_2O$. The pH should be adjusted to between 7.2 and 7.4. This solution should be filter-sterilized rather than autoclaved.
3. **PBS + Ca^{2+}/Mg^{2+}** — Dilute the 100X Ca^{2+}/Mg^{2+} solution 100-fold in PBS. Shake the solution until it is clear. Do not use this mix if the solution remains cloudy.
4. **PBS/HSA** — Filter sterilize PBS + Ca^{2+}/Mg^{2+} + 0.5% HSA solution. Store at 4°C.
5. **Hank's Balance Salt Solution (HBSS) supplemented with 0.2% sodium azide** — 5.4 mM KCl, 0.3 mM Na_2HPO_4, 0.4 mM KH_2PO_4, 4.2 mM $NaHCO_3$, 137 mM NaCl, 5.6 mM Dextrose, 0.2% NaN_3. The pH of the HBSS should be adjusted to between 7.2 and 7.4. Store at 4°C.
6. **FACS media** — HBSS supplemented with 1% bovine calf serum (Hyclone) and 0.2% sodium azide.
7. **Versene** — PBS supplemented with 0.1 g EDTA/500 ml PBS. This solution does not have to be kept sterile.

V. Summary and Future Directions

The assay described in this chapter utilizes flow cytometry to analyze the adhesive behavior of a phenotypically distinct subset of cells without having to physically separate this subset before performing the adhesion assay. The example that we have outlined utilizes GFP as the marker for discriminating this cellular population by flow cytometry. The autofluorescent properties of the GFP tag have allowed us to demonstrate inhibition of activation-dependent regulation of β1 integrin function by a dominant-negative form of the p85 subunit of PI 3-K,[25,27] and by a mutant form of the Cbl adaptor protein.[26] However, it should be noted that the basic aspects of this assay can be generally employed to analyze the adhesive behavior of a small subset of cells without further purification, provided that there is an available marker that can distinguish the cells of interest from other cells by flow cytometry. For example, integrin-dependent adhesion of distinct thymocyte subsets could be analyzed by staining the adherent cells for expression of the CD4 and CD8 antigens using appropriate FITC- and PE-conjugated antibodies. Since CD4 and CD8 distinguish various stages of thymocyte development, the adhesive behavior of these various subsets could be determined without actually physically purifying each subset. The recent advent of additional variants of GFP, such as blue fluorescent protein (BFP),[29] now makes it possible to use this assay to determine the effects of co-transfection of two different tagged constructs on integrin-mediated adhesion. Thus, the multi-color capabilities of the flow cytometer can be exploited in many ways to enhance the analytical power of this assay.

There are several other important aspects of this assay that deserve discussion. First, the assay is rapid. The adhesive behavior of transiently transfected cells can be assessed in 2 to 3 h. Second, the assay requires minimal manipulation of the cell population in order to quantitate the adhesion of the GFP^+ transfectants. Third, it is important to emphasize that each sample that is analyzed contains a small population of GFP^+ cells and a large population of cells that have been electroporated but do not express GFP. The presence of a large population of GFP^- cells that have been electroporated and manipulated in the same manner as the GFP^+ transfectants is a very important internal control, since adhesion of both subpopulations under various activation conditions can be quantitated and directly compared. Fourth, the sensitivity of the flow cytometer can be used to track and quantitate the adhesive behavior of GFP^+ cells, even if they represent a minor fraction of the total cell population. Using known amounts of HL-60 cells, we have found that we can accurately quantitate samples containing less than 1000 cells (Figure 1), which represent 0.2% of the starting population of 480,000 cells/sample in the adhesion assay shown in Figure 2. Since we routinely obtain expression of GFP constructs in HL-60 cells and T cells in at least 5 to 10% of the electroporated cells, this is above the lower level of sensitivity in this assay. Since the accuracy of the assay is dependent to some extent on the number of events that are acquired, a large number of events should be acquired if small subpopulations of cells are being analyzed. In general, we have found that a minimum of 100 events in any given gate is sufficient for subsequent data analysis. Finally, gating can be used to assess dose-dependent effects of dominant-negative and constitutively active proteins on cell adhesion, as shown in Figure 2. In this experiment, there is dramatically reduced adhesion to FN of $CD2^+$ HL-60 cells expressing high levels of GFP-Δp85 (the GFP^{++} gate in Figure 2) when compared to cells in the same sample expressing lower levels of GFP-Δp85 (GFP^+ gate) or no GFP-Δp85 (GFP^- gate). Thus, this assay can assess the effects of differing levels of expression of proteins encoded by transiently transfected cDNA clones on integrin-mediated cell adhesion.

In summary, the technique that we have outlined in this chapter combines a static adhesion assay with data acquisition and analysis using flow cytometric cell counting and phenotypic discrimination of cell subsets by fluorescence. Although we have used this approach extensively with GFP-tagged proteins to dissect signaling pathways that regulate integrin-mediated cell adhesion, this assay has general applicability to the analysis of the adhesive behavior of any mixed population of cells that can be discriminated by multi-parameter flow cytometry.

Acknowledgments

We thank members of the laboratory for their critical comments during the writing of this chapter. The work in the authors' laboratory is supported by NIH Grants AI31126 and AI38474, and U.S. Army Medical Research and Material Command Grant BC961925. W.J.K. is supported by NIH postdoctoral fellowship F32-AR09438. Y.S. is the Harry Kay Professor of Cancer Research at the University of Minnesota.

References

1. Dustin, M. L. and Springer, T. A., T-cell receptor cross-linking transiently stimulates adhesiveness through LFA-1, *Nature*, 341, 619, 1989.
2. van Kooyk, Y., van de Wiel-van Kemenade, P., Weder, P., Kuijpers, T. W., and Figdor, C. G., Enhancement of LFA-1-mediated cell adhesion by triggering through CD2 or CD3 on T lymphocytes, *Nature*, 342, 811, 1989.
3. Shimizu, Y., van Seventer, G. A., Horgan, K. J., and Shaw, S., Regulated expression and binding of three VLA (β1) integrin receptors on T cells, *Nature*, 345, 250, 1990.
4. Campbell, J. J., Hedrick, J., Zlotnik, A., Siani, M. A., Thompson, D. A., and Butcher, E. C., Chemokines and the arrest of lymphocytes rolling under flow conditions, *Science*, 279, 381, 1998.
5. Springer, T. A., Traffic signals for lymphocyte recirculation and leukocyte emigration: the multistep paradigm, *Cell*, 76, 301, 1994.
6. Shimizu, Y., van Seventer, G. A., Ennis, E., Newman, W., Horgan, K. J., and Shaw, S., Crosslinking of the T cell-specific accessory molecules CD7 and CD28 modulates T cell adhesion, *J. Exp. Med.*, 175, 577, 1992.
7. Campanero, M. R., Arroyo, A. G., Pulido, R., Ursa, A., de Matias, M. S., Sanchez-Mateos, P., Kassner, P. D., Chan, B. M. C., Hemler, M. E., De Landazuri, M. O., and Sanchez-Madrid, F., Functional role of α2/β1 and α4/β1 integrins in leukocyte intercellular adhesion induced through the common β1 subunit, *Eur. J. Immunol.*, 22, 3111, 1992.
8. Gailit, J. and Ruoslahti, E., Regulation of the fibronectin receptor affinity by divalent cations, *J. Biol. Chem.*, 263, 12927, 1988.
9. Shimizu, Y., Mobley, J. L., Finkelstein, L. D., and Chan, A. S. H., A role for phosphatidylinositol 3-kinase in the regulation of β1 integrin activity by the CD2 antigen, *J. Cell Biol.*, 131, 1867, 1995.
10. Shimizu, Y. and Hunt, S. W., III, Regulating integrin-mediated adhesion: one more function for PI 3-kinase? *Immunol. Today*, 17, 565, 1996.

11. Nakanishi, S., Catt, K. J., and Balla, T., A wortmannin-sensitive phosphatidylinositol 4-kinase that regulates hormone-sensitive pools of inositolphospholipids, *Proc. Natl. Acad. Sci. USA,* 92, 5317, 1995.
12. Cross, M. J., Stewart, A., Hodgkin, M. N., Kerr, D. J., and Wakelam, M. J. O., Wortmannin and its structural analogue demethoxyviridin inhibit stimulated phospholipase A_2 activity in Swiss 3T3 cells – wortmannin is not a specific inhibitor of phosphatidylinositol 3-kinase, *J. Biol. Chem.*, 270, 25352, 1995.
13. Mansour, S. J., Matten, W. T., Hermann, A. P., Candia, J. M., Rong, S., Fukasawa, K., Vande Woude, G. F., and Ahn, N. G., Transformation of mammalian cells by constitutively active MAP kinase kinase, *Science*, 265, 966, 1994.
14. Qian, D. P., Mollenauer, M. N., and Weiss, A., Dominant-negative ζ-associated protein 70 inhibits T cell antigen receptor signaling, *J. Exp. Med.*, 183, 611, 1996.
15. Ota, Y. and Samelson, L. E., The product of the proto-oncogene c-*cbl*: a negative regulator of the Syk tyrosine kinase, *Science*, 276, 418, 1997.
16. Keely, P. J., Westwick, J. K., Whitehead, I. P., Der, C. J., and Parise, L. V., Cdc42 and Rac1 induce integrin-mediated cell motility and invasiveness through PI(3)K, *Nature*, 390, 632, 1997.
17. Cary, L. A., Han, D. C., Polte, T. R., Hanks, S. K., and Guan, J. L., Identification of $p130^{Cas}$ as a mediator of focal adhesion kinase-promoted cell migration, *J. Cell Biol.*, 140, 211, 1998.
18. Klemke, R. L., Leng, J., Molander, R., Brooks, P. C., Vuori, K., and Cheresh, D. A., CAS/Crk coupling serves as a "molecular switch" for induction of cell migration, *J. Cell Biol.*, 140, 961, 1998.
19. King, W. G., Mattaliano, M. D., Chan, T. O., Tsichlis, P. N., and Brugge, J. S., Phosphatidylinositol 3-kinase is required for integrin-stimulated AKT and Raf-1/mitogen-activated protein kinase pathway activation, *Mol. Cell. Biol.*, 17, 4406, 1997.
20. Perlmutter, R. M. and Alberola-Ila, J., The use of dominant-negative mutations to elucidate signal transduction pathways in lymphocytes, *Curr. Opin. Immunol.*, 8, 285, 1996.
21. Cubitt, A. B., Heim, R., Adams, S. R., Boyd, A. E., Gross, L. A., and Tsien, R. Y., Understanding, improving and using green fluorescent proteins, *Trends Biochem. Sci.*, 20, 448, 1995.
22. Stearns, T., The green revolution, *Curr. Biol.*, 5, 262, 1995.
23. Gerdes, H. H. and Kaether, C., Green fluorescent protein: applications in cell biology, *FEBS Lett.*, 389, 44, 1996.
24. Tamura, M., Gu, J., Matusmoto, K., Aota, S., Parsons, R., and Yamada, K. M., Inhibition of cell migration, spreading, and focal adhesions by tumor suppressor PTEN, *Science*, 280, 1614, 1998.
25. Chan, A. S. H., Mobley, J. L., Fields, G. B., and Shimizu, Y., CD7-mediated regulation of integrin adhesiveness on human T cells involves tyrosine phosphorylation-dependent activation of phosphatidylinositol 3-kinase, *J. Immunol.*, 159, 934, 1997.
26. Zell, T., Warden, C. S., Chan, A. S. H., Cook, M. E., Dell, C. L., Hunt, S. W., III, and Shimizu, Y., Regulation of β1 integrin-mediated adhesion by the Cbl adapter protein, *Curr. Biol.*, 8, 814, 1998.

27. Kivens, W. J., Hunt, S. W., III, Mobley, J. L., Zell, T., Dell, C. L., Bierer, B. E., and Shimizu, Y., Identification of a proline-rich sequence in the CD2 cytoplasmic domain critical for regulation of integrin-mediated adhesion and activation of phosphoinositide 3-kinase, *Mol. Cell. Biol.*, 18, 5291, 1998.
28. Mobley, J. L. and Shimizu, Y., Measurement of cellular adhesion under static conditions, in *Current protocols in immunology*, Coligan, J. E., Kruisbeek, A. M., Margulies, D. H., Shevach, E. M., and Strober, W., Eds., Greene Publishing Associates, Brooklyn, 7.28.1-7.28.22.
29. Heim, R. and Tsien, R. Y., Engineering green fluorescent protein for improved brightness, longer wavelengths and fluorescence energy transfer, *Curr. Biol.*, 6, 178, 1996.

Chapter 16

Expression Cloning of Proteins That Modify Integrin Activation

Csilla A. Fenczik, Joe W. Ramos, and Mark H. Ginsberg

Contents

I. Introduction 235
II. General Considerations 237
III. Specific Protocols 238
A. Cell Sort 238
B. Hirt Supernatants 240
C. Analysis of Clones 241
D. Materials and Equipment 241
Acknowledgments 242
References 242

I. Introduction

Integrins transmit information in both directions across the plasma membrane. Integrin affinity for ligands ("activation") can be modulated by intracellular signals (inside-out signaling).[1] Conversely, integrin ligand-binding can modify cell shape, growth, survival, and gene expression (outside-in signaling).[2] The expression cloning strategy we describe uses inside-out signaling as a selective marker. The dynamic regulation of integrin adhesion by inside-out signaling is a crucial aspect of integrin function that is important in hemostasis, leukocyte extravasation, cell migration, and fibronectin matrix assembly. Inside-out signaling is energy dependent, cell-type

0-8493-3385-7/99/$0.00+$.50

specific, and requires the cytoplasmic domains of both the α and β subunits.[3] The proteins that mediate these cytoplasmic signals have not been fully defined.

One approach to identifying the proteins involved in inside-out signaling is to find proteins that directly interact with the integrin cytoplasmic domains. This has been done using both biochemical approaches and yeast two-hybrid screens. Several proteins have been shown to interact with integrin cytoplasmic domains by one or both of these methods including talin, α-actinin, filamin, β_3-endonexin, and Rack1.[4-9] However, the functional relevance of these proteins in regulating inside-out integrin signaling remains unclear.

Genetic analysis has been a successful method for mapping intracellular signal transduction pathways *in vivo.*[10] In general, these approaches have involved the isolation and characterization of mutants that perturb functions by disrupting signaling cascades. Recently, methods have been developed to isolate proteins that complement overexpressed signaling mutants in cultured cells.[11-13] We have used concepts from these techniques to develop expression cloning strategies to identify potential integrin regulatory proteins in a cell culture system. The advantage of this approach is that it does not rely on a physical interaction between integrins and possible regulatory molecules, but rather on a functional relationship. A second important advantage is that all proteins expressed by the library are tested without inherent bias. The major disadvantage is that there are certain functional interactions that are not likely to be uncovered by such an approach. For instance, if a multiprotein complex is required for the rescue of integrin suppression, it may be impossible to identify it by this method.

We have developed two screens to isolate proteins that regulate integrin activation. The first screen relies on suppression of integrin activation by overexpression of isolated integrin β_1 cytoplasmic domains in the form of chimeras with the Tac subunit of the IL_2 receptor (Tac-β_1).[14] This inhibition of integrin signaling is structurally specific as chimeric molecules with either integrin α cytoplasmic domains or certain β cytoplasmic domain point mutants lack inhibitory activity.[15] Suppression of integrin activation by Tac-β_1 is also cell autonomous and concentration dependent. Hence, the free β_1 tails may act as competitive inhibitors of activation by titrating out proteins necessary for integrin signaling. We reasoned, therefore, that if we overexpressed these proteins we would overcome Tac-β_1 suppression. Our cloning strategy was to co-express Tac-β_1 and a cDNA library and isolate cDNAs that prevent Tac-β_1 suppression. CD98, an early T-cell activation antigen, was identified as an integrin regulatory protein using this approach.[16]

A second screen relied on the observation that overexpression of activated forms of the small GTP-binding protein H-Ras results in the suppression of integrin activation.[17] H-Ras mediated integrin suppression correlates with the activation of ERK MAP kinase and does not require mRNA transcription or protein synthesis. The identity of all the molecules involved in H-Ras signaling to integrins are unknown. Hence, we devised a screen to isolate cDNAs encoding proteins that act as suppressors of this novel H-Ras function. In this strategy we co-expressed H-Ras, a cDNA library, and a transfection marker and isolated cDNAs from cells in which H-Ras suppression was prevented.

II. General Considerations

When designing a similar expression cloning strategy, there are several factors that need to be considered. First, one needs to have a system in which integrin activation can be easily measured. We have created a CHO cell line ($\alpha\beta$py cells) that contains integrated copies of the chimeric integrin, $\alpha_{IIb}\alpha_{6A}$ and $\beta_3\beta_1$.[17] This chimeric integrin contains the extracellular domains of the platelet integrin, $\alpha_{IIb}\beta_3$ and the cytoplasmic domains of $\alpha_{6A}\beta_1$.Thus, it has the binding properties of $\alpha_{IIb}\beta_3$ and its activation state is controlled through the $\alpha_{6A}\beta_1$ cytoplasmic domains. Consequently, flow cytometry (FACS) can be used to follow the activation of the chimeric integrin by measuring the binding of fibrinogen or the ligand mimetic monoclonal antibody, PAC1.[18] Both of these ligands only bind the chimeric integrin when it is in the activated state.

The chosen cell line must also be easily transfectable, as the strategy relies on transient expression of a cDNA library. We typically get above 60% of the $\alpha\beta$py cells transfected. This strategy also requires autonomous replication driven by elements such as nuclear antigens from SV40, polyoma, or Epstein-Barr virus. Replication of the cDNAs in the mammalian cells serves several purposes:

1. Because of the increased number of each plasmid in the transfected cells, cDNAs are more easily recovered from the selected positive cells.
2. The increased copy number of each plasmid results in a corresponding increase in the protein level encoded by the inserted cDNA.
3. Replication of the cDNA library in mammalian cells allows us to distinguish between the transfected library cDNA and the non-replicated suppressive plasmid, by selection with the methylation specific enzyme, DpnI.

A critical parameter to consider when performing expression cloning strategies is the cDNA library. A number of cDNA libraries in pcDNA1 are commercially available (Invitrogen, Clontech). We chose the pcDNA1 vector because it contains the CMV promoter for high-level constitutive expression in mammalian cells, the SV40 and polyoma origin of replication for episomal replication, the ColE1 origin for growth in *E. Coli,* and the supF suppressor tRNA for selection in *E Coli* strains that carry the P3 episome. The quality of the cDNA library is another important factor. Ideally, the library should not have been amplified or at worst amplified no more than once. This is because during the amplification process very rare cDNAs and cDNAs that affect bacterial growth may be lost, thus diminishing the complexity of the library. The cDNA library we used had 1.8×10^7 primary recombinants and was size-selected for clones that were above 0.5 kb. In order to ensure the presence of full-length clones in the library, we chose a library that had been oligo dT primed. Random primed libraries, which can contain antisense and protein fragments, may also be useful for some strategies.

We used two-color flow cytometry to isolate cells that were able to bind to the activation-specific antibody PAC-1 despite high levels of Tac-β_1. Because of the sensitivity of this technique, we were able to screen millions of cells and exhaustively test the cDNA library. After cell sorting, the cDNAs are recovered by generating

Hirt supernatants with the selected cells.[19] The isolated DNA is then transformed into bacteria to be amplified. We transfected the αβpy cells with Lipofectamine® (Gibco, BRL), which in our hands gives very high levels of transfection in CHO cells, essential in this type of cloning scheme. A disadvantage of lipofectamine transfection is that large numbers of plasmids are delivered to each recipient cell. Because of this, it may be impossible to further enrich for a single positive cDNA by doing further selections with recovered, pooled cDNAs. One way to overcome this problem is to use alternative DNA transfection procedures in the second round of selection that deliver only one plasmid per cell, such as protoplast fusion or electroporation.[20] However, the way we chose to overcome this problem was to do only one round of initial transfection and selection. From this, we generated pools of plasmids and maintained them in bacteria. The pools were then screened for their ability to rescue the suppression of integrin activation. The positive pools were subdivided and rescreened until a single clone was obtained.

We include several controls in our analysis. In order to test the efficiency of transfection and the integrity of the assay for each round of selection, we include cells that have been transfected with Tac-α_5 alone in our assay. Cells that have been transfected with Tac-β_1 alone are also included in each round of selection to be sure that the suppressive molecule is working as expected. The difference between PAC1 binding of cells highly transfected with Tac-α_5 and Tac-β_1 is also used to draw the gate in which the positive cells are selected (Figure 1). In order to test whether the cells have been replicating plasmids, we take advantage of the fact that DNA replicated in bacterial cells is methylated, while DNA replicated in mammalian cells is not. We isolate DNA from the control plate transfected with Tac-α_5 and the cDNA library (see Section III.B.). Half of the DNA is then digested with the methylation specific enzyme DpnI. Equal amounts of the digested DNA and non-digested DNA are transformed into bacteria and plated. A comparison of the number of colonies on each plate will reveal whether or not the cells have been replicating plasmid.

III. Specific Protocols

A. Cell Sort

1. Plate cells at 2×10^6 cells per plate 24 h prior to transfection. We use between 12 to 18 plates.
2. Transfect cells as follows: Control Plate 1:Tac-α_5 (2 μg) and CHO Library (5 μg); Control Plate 2: Tac-α_5 (2 μg) and pcDNA1 (5 μg); Control Plate 3: Tac-β_1 (2 μg) and pcDNA1 (5 μg); and Sort Plates: Tac-β_1 (2 μg) and CHO Library (5 μg).
3. 24 h after the start of the transfection, wash plates with complete DMEM media.
4. 48 h after the start of transfection, remove the media and add 3 mls of Cell Dissociation buffer to the plates. Let this sit 5 min. Use a transfer pipette to collect the cells and put them into the appropriately labeled centrifuge tube. At this point, combine the cells to be sorted (the Tac-β_1+lib cells). Set aside Control Plate 1 (Tac-α_5 + cDNA library), to be used later as a Hirt supernatant control.

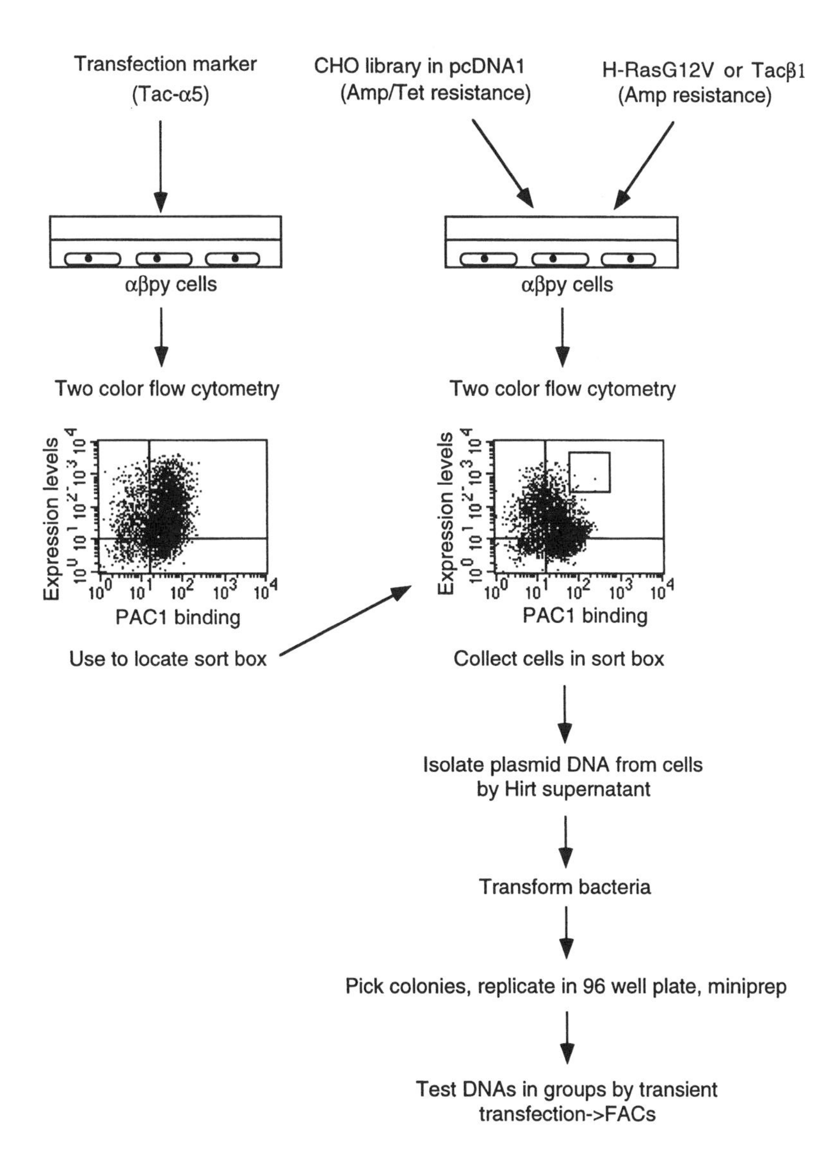

FIGURE 1
Flow chart summarizing the expression cloning strategy presented.

5. Spin cells down at 1000 RPM in tabletop centrifuge at room temperature. Aspirate off supernatant.
6. Resuspend the cells to be sorted (Tac-β_1 + cDNA library) in 1.4 ml complete DMEM total. Divide the sort cells into six tubes (Falcon 2054) of 225 µl each. Add 25 µl of diluted PAC1 ascites (PAC1 should be titered in each lab and the appropriate dilution determined). Resuspend cells transfected with Tac-β_1+pcDNA1 in 45 µl and add 5 µl of the diluted PAC1 ascites. Resuspend the Tac-α_5 control (Tac-α_5+pcDNA1) in 200 µl complete DMEM, aliquot into four separate tubes labeled 1, A, B, and C. Add 5 µl of

diluted PAC1 to tubes 1 and B, and nothing to the other two. The cells in tube 1 will be stained with both PAC1 and 7G7B6 to measure transfection levels and to determine the sort box. The cells in the next three tubes will be used as compensation controls for the FACS: Tube A is the no antibody control; B, the FITC only control; C, the PE only control. Incubate at room temperature for 30 min.

7. Add 2 ml complete DMEM to each tube, centrifuge 5 min at 1000 RPM. Decant liquid and resuspend cells in 250 μl DMEM/4% (v/v) biotinylated-7G7B6. Incubate on ice for 30 min.
8. Wash cells with 2 ml complete DMEM, centrifuge 5 min, and decant liquid. Resuspend cells in 250 μl DMEM/4% streptavidin-phychoerythrin/4% FITC-labeled anti-mouse-IgM μ chain antibodies. Incubate on ice in the dark for 30 min.
9. Add propidium iodide (0.1 μg to 0.5 μg/1 × 10^5 cells) during the last 5 min of the PE/FITC antibody incubation.
10. Wash cells with 2 mls of cold 1 × PBS. Spin cells down at 1000 RPM, resuspend, and combine cells to be sorted in 4 mls of 1 × PBS, resuspend other controls in 500 μl 1 × PBS each.
11. Use FACs to collect PE/FITC positive cells. Collect cells into 3 mls complete DMEM. We set our collection window as shown in Figure 1. Generally, we collected at least 50,000 cells. Note the number of cells sorted and collected.

B. Hirt Supernatants

1. Pipette 1.5 ml of cells collected from the sort into an Eppendorf tube and spin down for 5 min at 1000 × *g* in a centrifuge. Remove supernatant and pipette remaining 1.5 mls into the same tube and spin the cells onto the previous cell pellet. Remove supernatant. Now all cells should be collected at the bottom of the Eppendorf tube.
2. Add 400 μl of 0.6% SDS/10 mM EDTA and let sit at room temperature for 20 min. At the same time remove the DMEM from the Tac-α_5+lib plate, wash with PBS, and add 800 μl of 0.6% SDS/10 mM EDTA to the saved plate. Also let this sit at room temperature for 20 min. Collect lysate from plate into Eppendorf tube with cell scraper.
3. Add 100 μl (or 200 μl to plate lysate) of 5 M NaCl, mix by inversion, leave at 4°C overnight.
4. Spin in microfuge on high for 5 min. Carefully remove supernatant to a new tube.
5. Phenol:chloroform extract the supernatant twice. Transfer the supernatant to new tube and add 1 μl of glycogen (MolBiol grade). Fill the tube to the top with 100% ethanol to precipitate DNA. Spin down for 5 min at full speed in microfuge. Remove supernatant. Resuspend pellet in 100 μl of ddH_2O.
6. Add 10 μl of 3M sodium acetate and 300 μl of 100% ethanol and repeat precipitation spinning down in a microfuge for 20 min in a cold room. Wash pellet with 70% ethanol. Resuspend the pellet in 8 μl H_2O. Resuspend the control Tac-α_5+lib pellet in 35 μl H_2O.
7. Digest the DNAs overnight at 37°C with DpnI.
8. Incubate each entire 10 μl digestion with 100 μl of MC10B/P3 cells on ice for 20 min in falcon 2059 tubes. Transfer tubes to 42°C bath for 1 min, return to ice. Add 1 ml SOC media and incubate on the shaker at 37°C for 1 h. Plate entire transformation of each on Amp (12.5 μg/ml)/Tet (7.5 μg/ml) plates. Leave overnight at 37°C.

C. Analysis of Clones

1. Count the number of colonies that grow.
2. Pick the colonies from the sort plate and inoculate a 2 ml miniprep grow for each. Let grow overnight at 37°C.
3. Prepare 96-well microtiter plates with 50 µl of 60% glycerol in each well. Prepare a frozen glycerol stock of each miniprep by adding 150 µl of the miniprep to a numbered microtiter plate. The coordinates of the microtiter plate will serve as the name of each colony/clone (i.e., plate 1 A1 or 1A1 for the first clone picked, etc.). Seal the microtiter plates with parafilm and store at –80°C for later use.
4. After preparing the glycerol stock, pool the minipreps in groups of no more than 20. Carefully note the plate and coordinates corresponding to each set of pooled minipreps. Prepare DNA from each of the pools by Qiagen prep or CsCl preparation according to standard protocols.
5. To identify pools containing cDNAs that prevent Tac-β_1 or H-Ras suppression, co-transfect 8 µg of each pool of DNA into cells with the appropriate test plasmids (e.g., Tac-β_1). Look for rescue by analytical FACs staining with PAC1 and 7G7B6. The rescue can be very subtle at this point. Pay particular attention to any increase in cell number in the upper right hand quadrant.
6. Once a candidate batch has been identified, go back to the saved microtiter plate corresponding to that pool and start a miniprep grow from each colony in the pool. Prepare DNA by standard methods.
7. To test DNAs from the pool, combine them into groups of 4 to 5 and co-transfect them with the appropriate test plasmids as above. Look for rescue by analytical FACs. The rescue should be more obvious at this step. Next, test each DNA from the candidate pool individually by analytical FACs.
8. Sequence the positive DNA and begin characterization of the protein function.

D. Materials and Equipment

Cell line: αβpy cells, a CHO-K1 cell line stably transfected with the polyoma large T antigen and $\alpha_{IIb}\alpha_{6A}$ and $\beta_3\beta_1$ (as described in Hughes, et al.).

Cell Dissociation Buffer (Gibco BRL #13151-014)

cDNA library: CHO cell library directionally cloned into the mammalian expression vector, pcDNA1 (Invitrogen).

Primary and secondary antibodies PAC1 IgM ascites ([18]; a gift from Sandy Shattil); 7G7B6 (ATCC #HB 8784) biotinylated with biotin-N-hydroxysuccinimide (Pierce) according to manufacturer's instructions; FITC-anti-mouse IgM Mu chain (BioSource).

Streptavidin-phycoerthyrin (Molecular Probes)

Ro43 (inhibits IIb/IIIa binding; diluted to 10 µM)

Lipofectamine (Gibco BRL #13151-014)

Propidium iodide (Sigma)

Phosphate buffered saline (PBS)

0.6% SDS/10 mM EDTA
5 M NaCl
Phenol:chloroform
Glycogen
3 M sodium acetate
DpnI
MC1061/P3 cells (Invitrogen)
60% Glycerol
96-well microtiter plates
Flow Cytometer (FACScalibur, Becton Dickenson)
Fluorescence activated cell sorter (FACSTAR, Becton Dickenson)

Acknowledgments

Correspondence should be addressed to M.H.G. (ginsberg@scripps.edu). This work was supported by grants from the NIH and Cor Therapeutics. C.A.F. is supported by a Department of Defense Breast Cancer Research Award. J.W.R. is a fellow of the Leukemia Society of America. Manuscript #11950-VB from T.S.R.I.

References

1. Hughes, P. E. and Pfaff, M., Integrin affinity modulation, *Trends Cell Biol.*, 8, 359, 1998.
2. Schwartz, M. A., Schaller, M. D., and Ginsberg, M. H., Integrins: emerging paradigms of signal transduction, *Annu. Rev. Cell Biol.*, 11, 549, 1995.
3. Williams, M. J., Hughes, P. E., O'Toole, T. E., and Ginsberg, M. H., The inner world of cell adhesion: integrin cytoplasmic domains, *Trends Cell Biol.*, 4, 109, 1994.
4. Horwitz, A., Duggan, K., Buck, C., Beckerle, M., and Burridge, K., Interaction of plasma fibronectin receptor with talin a transmembrane linkage,*Nature*, 320, 531, 1986.
5. Otey, C. A., Pavalko, F. M., and Burridge, K., An interaction between alpha-actinin and the beta 1 integrin subunit *in vitro*, *J. Cell Biol.*, 111, 721, 1990.
6. Pfaff, M., Liu, S., Erle, D. J., and Ginsberg, M. H., Integrin beta cytoplasmic domains differentially bind to cytoskeletal proteins, *J. Biol. Chem.*, 273, 6104, 1998.
7. Sharma, C. P., Ezzel, R. M., and Arnaout, M. A., Direct interaction with filamin (ABP-280) with the beta 2-integrin subunit CD18, *J. Immunol.*, 154, 3461, 1995.
8. Shattil, S. J., O'Toole, T., Eigenthaler, M., et al., Beta 3-endonexin, a novel polypeptide that interacts specifically with the cytoplasmic tail of the integrin beta 3 subunit, *J. Cell Biol.*, 131, 807, 1995.
9. Lilental, J. and Chang, D. D., Rack1, a receptor for activated protein kinase C, interacts with integrin beta subunit, *J. Biol. Chem.*, 273, 2379, 1998.
10. Perrimon, N. and Desplan, C., Signal transduction in the early Drosophila embryo: when genetics meets biochemistry, *Trends Biochem. Sci.*, 19, 509, 1994.
11. Han, L. and Colicelli, J., A human protein selected for interference with Ras function interacts directly with Ras and competes with Raf1, *Mol. Cell. Biol.*, 15, 1318, 1995.

12. Akiyama, Y. and Koreaki, I., A new Escherichia coli gene, fdrA, identified by suppression analysis of dominant negative FtsH mutations, *Mol. Gen. Genet.*, 249, 202, 1995.
13. Simonsen, H. and Lodish, H., Cloning by function: expression cloning in mammalian cells, *Trends Prot. Sci.*, 15, 437, 1994.
14. LaFlamme, S. E., Akiyama, S. K., and Yamada, K. M., Regulation of fibronectin receptor distribution, *J. Cell Biol.*, 117, 437, 1992.
15. Chen, Y., O'Toole, T., Shipley, T., et al., "Inside-Out" signal transduction inhibited by isolated integrin cytoplasmic domains, *J. Biol. Chem.*, 269, 18307, 1994.
16. Fenczik, C., Sethi, T., Ramos, J. W., Hughes, P. E., and Ginsberg, M. H., Complementation of dominant suppression implicates CD98 in integrin activation, *Nature*, 390, 81, 1997.
17. Hughes, P., Renshaw, M. W., Pfaff, M., et al., Suppression of integrin activation: A novel function of a Ras/Raf-initiated MAP kinase pathway, *Cell*, 88, 521, 1997.
18. Shattil, S. J., Hoxie, J. A., Cunningham, M., and Brass, L. F., Changes in the platelet membrane glycoprotein IIb-IIIa complex during platelet activation, *J. Biol. Chem.*, 260, 11107, 1985.
19. Hirt, B., Selective extraction of polyoma DNA from infected mouse cell cultures, *J. Mol. Biol.*, 26, 365, 1967.
20. Hollenbaugh, D. and Aruffo, A., Specialized strategies for screening libraries, in *Current Protocols in Molecular Biology*, Ausubel, F. M., Brent, R., Kingston, R., Moore, D., Seidman, J., Smith, J., and Struhl, K., Eds., John Wiley & Sons, 1998 Vol 1. Suppl. 23, pp. 6.11.1–6.11.16.

Chapter

Regulation of Cell Contractility by RhoA: Stress Fiber and Focal Adhesion Assembly

Betty P. Liu, Magdalena Chrzanowska-Wodnicka, and Keith Burridge

Contents

I. Introduction ... 246
 A. Regulation of Nonmuscle Contractility ... 246
 B. Contractility and Focal Adhesion Assembly ... 249
II. Microinjection Protocols ... 250
 A. Purification of Recombinant Proteins ^{14}VRhoA and C3 Exotransferase ... 250
 B. Microinjection ... 252
 C. Cell Culture ... 253
III. Analyses of Microinjected Cells ... 253
 A. Immunofluorescence ... 253
 B. Silicone Rubber Substrata ... 254
 C. Cell Labeling, Myosin Immunoprecipitation, and Quantitation of Myosin Light Chain Phosphorylation ... 254
IV. Future Directions ... 255
Acknowledgments ... 256
References ... 256

0-8493-3385-7/99/$0.00+$.50

I. Introduction

Many cells in culture adhere to the underlying substratum via specialized regions, known as focal adhesions. These are areas where integrins, receptors for extracellular matrix (ECM) proteins, are clustered and adhesion is made to ECM components adsorbed to the surface on which the cells are growing (for reviews see References 1 and 2). Focal adhesions serve to link the extracellular matrix on the outside of the cell to the actin cytoskeleton on the inside. At their cytoplasmic face, the clustered integrins within focal adhesions anchor bundles of actin filaments (stress fibers) to the membrane. Several structural proteins (talin, vinculin, and alpha-actinin) mediate this attachment of stress fibers to the cytoplasmic domains of integrins. In addition to a structural role, focal adhesions are also sites of signal transduction. Many components involved in signal transduction have been identified in focal adhesions, and signaling cascades emanate from focal adhesions in response to integrin clustering and engagement.[1-3] Because focal adhesions are easily visualized and studied, they provide a model for studying cell adhesion to the extracellular matrix mediated by integrins.

The assembly of focal adhesions and their associated bundles of actin filaments is regulated by the GTPase RhoA.[4] How RhoA regulates the assembly of these structures has been the subject of intense investigation for several years. It was suggested many years ago that stress fibers were under isometric tension[5,6] and that the isometric tension was itself responsible for their formation.[6] This led us to investigate whether the assembly of stress fibers and focal adhesions induced by RhoA might be the result of RhoA stimulating contractility and generating tension. This is indeed what we found,[7] which has led us to propose a model for how tension may lead to the assembly of focal adhesions by clustering integrins.[2,7,8] In this chapter, after a brief review of nonmuscle contractility and its relationship to focal adhesion assembly, we describe methods for studying RhoA by microinjection, and describe assays for the contractility of cultured cells.

A. Regulation of Nonmuscle Contractility

Contractility in nonmuscle cells is generated by the interaction of myosin II with actin filaments. This interaction is regulated at multiple levels (for reviews, see References 2 and 9). At the immediate level of myosin binding to actin, there is regulation both on the actin filaments by proteins such as caldesmon and calponin, and on the myosin by phosphorylation of the regulatory light chain in the myosin head. Currently, more is known about the factors that affect myosin light chain phosphorylation and we will focus on these. Phosphorylation of the myosin light chain occurs at multiple sites and is catalyzed by several distinct protein kinases. Specific phosphorylations can either inhibit or stimulate the interaction of myosin with actin filaments.[9] Two kinases responsible for inhibitory phosphorylations are protein kinase C (PKC) and the mitotic kinase, $p34^{cdc2}$. Both kinases phosphorylate Ser-1 and Ser-2 in the regulatory myosin light chain. In addition, PKC also phosphorylates Thr-9. The phosphorylation of these sites decreases the actin-activated ATPase of the myosin and decreases the stability of myosin filaments.[9] In contrast, phosphorylation on Ser-19

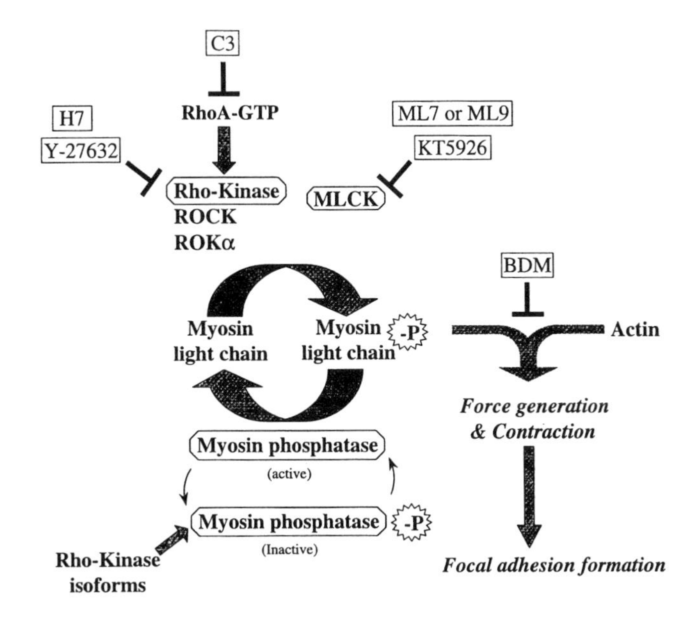

FIGURE 1

Regulation of cell contractility by myosin light chain phosphorylation. Some of the pathways regulating myosin light chain phosphorylation on serine-19 and threonine-18 are illustrated. Inhibitors that have been used to block specific steps in these pathways are indicated.

and/or Thr-18 is associated with a stimulation of the actin-activated myosin ATPase activity and with a conformational change in the myosin monomer that promotes assembly of myosin filaments.[10] Several kinases have been identified that phosphorylate the regulatory light chain at these sites. The most studied has been the myosin light chain kinase (MLCK), an enzyme that is itself regulated by Ca^{2+}/calmodulin binding.[9] In the absence of calcium, calmodulin does not bind MLCK and MLCK is inhibited catalytically by an intramolecular inhibitory domain. Upon elevation of intracellular calcium, the Ca^{2+}/calmodulin complex binds to MLCK, inducing a conformational change in MLCK that relieves the inhibition, thereby activating the kinase. MLCK is also a target for phosphorylation by PKA, PKG, and CaM kinase II, all of which phosphorylate a site that decreases the affinity of MLCK for Ca^{2+}/calmodulin.[11,12] This inhibits MLCK activation, decreases myosin light chain phosphorylation, and results in decreased contractility.

Many agents that activate the GTPase RhoA stimulate cell contractility[7,13-18] and this has been correlated with an increase in myosin light chain phosphorylation.[7,16,17] One family of RhoA-activated kinases includes Rho-kinase,[19] p160^{ROCK},[20] and ROKα,[21] which are identical or closely related kinases identified by different groups. Members of this kinase family phosphorylate the regulatory myosin light chain on Ser-19,[22] thus mimicking the action of MLCK. In addition, these kinases phosphorylate the myosin phosphatase that removes phosphates from Ser-19 and Thr-18.[23] The phosphorylation of the myosin phosphatase inhibits its activity, again resulting in elevated myosin light chain phosphorylation and enhanced contractility. Pathways regulating myosin light chain phosphorylation and contractility are illustrated in Figure 1. This figure, as well as Table 1, includes various pharmacological reagents

TABLE 1
Contractility Inhibitors

Inhibitor	Intracellular Target	Effective Concentrations	Distributor	References
2,3-butanedione 2-monoxime (BDM)	Actin-myosin interaction. Muscle myosin ATPase activity and nonmuscle myosin II.	20 mM	Sigma	28, 40-43
C3	RhoA	25–50 μg/ml(media) 50–250 μg/ml (microinjection)	Calbiochem	44
1-(5-isoquinolinylsulfonyl)-2-methyylpiperazine (H7)	p160ROCK(Rho–Kinase)	300 μM (Swiss and Balb3T3) 150 μM (MCF10A)	Biomol and Alexis Biochemical	7, 29, 30
KT5926	Myosin Light Chain Kinase	15 μM (Swiss and Balb3T3) 60 μM (MCF10A)	Sigma	45
ML7 and ML9	Myosin Light Chain Kinase	25 μM	Biomol	46
Y-27632	p160ROCK (Rho–Kinase)	10 μM	Narumiya, S.	30

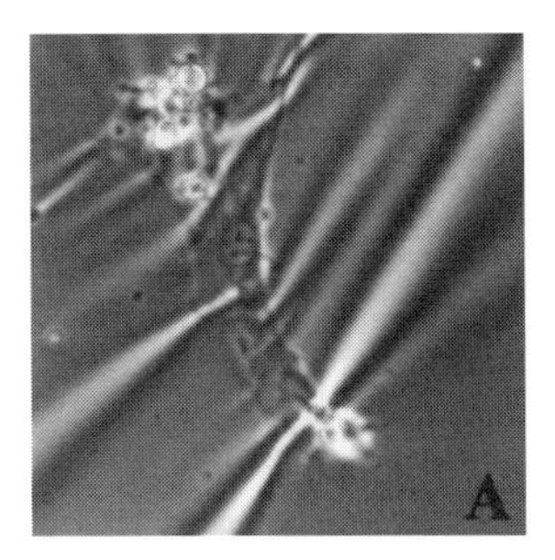

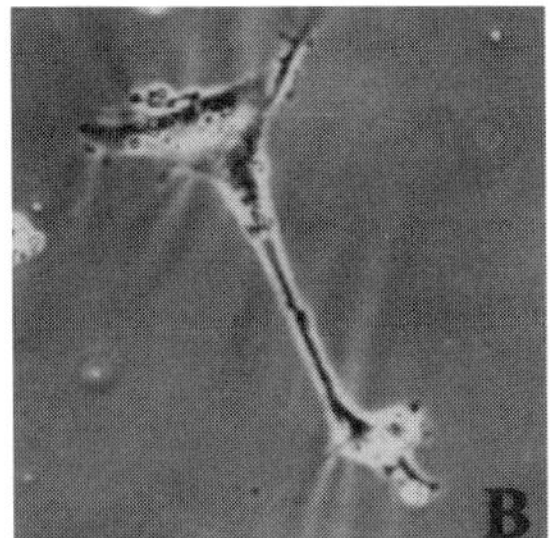

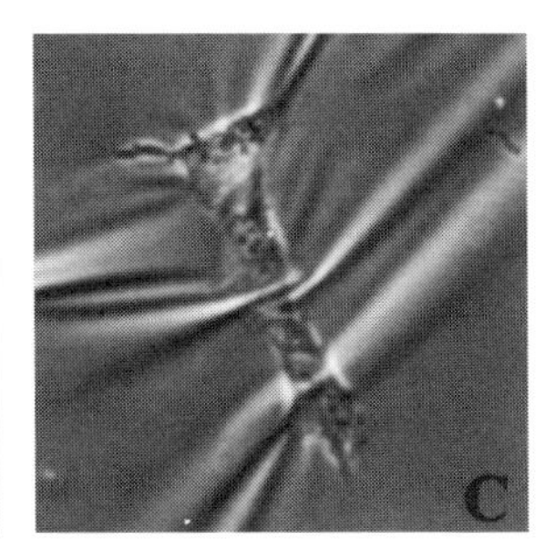

FIGURE 2
Serum stimulates fibroblast contractility. Balb/c 3T3 fibroblasts were plated on silicon rubber substrata. In the presence of serum, these cells generate tension on the underlying substratum that is revealed by wrinkles in the rubber (*A*). Serum starvation for 4 h leads to relaxation of tension and a decrease in the wrinkles (*B*). Replacement of serum leads to the activation of RhoA and wrinkles are seen prominently in the rubber, indicating stimulation of contractility (*C*). Each panel represents the same field of view. *Panel C* was photographed approximately 30 min after replacement of medium containing serum.

that have been used to inhibit myosin light chain phosphorylation and other steps in nonmuscle contractility.

B. Contractility and Focal Adhesion Assembly

Many experimental observations support the idea that tension is important for both the assembly and maintenance of focal adhesions and stress fibers (reviewed in References 2 and 8). These two structures are closely correlated: focal adhesions are rarely, if ever, seen in the absence of an attached bundle of microfilaments (stress fiber). The observation that RhoA regulates the formation of both stress fibers and focal adhesions led to the suggestion that RhoA may exert its effect on these structures by controlling cell contractility.[24] This was tested by using several of the inhibitors in Figure 1 and Table 1 to block contractility in cells in which RhoA was activated.[7] Contractility was assayed using the technique developed by Harris and his colleagues of culturing cells on flexible rubber substrata. In this assay, tension exerted by cells on the substratum is detected as wrinkles in the rubber.[25] This technique (described in detail below) is non-quantitative, but allows changes in the contractile state of cells to be easily visualized. Many of the studies on the action of RhoA have employed cells, such as Swiss 3T3 or Balb/c 3T3 cells, in which RhoA becomes largely inactive in response to serum starvation. Under these conditions of quiescence, there is a loss of stress fibers and focal adhesions. These can be rapidly induced to form by stimulating with a number of agents that activate RhoA (e.g., serum, lysophosphatidic acid, sphingosine-1-phosphate, bombesin, etc.). Quiescence is accompanied by a decrease in contractility of the cells and agents that activate RhoA and restore contractility (illustrated in Figure 2) in parallel with the restoration of stress fibers and focal adhesions.[7] The importance of RhoA in the induction of stress fibers, focal adhesions, and contractility has been demonstrated by blocking RhoA with the Botulinum toxin, C3 exotransferase.[26,27] This ADP-ribosylates RhoA, inhibiting its interactions with downstream effectors. Direct introduction into quiescent cells of constitutively activated forms of RhoA, such as ^{14}VRhoA, also rapidly stimulates the reassembly of stress fibers and focal adhesions.[4]

Few direct inhibitors of actin-myosin interaction have been identified. One low affinity inhibitor of myosin ATPase activity is BDM.[28] Treatment of cells with BDM inhibits the development of tension, and blocks RhoA-induced assembly of stress fibers and focal adhesions.[7] Similarly, inhibitors of MLCK decrease the formation of these structures in response to RhoA activation, although generally we have found them to be less effective than direct inhibitors of RhoA itself (e.g., C3 toxin) or its downstream kinases. One potent inhibitor of cell contractility that very effectively blocks RhoA-induced focal adhesion and stress fiber formation is the compound H7.[7,29] H7 leads to the rapid disassembly of stress fibers and focal adhesions in normal cells.[29] H7 is a broad specificity serine/threonine kinase inhibitor that was originally used to inhibit PKC. PKC is not involved in this response to H7 because other, more specific inhibitors of PKC, had no effect on contractility or stress fibers and focal adhesions.[29] Recently, it has been shown that H7 inhibits the p160^{ROCK} family of RhoA-activated kinases at significantly lower concentrations than those at which it acts on PKC.[30] The action of H7 on RhoA-activated kinases accounts for its potent inhibition of RhoA-stimulated contractility and stress fiber and focal adhesion assembly.

Experimental depolymerization of microtubules was shown to stimulate cell contractility and enhance the assembly of stress fibers.[31] This observation was extended to show that microtubule depolymerization also leads to the formation of focal adhesions and to enhanced tyrosine phosphorylation of several focal adhesion components.[32] The assembly of stress fibers, focal adhesions, and tyrosine phosphorylation of focal adhesion components in response to microtubule depolymerization is blocked by inhibiting RhoA with C3 toxin.[33-35] This is illustrated in Figure 3. Several groups have made this observation and presented evidence that microtubule depolymerization leads in some way to RhoA activation.[33-35]

II. Microinjection Protocols

A. Purification of Recombinant Proteins ^{14}VRhoAA and C3 Exotransferase

Plasmid constructs containing glutathione-S-transferase (GST) fused to ^{14}VRhoA (constitutively active RhoA) or C3 exotransferase (gifts from A. Hall, University College, London, U.K. and L. Feig, Tufts University, Boston, MA, respectively) are expressed in DH5 and JM101 Escherichia coli strains. For both fusion proteins, a 100 ml of freshly saturated, overnight culture are diluted 1:10 in Luria Broth (LB) and allowed to grow for 1 h at 37°C. Expression of ^{14}VRhoA is induced by 0.1 μM final concentration of isopropyl-β-D-thiogalactopyranoside (Boehringer, Mannheim, Germany) at 37°C for 3 h, while C3 exotransferase is induced at room temperature overnight in order to obtain a high yield of the fusion protein. Bacteria are then pelleted by centrifugation for 10 min at 5000 × *g* at 4°C. All manipulations from

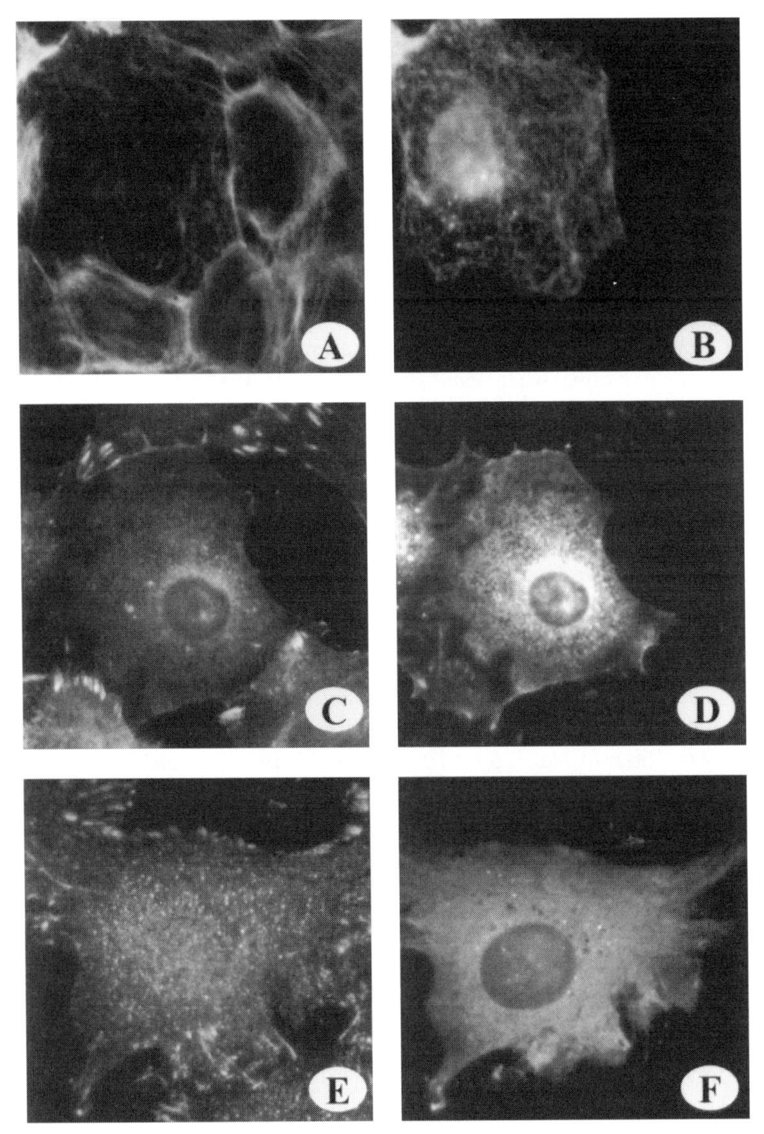

FIGURE 3

The induction of stress fibers, focal adhesions and tyrosine phosphorylation, by microtubule depolymerization is inhibited by C3 toxin. Quiescent Balb/c 3T3 cells were microinjected with recombinant C3 exotransferase (*A-D*) prior to microtubule depolymerization with nocodazole at 5 μg/ml for 30 min. Cells were stained for actin (*A*), and vinculin (*C, E*). C3-injected cells were visualized by co-injection with FITC-labeled IgG (*B, D*). Uninjected cells contain prominent actin stress fibers (*A*) and vinculin within focal adhesions (*C*). Microinjection of FITC-IgG alone did not affect the induction of focal adhesions by microtubule depolymerization (*E, F*).

this point on are done on ice. GST-^{14}VRhoA expressing cells are resuspended in 50 mM Tris-HCl pH 7.6, 50 mM NaCl, 5 mM $MgCl_2$, 5 mM DTT, 1 mM PMSF, and 20 μg/ml aprotinin (resuspension buffer), while GST-C3 exotransferase expressing cells are resuspended in 1% Triton X-100, 1 mM EDTA, 5 mM DTT, 1 mM PMSF, and 20 μg/ml aprotinin. The suspensions are aliquoted into multiple 1.5 ml microfuge tubes and sonicated on ice, three times for 10 sec each, with a Vibra Cell™ (Danbury, CT) probe-tip set at microtip limit of 4. Lysates are clarified by centrifugation at 12,000 × *g* to pellet the insoluble fractions. The supernatants are poured over glutathione agarose columns (Pharmacia, Piscataway, NJ) with a binding capacity of 5 mg of GST fusion protein per 1 ml bed volume. The columns are washed with 5 to 10 bed volumes of the respective resuspension buffer and the GST constructs are released from the column by elution with 100 mM glutathione in 50 mM Tris-HCl, pH 8.0. For many experiments, the GST constructs can be used without cleavage of the GST moiety. However, in some circumstances we remove the GST; for example, when incubating cells with C3 in the medium. To release ^{14}VRhoA or C3 from the GST, we wash the columns with 5 to 10 bed volumes of 50 mM Tris-HCl pH 8, 150 mM NaCl, 5 mM $MgCl_2$, 2.5 mM $CaCl_2$, and 5 mM DTT (thrombin buffer). We make a cleavage reaction solution by adding thrombin (Sigma, St. Louis, MO, or Haematologic Technologies Inc., Essex Junction, VT) to 1 bed volume of the thrombin buffer and load the mixture onto the column. We allow thrombin to cleave the fusion protein for 30 min to 1 h at 4°C by plugging the bottom of the column. This cleavage reaction can be repeated with fresh thrombin if there is fusion protein remaining on the column after the first cleavage. ^{14}VRhoA and C3 are retrieved by washing the column with a bed volume of thrombin buffer without $CaCl_2$. To remove thrombin, the eluate is incubated with 3 batches of 50% p-aminobenzamide agarose beads (Sigma) previously equilibrated in 50 mM Tris-HCl pH 7.6, 150 mM NaCl, 5 mM $MgCl_2$ in 4°C for 1 h. Beads are pelleted at 500 × *g* and the supernatant is dialyzed against microinjection buffer consisting of 10 mM Tris-HCl pH 7.6, 150 mM NaCl, 2 mM $MgCl_2$ and 0.1 mM DTT at 4°C. Proteins that are too dilute are concentrated by centrifugation at 5000 × *g* in a Centricon 10 concentrator (Amicon, Beverly, MA). Constitutively active RhoA is microinjected at a final concentration of 1 to 2 mg/ml. We have found C3 to be toxic to the cells when microinjected at more than 0.5 mg/ml. C3 microinjected at 50 μg/ml is sufficient to block stress fibers and focal adhesions induced by LPA, serum, and nocodazole.

B. Microinjection

Purified recombinant proteins are mixed with marker IgG (for example, FITC-conjugated goat-anti-human IgG fraction; Cappel, Organon-Technika, West Chester, PA) at a 4:1 ratio and spun in a Beckman TL100 Ultracentrifuge at 30,000 rpm for 10 min to ensure no particulate material clogs the microinjection needles. Proteins are then loaded by capillary action into needles (Narishige, Tokyo, Japan) pulled on a Narishige Model PC-10 needle puller, with the heater level set at 67. Needle tip diameter and tapering is affected by the temperature at which the needle is pulled. Lower heat settings generate more blunt needle points that do not clog as often as needles pulled with higher heat settings. For microinjection, we use a Leitz micromanipulator to position the needle relative to the cells. Pressure is applied to the contents of the needle using a 10 ml syringe attached to the needle holder by flexible, air-tight tubing. We try to keep the microinjection time for a particular coverslip limited between 15 to 20 min at room temperature and allow the cells to recover at 37°C in an atmosphere of 10% CO_2 for 15 to 30 min or longer depending on the experiment.

C. Cell Culture

Balb/c 3T3 and Swiss 3T3 fibroblasts (ATCC, Rockville, MD) are cultured in Dulbecco's modified Eagle's medium (DMEM) (Life Technologies, Inc., Grand Island, NY), supplemented with 10% calf serum, 50 units/ml penicillin, and 50 µg/ml streptomycin at 37°C with 10% CO_2. We passage both cell lines by diluting cultures 1:5 to 1:10 once they reach 50 to 60% confluency. If cultures are allowed to become confluent, they rapidly lose the ability to respond to serum starvation by loss of stress fibers and focal adhesions. In addition, we have typically found that cells kept beyond the 10th to 12th passage after we have obtained them lose their ability to respond to serum starvation. Consequently, we have found that it is necessary to frequently return to frozen stocks of early passages.

To obtain quiescence with the Swiss 3T3 cells, they are seeded in sterilized 12 mm circular glass coverslips (Carolina Biologicals, Burlington, NC) and allowed to reach 90 to 100% confluency. They are then rinsed twice with serum-free DMEM containing 50 Units/ml penicillin, and 50 µg/ml streptomycin, and kept in this serum-free medium for 16 to 20 h at 37°C with 10% CO_2. The Balb/c 3T3 cells require less time to become quiescent, as judged by the loss of stress fibers and focal adhesions. These cells are incubated in serum-free medium for just 1 to 4 h to induce quiescence.

Quiescent Swiss and Balb/c 3T3 fibroblasts assemble stress fibers and focal adhesions within 5 to 20 min after addition of 10% serum or 5 µg/ml of nocodazole. To block the role of RhoA in these responses, we have microinjected the C3 toxin. Alternatively, we have modified a method previously described by Yamamoto and colleagues,[36] in which cells are incubated with C3 in the medium and C3 is spontaneously taken up into the cells. Swiss 3T3 cells can be serum-starved for 16 to 20 h as described above, but with the addition of 25 to 50 µg/ml of C3 in the serum-free medium. We have found that this treatment results in blocking the responsiveness of 50 to 70% of the cells to subsequent activation of RhoA by serum or lysophosphatidic acid. Longer incubation times in C3 or higher concentrations will induce cell rounding and death in the same amount of time. This approach has the advantage that large numbers of cells can be exposed to C3 for biochemical analysis, in contrast to microinjection where only a few cells can be analyzed.

III. Analyses of Microinjected Cells

A. Immunofluorescence

Balb/c 3T3 and Swiss 3T3 fibroblasts are fixed in 3.7% formaldehyde in Dulbecco's phosphate buffer saline (PBS) for 10 min, rinsed in tris buffered saline (TBS) at pH

7.6 for 1 min, and permeabilized for 1 to 5 min in TBS containing 0.5% Triton X-100, before staining with primary and secondary antibodies. We dilute our primary and secondary antibodies in TBS containing 0.1% azide. Anti-mouse and anti-rabbit antibodies conjugated to fluorescein or rhodamine are available from a number of commercial sources. We have used secondary antibodies from Chemicon, Cappel, and Jackson Labs. Incubations with primary and secondary antibodies are performed in humidified chambers for 45 min at 37°C. Between incubations, coverslips are washed for 5 to 10 min in TBS. For visualization of phosphotyrosine, we use either a polyclonal or monoclonal (PY20) anti-phosphotyrosine antibody (Transduction Labs, Lexington, KY). For visualization of focal adhesions, many different antibodies can be used. We recommend monoclonal antibodies against vinculin (VIN 11-5) or talin (8d4) from Sigma, or against the focal adhesion kinase (FAK) (clone 77) from Transduction Labs. For visualizing filamentous actin, cells are incubated with rhodamine-conjugated phalloidin (Molecular Probes, Eugene, OR) diluted to 1:500. This can be included with the second antibody.

B. Silicone Rubber Substrata

We modified the protocol of Harris et al.[25] for visualization of contractility on rubber substrata.[7,37] Our protocol yields flexible rubber substrata coated with a thin layer of gold-palladium. Such surface treatment promotes cell adhesion and spreading by decreasing the natural hydrophobic nature of the silicone. 1 ml of silicone rubber (Dimethylpolysiloxane, 60,000 centistokes; Sigma) is dispensed from a 60 ml syringe into 35 mm tissue culture dishes and allowed to spread for 3 to 4 h. The silicone rubber is then coated with a thin layer of palladium-gold, using a scanning electron microscope coating Unit E5100 (Polar Instruments, Inc.) for 16 sec at 20 milliamps in a >99% argon-filled chamber. The UV glow discharge that occurs during the coating process serves to polymerize the top surface of the silicone rubber. When plating cells on the rubber substrate, care should be taken not to disturb it with the flow of media or to damage it with a pipet tip. Polymerized rubber substrates can be stored for up to 1 month at room temperature in the dark, on a level surface, and under sterile conditions.

C. Cell Labeling, Myosin Immunoprecipitation, and Quantitation of Myosin Light Chain Phosphorylation

To analyze phosphorylation of the myosin light chains, we adapted the protocol from Goekeler and Wysolmerski.[16] In order to quantitate the amount of phosphorylation on light chains relative to the amount of myosin precipitated, we double labeled cells with Tran-^{35}S-label (70% L-Methionine (^{35}S) + 30% L-Cysteine (^{35}S) (ICN Biochemicals, Costa Mesa, CA) and ^{32}P-orthophosphoric acid (ICN Biochemicals). Cells are grown as described above in 24-well Falcon dishes (Becton Dickinson, RTP, NC). After confluency is reached, cells are washed in phosphate-free Eagle's medium (Life Technologies, Inc.) and labeled with 100 μCi/ml Tran-^{35}S-label and 125 μCi/ml ^{32}P-orthophosphoric acid in tissue culture phosphate-free media containing 2% calf serum for 16 to 40 h. Prior to stimulation with LPA with or without inhibitors, cells are incubated in serum-free (phosphate-free) media to obtain quiescence. Medium is aspirated and cells are scraped off the dish in 150 μl of ice cold buffer A, (25 mM Tris-HCl pH 8.0, 300 mM NaCl, 100 mM $Na_4P_2O_7$, 75 mM NaF, 5 mM EGTA, 5 mM EDTA, 10 μg aprotinin and 10 μg leupeptin). The dishes are rinsed once with 50 μl of buffer A, which is combined with the lysate. The insoluble fraction is separated from the lysate by centrifugation at 132,000 × *g* for 10 min in a Beckman TL-100 centrifuge. Myosin is immunoprecipitated from the supernatants using 3 μl of a polyclonal rabbit anti-platelet myosin IgG fraction (a gift from Dr. R. S. Adelstein, NIH, Bethesda, MD) and 100 μl of a 10% suspension of protein A-sepharose in buffer A. The beads are washed once in buffer A and twice in buffer A in which the concentration of NaCl is decreased to 100 mM. Samples are prepared by boiling the beads in SDS gel sample buffer and separated by electrophoresis on 15% polyacrylamide-SDS gels. These are stained with Coomasie blue, destained, dried, and exposed to a phosphor imaging screen (Molecular Dynamics, Sunnyvale, CA) or Kodak X-ray film.

For quantitation of ^{32}P incorporation into myosin light chains, we have used a phosphor imager (Molecular Dynamics), and have adapted the protocol of Johnston et al.[38] Because the amount of myosin immunoprecipitated from different samples is difficult to control, we have compared the amount of ^{32}P incorporated into myosin light chains with the amount of ^{35}S incorporated into the myosin heavy chain. This has allowed us to compare the relative amount of myosin light chain phosphorylation even when different amounts of myosin have been precipitated from different samples. First, the position of bands corresponding to myosin light chains and heavy chains is established by running a sample of purified smooth muscle myosin along with the labeled samples. The gel is then directly exposed to phosphor imaging screens and the total amount of radioactivity (^{35}S + ^{32}P) is quantitated for myosin heavy chains and myosin light chains. To quantitate the contribution of ^{32}P, a second exposure is obtained with a filter of a triple layer heavy-duty Reynolds aluminum foil, to block ^{35}S radiation. This attenuates the detection of ^{35}S to less than 1%, while detection of ^{32}P remains at greater than 97%. The amount of ^{35}S is calculated for each myosin band by subtracting the number of counts with the filter (^{32}P contribution) from the number of total counts obtained without the filter (^{32}P + ^{35}S). Incorporation of ^{32}P into light chains is calculated relative to the incorporation of ^{35}S into the corresponding heavy chains (a measure of the amount of myosin in the sample) by dividing the number of ^{32}P counts from each light chain sample by the number of ^{35}S counts from the corresponding heavy chain.

IV. Future Directions

A few years ago, it was presumed that the contractile state of nonmuscle cells was largely controlled by the MLCK, which is regulated, in turn, by the level of intracellular free calcium. However, it is now recognized that nonmuscle contractility is also regulated by RhoA and its downstream kinases. The interplay between these two regulatory pathways will be important to unravel. Many factors are being identified that stimulate contractility by activating RhoA. In many circumstances, the regulation of contractility by RhoA may be the dominant mechanism.

We have considered here the role of contractility and isometric tension in mediating the assembly of stress fibers and focal adhesions. Contractility is also very important in regulating the integrity of cadherin-mediated cell-cell junctions. Here, however, tension generated within cells contributes to the disassembly of these junctions. This was first identified in studies of epithelial cell-cell junctions that were disassembling due to removal of extracellular calcium. The disassembly of these junctions could be blocked by inhibitors of contractility.[29,39] Tension generated within the cells contributed to the disassembly of the junctions that were weakened by calcium removal. Similarly, epithelial junctions weakened by activation of oncogenes could be restored to a more normal organization by inhibiting contractility with many of the agents discussed here.[37] Epithelial and endothelial junctions are affected in many physiological and pathological conditions. Pharmacological reagents that inhibit the contractile machinery of nonmuscle cells may have clinical significance in therapies aimed at maintaining normal junctional integrity.

Acknowledgments

Work in the authors' laboratory was supported by NIH Grants GM29860 and HL45100.

References

1. Jockusch, B. M., Bubeck, P., Giehl, K., Kroemker, M., Moschner, J., Rothkegel, M., Rudiger, M., Schluter, K., Stanke, G., and Winkler, J., The molecular architecture of focal adhesions, *Annu. Rev. Cell Biol.*, 11, 379, 1995.
2. Burridge, K. and Chrzanowska-Wodnicka, M., Focal adhesions, contractility and signaling, *Annu. Rev. Cell Dev. Biol.*, 12, 463, 1996.
3. Schwartz, M. A., Schaller, M. D., and Ginsberg, M. H., INTEGRINS: Emerging paradigms of signal transduction, *Annu. Rev. Cell Biol.*, 11, 549, 1995.
4. Ridley, A. J. and Hall, A., The small GTP-binding protein rho regulates the assembly of focal adhesions and actin stress fibers in response to growth factors, *Cell*, 70, 389, 1992.
5. Heath, J. P. and Dunn, G. A., Cell to substratum contacts of chick fibroblasts and their relation to the microfilament system. A correlated interference-reflexion and high-voltage electron-microscope study, *J. Cell Sci.*, 29, 197, 1978.
6. Burridge, K., Are stress fibres contractile? *Nature*, 294, 691, 1981.
7. Chrzanowska-Wodnicka, M., and Burridge, K., Rho-stimulated contractility drives the formation of stress fibers and focal adhesions, *J. Cell Biol.*, 133, 1403, 1996.
8. Burridge, K., Chrzanowska-Wodnicka, M., and Zhong, C., Focal adhesion assembly, *Trends Cell Biol.*, 7, 342, 1997.

9. Tan, J. L., Ravid, S., and Spudich, J. A., Control of nonmuscle myosins by phosphorylation, *Annu. Rev. Biochem.*, 61, 721, 1992.
10. Citi, S. and Kendrick-Jones, J., Regulation of non-muscle myosin structure and function, *Bioessays*, 7, 155, 1987.
11. Nishikawa, M., de Lanerolle, P., Lincoln, T. M., and Adelstein, R. S., Phosphorylation of mammalian myosin light chain kinases by the catalytic subunit of cyclic AMP-dependent protein kinase and by cyclic GMP-dependent protein kinase, *J. Biol. Chem.*, 259, 8429, 1984.
12. Tansey, M. G., Luby-Phelps, K., Kamm, K. E., and Stull, J. T., Ca(2+)-dependent phosphorylation of myosin light chain kinase decreases the Ca2+ sensitivity of light chain phosphorylation within smooth muscle cells, *J. Biol. Chem.*, 269, 9912, 1994.
13. Jalink, K. and Moolenaar, W. H., Thrombin receptor activation causes rapid neural cell rounding and neurite retraction independent of classic second messengers, *J. Cell Biol.*, 118, 411, 1992.
14. Suidan, H. S., Stone, S. R., Hemmings, B. A., and Monard, D., Thrombin causes neurite retraction in neuronal cells through activation of cell surface receptors, *Neuron*, 8, 363, 1992.
15. Giuliano, K. A. and Taylor, D. L., Formation, transport, contraction, and disassembly of stress fibers in fibroblasts, *Cell Motil. Cytoskeleton*, 16, 14, 1990.
16. Goeckeler, Z. M. and Wysolmerski, R. B., Myosin light chain kinase-regulated endothelial cell contraction: the relationship between isometric tension, actin polymerization, and myosin phosphorylation, *J. Cell Biol.*, 130, 613, 1995.
17. Kolodney, M. S. and Elson, E. L., Correlation of myosin light chain phosphorylation with isometric contraction of fibroblasts, *J. Biol. Chem.*, 268, 23850, 1993.
18. Jalink, K., van Corven, E. J., Hengeveld, T., Morii, N., Narumiya, S., and Moolenaar, W. H., Inhibition of lysophosphatidate- and thrombin-induced neurite retraction and neuronal cell rounding by ADP ribosylation of the small GTP-binding protein Rho, *J. Cell Biol.*, 126, 801, 1994.
19. Matsui, T., Amano, M., Yamamoto, T., Chihara, K., Nakafuku, M., Ito, M., Nakano, T., Okawa, K., Iwamatsu, A., and Kaibuchi, K., Rho-associated kinase, a novel serine/threonine kinase, as a putative target for the small GTP binding protein Rho, *EMBO J.*, 15, 2208, 1996.
20. Ishizaki, T., Maekawa, M., Fujisawa, K., Okawa, K., Iwamatsu, A., Fujita, A., Watanabe, N., Saito, Y., Kakizuka, A., Morii, N., and Narumiya, S., The small GTP-binding protein Rho binds to and activates a 160 kDa Ser/Thr protein kinase homologous to myotonic dystrophy kinase, *EMBO J.*, 15, 1885, 1996.
21. Leung, T., Manser, E., Tan, L., and Lim, L., A novel serine/threonine kinase binding the ras-related RhoA GTPase which translocates the kinase to peripheral membranes, *J. Biol. Chem.*, 270, 29051, 1995.
22. Amano, M., Ito, M., Kimura, K., Fukata, Y., Chihara, K., Nakano, T., Matsuura, Y., and Kaibuchi, K., Phosphorylation and activation of myosin by Rho-associated kinase (Rho-kinase), *J. Biol. Chem.*, 271, 20246, 1996.
23. Kimura, K., Ito, M., Amano, M., Chihara, K., Fukata, Y., Nakafuku, M., Yamamori, B., Feng, J. H., Nakano, T., Okawa, K., Iwamatsu, A., and Kaibuchi, K., Regulation of myosin phosphatase by Rho and Rho-associated kinase (Rho-kinase), *Science*, 273, 245, 1996.

24. Chrzanowska-Wodnicka, M. and Burridge, K., Rho, rac and the actin cytoskeleton, *Bioessays*, 14, 777, 1992.
25. Harris, A. K., Wild, P., and Stopak, D., Silicone rubber substrata: a new wrinkle in the study of cell locomotion, *Science*, 208, 177, 1980.
26. Narumiya, S., Sekine, A., and Fujiwara, M., Substrate for botulinum ADP-ribosyltransferase, Gb, has an amino acid sequence homologous to a putative rho gene product, *J. Biol. Chem.*, 263, 17255, 1988.
27. Aktories, K. and Hall, A., Botulinum ADP-ribosyltransferase C3: a new tool to study low molecular weight GTP-binding proteins, *Trends Pharmacol. Sci.*, 10, 415, 1989.
28. Cramer, L. P. and Mitchison, T. J., Myosin is involved in postmitotic cell spreading, *J. Cell Biol.*, 131, 179, 1995.
29. Volberg, T., Geiger, B., Citi, S., and Bershadsky, A. D., Effect of protein kinase inhibitor H-7 on the contractility, integrity, and membrane anchorage of the microfilament system, *Cell Motil. Cytoskeleton*, 29, 321, 1994.
30. Uehata, M., Ishizaki, T., Satoh, H., Ono, T., Kawahara, T., Morishita, T., Tamakawa, H., Yamagami, K., Inui, J., Maekawa, M., and Narumiya, S., Calcium sensitization of smooth muscle mediated by a Rho-associated protein kinase in hypertension, *Nature*, 389, 990, 1997.
31. Danowski, B. A., Fibroblast contractility and actin organization are stimulated by microtubule inhibitors, *J. Cell Sci.*, 93, 255, 1989.
32. Bershadsky, A., Chausovsky, A., Becker, E., Lyubimova, A., and Geiger, B., Involvement of microtubules in the control of adhesion-dependent signal transduction, *Curr. Biol.*, 6, 1279, 1996.
33. Liu, B. P., Chrzanowska-Wodnicka, M., and Burridge, K., Microtubule depolymerization induces stress fibers, focal adhesions, and DNA synthesis via the GTP-binding protein Rho, *Cell Adhesion and Commun.*, 5, 249, 1998.
34. Zhang, Q., Magnusson, K. E., and Mosher, D. F., Lysophosphatidic acid and microtubule-destabilizing agents stimulate fibronectin matrix assembly through rho-dependent actin stress fiber formation and cell contraction, *Mol. Biol. Cell*, 8, 1415, 1997.
35. Enomoto, T., Microtubule disruption induces the formation of actin stress fibers and focal adhesions in cultured cells: possible involvement of the Rho signal cascade, *Cell Struct. Funct.*, 21, 317, 1996.
36. Yamamoto, M., Marui, N., Sakai, T., Morii, N., Kozaki, S., Ikai, K., Imamura, S., and Narumiya, S., ADP-ribosylation of the rhoA gene product by botulinum C3 exoenzyme causes Swiss 3T3 cells to accumulate in the G1 phase of the cell cycle, *Oncogene*, 8, 1449, 1993.
37. Zhong, C., Kinch, M. S., and Burridge, K., Rho-stimulated contractility contributes to the fibroblastic phenotype of Ras-transformed epithelial cells, *Mol. Biol. Cell*, 8, 2329, 1997.
38. Johnston, R. F., Pickett, S. C., and Barker, D. L., Double-label image analysis using storage phosphor technology, *Comp. Methods Enzymol.*, 3, 128, 1991.
39. Citi, S., Volberg, T., Bershadsky, A. D., Denisenko, N., and Geiger, B., Cytoskeletal involvement in the modulation of cell-cell junctions by the protein kinase inhibitor H-7, *J. Cell Sci.*, 107, 683, 1994.

40. Higuchi, H. and Takemori, S., Butanedione monoxime suppresses contraction and ATPase activity of rabbit skeletal muscle, *J. Biochem. (Tokyo)*, 105, 638, 1989.
41. Herrmann, C., Wray, J., Travers, F., and Barman, T., The effects of 2,3 butanedione monoxime on myosin and myofibrillar ATPases. An example of an uncompetitive inhibitor, *Biochem.*, 31, 12227, 1993.
42. Osterman, A., Arner, A., and Malmqvist, U., Effects of 2,3-butanedione monoxime on activation of contraction and crossbridge kinetics in intact and chemically skinned smooth muscle fibers from guinea pig taenia coli, *J. Muscle Res. Cell Motil.*, 14, 186, 1993.
43. McKillop, D. F., Fortune, N. S., Ranatunga, K. W., and Geeves, M. A., The influence of 2,3-butanedione 2-monoxime (BDM) on the interaction between actin and myosin in solution and in skinned muscle fibres, *J. Muscle Res. Cell Motil.*, 15, 309, 1994.
44. Dillon, S. T. and Feig, L. A., Purification and assay of recombinant C3 transferase, *Methods Enzymol.*, 256, 174, 1995.
45. Nakanishi, S., Yamada, K., Iwanashi, K., Kuroda, K., and Kase, H., KT5926, a potent and selective inhibitor of myosin light chain kinase, *Mol. Pharmacol.*, 37, 482, 1990.
46. Saitoh, M., Ishikawa, T., Matsushima, S., Naka, M., and Hidaka, H., Selective inhibition of catalytic activity of smooth muscle myosin light chain kinase, *J. Biol. Chem.*, 262, 7796, 1987.

SECTION IV

General Methods for Signaling Studies of Cell Adhesion Molecules

Chapter 18

Intra- and Intercellular Localization of Proteins in Tissue *in situ*

Peter A. Piepenhagen and W. James Nelson

Contents

I. Introduction 264
II. General Considerations 264
A. Fixation 264
B. Embedding/Sectioning 266
C. Permeablization 268
D. Blocking Non-Specific Background 268
E. Primary Antisera 269
F. Secondary Antisera/Detection Systems 270
III. Protocol 272
A. Fixing and Sectioning Tissue 272
B. Immunofluorescent Staining of Tissue Sections 274
IV. Preparation of Buffers 275
A. CSK Buffer 275
B. Blocking Solution 275
C. Antibody Buffer 276
D. Preparation of Paraformaldehyde-Lysine-Peroxidate (PLP) Fixative 276
E. Protocol for Preparing "Subbed" Slides 276
V. Concluding Remarks 277
References 277

0-8493-3385-7/99/$0.00+$.50

I. Introduction

Understanding signaling pathways in complex tissues requires a combination of experimental approaches. One approach widely used to examine cellular and subcellular distributions of signaling molecules is immuno-localization of proteins within cells and tissues. This technique is of particular importance in the study of signaling cascades in complex organs and tissues, such as the kidney, in which multiple cell types, each with a different structure and function, are present. Clues of a given protein's function are provided by knowing: 1) which cell types within a complex multicellular tissue express the protein; 2) when during development the protein is expressed; and 3) with which intracellular compartment(s) it associates. In addition, co-localization of two or more proteins provides supporting evidence that they may directly or indirectly interact.

It is of considerable interest to understand how best to obtain high quality immuno-localization data. Because of their relatively large size and structural and functional polarity, renal epithelial cells lend themselves to studies of subcellular localization of proteins. While immuno-localization of proteins within cultured renal epithelial cells is relatively straightforward and has been exhaustively described elsewhere, immuno-localization of proteins within renal epithelial cells *in situ* is more problematic due to the complex three-dimensional structure of kidneys and to the occurrence of potentially cross-reacting proteins not present within cell cultures. In this chapter, we present general techniques for conducting immuno-localization of proteins within embryonic and adult kidney. We discuss the parameters that must be considered when devising an immuno-localization protocol, potential problems that may be encountered, and strategies by which these problems may be overcome. Finally, we present a protocol for conducting immunofluorescence localization of proteins within kidney tissue *in situ*. This protocol has been used extensively within our laboratory to study the subcellular localization of cytoskeletal, cell adhesion, and ion transporting proteins within developing and adult mouse kidney.[1,2]

II. General Considerations

A. Fixation

The first issue which must be considered when designing an immunohistochemical protocol is whether tissue is to be fixed and, if so, how. One option involves conducting immunohistochemistry with unfixed tissue which has been snap-frozen and cryosectioned immediately after removal from the animal.[3-8] While this approach circumvents potential artifacts introduced by fixation, it does not preserve tissue architecture as well as the use of fixed tissue, and opens up the possibility that protein localization and conformation may change during the immuno-staining procedure. Because of this, unfixed frozen tissue sections are better used as a benchmark by which to evaluate the antigenicity of proteins in fixed tissue.[6,7] A large number of different fixatives useful for cell and tissue fixation have been described in the

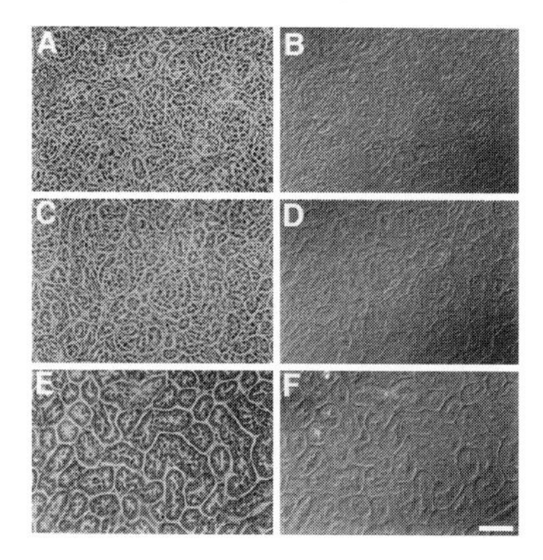

FIGURE 1
Morphological preservation of kidney tissue by post-mortem fixation in PLP. Kidneys from mice at embryonic day 16 (panels A and B), birth (panels C and D), and 10 weeks after birth (panels E and F) were fixed, cryoprotected, and cryosectioned at 5 μm according to the protocol described in the text. The degree of morphological preservation is revealed by phase contrast (panels A, C, and E) and differential interference contrast (panels B, D, and F) microscopy. Scale bar in panel F applies to all panels and is 50 μm.

literature.[9,10] Those commonly used to fix kidney tissue for immunohistology include 2 to 4% paraformaldehyde (4°C),[6,11-15] 95% ethanol (–20°C or 4°C),[16-20] 100% methanol (–20°C),[21-24] and acetone (–20°C).[7,16,17,25-29] Formaldehyde,[8,30] glutaraldehyde,[6] and Bouin's fixative[31,32] have also been used on occasion to fix kidney tissue for immunohistology. While these fixatives do an excellent job of preserving tissue morphology and cellular ultrastructure, they tend to reduce antigenicity of tissue. For this reason, they are generally considered too harsh for immuno-staining.

The best fixative must be empirically determined and should be evaluated for its ability to preserve immunological reactivity (in comparison to unfixed frozen sections) and important morphological features of the organ or tissue under study. While no one fixative will prove ideal for all purposes, we have found the paraformaldehyde-lysine-periodate (PLP) fixative formulated by McClean and Nakane (1974)[33] to be the most generally useful. This fixative appears to preserve both morphology (Figure 1) and immunological reactivity (see Figures 3 and 4) well in adult and embryonic kidney following post-mortem fixation, cryoprotection, and frozen sectioning (see below). This fixative employs a low concentration of paraformaldehyde (2%) and achieves an acceptable level of tissue preservation by also cross-linking carbohydrate moieties present within the tissue. This strategy decreases cross-linking through protein backbones where most antibody epitopes are located, thereby preventing masking or destruction of antigenic epitopes.

The method of fixation, pre- or post-mortem, will depend upon the tissue to be analyzed and the morphological requirements of the investigator. Pre-mortem fixation will more faithfully preserve morphological features of the tissue found in the living animal. In kidney, for instance, lumens of renal tubules rapidly collapse after death of the animal and removal of the kidneys due to ischemic damage.[11] This phenomenon can be avoided by pre-mortem perfusion fixation of kidneys following anesthesia. Pre-mortem fixation usually involves perfusion with Ringer's solution or

PBS to flush blood from the kidney, followed by perfusion with 4°C fixative for 5 to 10 min. After perfusion fixation within the animal, kidneys are usually removed and further fixed at 4°C for 1 to 12 h. The fixatives most commonly employed for pre-mortem fixation are 2 to 4% paraformaldehyde[11,12] and PLP.[34] While undoubtedly the best approach, pre-mortem fixation requires minor surgery and is difficult to perform in small species such as the mouse (but see Peters et al., 1995).[32]

For this reason, we have employed post-mortem fixation in our studies of mouse kidney. We find that if kidneys are removed rapidly after death of the animal and placed immediately into fixative, morphological perturbation is minimal (see Figure 1). Tissue should be cut into smaller pieces, if required, such that the tissue pieces are not more than 5 mm thick. This permits rapid penetration of the fixative. Unless there are compelling reasons to do otherwise, fixation should always be carried out at 4°C to prevent proteolysis of the tissue during fixation. While some investigators fix tissue for prolonged periods,[11,13,32,34] we have observed that fixation for periods longer than 30 to 45 min does not provide any additional morphological preservation. If cryosectioning is to be employed (see below), another parameter to consider is whether to fix tissue before or after sectioning; if paraffin or plastic sectioning is used, tissue must necessarily be fixed prior to embedding and sectioning. While some other investigators have used a post-sectioning fixation protocol,[6,7,16,17,26-29] there is no obvious advantage to fixing after rather than prior to sectioning, and our experience has been that morphology of kidney tissue is much better preserved if tissue is fixed and cryoprotected prior to sectioning.

B. Embedding/Sectioning

The size and thickness of kidney, even in small species such as the mouse, requires that tissue be sectioned to obtain adequate spatial resolution for the examination of subcellular localization of proteins. Two common approaches are mechanical sectioning followed by immuno-staining, and whole mount immuno-staining followed by optical sectioning using confocal or deconvolution microscopy. Whole mount staining is often used in cases where samples are comprised of a small number of cells or are only a few cell layers thick. Examples include *Xenopus* and *Drosophila* embryos, imagninal discs from *Drosophila*, and the nematode *Ceanorhabditis elegans*. This approach is not generally useful for examining protein expression and localization in kidney. At almost all developmental stages, expression and organization of extracellular matrix and basement membrane components are too extensive to allow antibodies to penetrate more than a few cell layers into the tissue; this is true even in the case of midgestational embryonic mouse kidneys. The exceptions to this rule are the pronephros, metanephros, and early metanephros in which whole mount immunohistochemistry has been used successfully.[21-24] For this reason, embedding and mechanical sectioning followed by immuno-staining is a more feasible approach to immuno-localization within kidney at most developmental stages.

Mechanical sectioning requires that tissue be embedded in some medium to stabilize it and hold it in place while sections are cut. The three types of embedding media commonly available and compatible with immunostaining are paraffin wax, various water-soluble plastics, and cryoembedding compounds. Paraffin wax and water-soluble plastics possess the advantage that tissue embeddded within them can be sectioned at room temperature using an inexpensive microtome. Their disadvantages, however, are that the processing required to embed tissue and to stain sections cut from embedded tissue may destroy the antigenicity of proteins in the tissue being examined.[6,35] While some investigators have reliably used paraffin-[6,8,18-20,30,32,35] and plastic-embedded tissue sections[14,15] for conducting immunohistochemistry, many investigators have reported inconsistent results with such sections.[10,36] Our experience with at least one water-soluble plastic embedding media, Immuno-bed (Polysciences), is that while it preserves tissue morphology well, only a small number of our antisera give robust and reproducible staining with plastic-embedded sections.

We have found that frozen tissue sections react consistently and strongly with all of our antisera (22 different antibodies from rabbit, rat, and mouse). While frozen sectioning is slightly more difficult than plastic or paraffin sectioning and does not preserve morphology quite as well as either plastic or paraffin, the morphological preservation is adequate (see Figure 1) and more than compensated for by the increased immunological reactivity. For these reasons, we believe that cryoembedding and sectioning are generally preferable to either paraffin or plastic embedding. Exceptions may occur in cases in which a cryostat is unavailable or in which morphological preservation is of greater importance. In these cases, staining patterns obtained with paraffin- or plastic-embedded samples should be compared to staining patterns obtained with unfixed or lightly fixed cryosections.[6,35]

While it is possible to freeze and cryosection unfixed tissue immediately after removal from the animal, we have found that morphological preservation is much better if tissue is fixed and cryoprotected prior to sectioning. Tissue can be cryoprotected in dimethylsulfoxide (DMSO),[34] glycerol, or sucrose.[1,2,11-13,37] Most often, fixed tissue is incubated in a graded series of DMSO, glycerol, or sucrose (5, 10, 25%) until the tissue has equilibrated with the final solution. This procedure decreases the concentration of water within the tissue so that as the tissue is frozen, formation of water crystals is minimized. Formation of large water crystals causes disruption of cell and tissue morphology by compromising the integrity of membranes and other intracellular structures. An alternative protocol to the graded DMSO, glycerol, or sucrose solutions which we have found convenient is to place tissue after fixation and washing into a solution of 2.5 M sucrose in PBS[1,2,37]; 10 volumes of this solution should be used for every volume of tissue to be processed. The tissue undergoes two incubations of 24 h each at 4°C in this solution and is then transferred to fresh 2.5 M sucrose solution, in which it is stored for up to several months at 4°C. As required, tissue is removed from storage at 4°C, frozen in cryoembedding medium, and sectioned using a cryostat.

Formation of water crystals can also be minimized if the sample is frozen rapidly. While direct immersion in liquid N_2 may freeze cyroprotected samples fast enough to prevent formation of large ice crystals,[1,2] the most effective way to freeze samples quickly is to immerse them in liquid N_2-cooled isopentane[3-5,11,12] or liquid N_2-cooled liquid Freon 22.[34] Unlike liquid N_2, isopentane and liquid Freon 22 do not vaporize when placed in contact with the warm sample and therefore conduct heat away from the sample more quickly. If samples are to be cryosectioned unfixed and non-cryoprotected, snap freezing in liquid N_2-cooled isopentane or liquid N_2-cooled liquid Freon 22 is strongly recommended. The appropriate temperature for cyrosectioning must be empirically determined, and will depend upon the type of tissue being sectioned and whether or not it is cyroprotected. In the case of non-cryoprotected kidney tissue, our experience is that –20°C to –25°C works well for kidney. In the case of cryoprotected tissue, freezing point depression due to the presence of the cryoprotectant requires that frozen sectioning be carried out at lower temperatures than is the case for non-cryoprotected tissue. In the case of tissue cyroprotected using 2.5 M sucrose, we have found that –40°C to –45°C is best. For cryoprotected tissue, the temperature of sectioning will be primarily determined by the concentration of cryoprotectant rather than by the fat or extracellular matrix content of the tissue itself, as is the case for non-cryoprotected tissue. If the refrigeration unit on the cryostat used is incapable of reaching these temperatures, it may be necessary to place some dry ice into the cryostat chamber to further lower the temperature. For cellular and subcellular localization of proteins, sections should be between 3 to 10 µm thick. Such frozen sections can be used immediately or stored at –20°C or –80°C in a slide box for at least one week, but in all cases should not be allowed to thaw until they are to be used.

C. Permeablization

Even though cells present within tissue sections have been cut open through the process of sectioning, we have found it useful to extract tissue sections with lipophilic agents before conducting immunohistochemical staining.[1,2] Extraction appears to improve the accessibility of antibodies to their cognate epitopes within the tissue sections. For this purpose, either methanol or detergents can be used. Slides with sections are incubated for 15 to 30 sec with either –20°C 100% methanol or with 4°C buffers containing 0.5% Triton X-100 or another non-ionic detergent such as NP-40 (the recipe for one such buffer used by our laboratory is given in the protocol below).

D. Blocking Non-Specific Background

One of the most problematic aspects of conducting immunohistochemistry on tissue sections is the non-specific background which is often observed. This is usually much less of a problem in cell culture because there are fewer potential cross-reacting proteins. The most severe background problems appear to occur when secondary antibodies directed against primary antibodies from the species under study are used (e.g., anti-mouse secondary antibodies are used to stain sections of mouse tissue). In mouse kidney, anti-mouse secondary antibodies react strongly with components of the basement membrane even in the absence of primary antibodies (Figure 2). Because of this, extra care must be taken to block non-specific antibody interactions in tissue sections. Our experience indicates that PBS containing 20% normal serum (from the species in which the secondary antibody was raised), 0.2% bovine serum albumin (BSA), 50 mM NH_4Cl, 25 mM lysine, and 25 mM glycine works well as

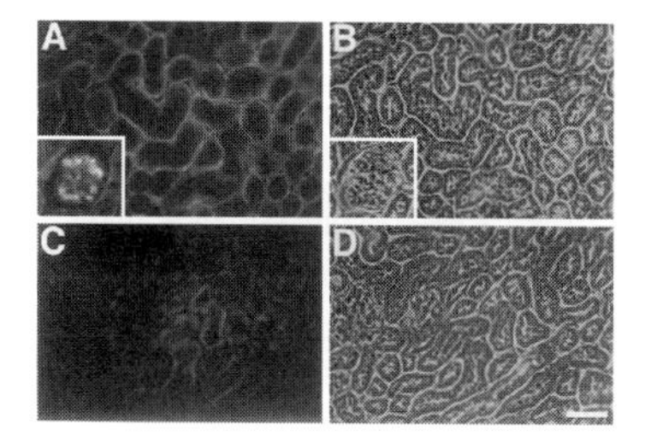

FIGURE 2

Blocking non-specific binding of secondary antibodies. Adult mouse kidneys were fixed, cryoprotected, and cryosectioned at 5 μm according to the protocol described in the text. The section shown in panels A and B was blocked using the standard blocking solution described in the text; it was then stained with goat anti-mouse FITC-conjugated secondary antibody in the absence of primary antibody. Note the prominent background staining of basement membranes in glomeruli (inset) and around renal tubules. The section shown in panels C and D was blocked with the standard blocking solution described in the text, supplemented with unlabeled goat anti-mouse IgG at a final concentration of 0.4 mg/ml; it was then stained with goat anti-mouse FITC-conjugated secondary antibody. Note that using this blocking regimen eliminates the background staining of basement membranes. Scale bar in panel D applies to all panels and is 50 μm.

a blocking buffer in almost all cases. Normal serum and BSA saturate non-specific binding sites for the primary and secondary antibodies, and the NH_4Cl, lysine, and glycine quench any unreacted aldehyde groups left from fixation so that these cannot interact with antibodies. If secondary antibodies directed against the species under study or a closely related species are to be used, unlabeled secondary antibody at a concentration of 0.4 mg/ml should also be included in the blocking solution. The efficacy of this addition is illustrated in Figure 2. We have found that blocking of non-specific binding is complete after 2 h at room temperature, but the incubation with blocking buffer may be carried out for longer periods if convenient; if incubation with blocking buffer is to be carried out overnight, it should be conducted at 4°C to assure preservation of samples.

E. Primary Antisera

While monoclonal or affinity-purified polyclonal antisera give the cleanest results when used to immuno-stain tissue sections, acceptable results can be obtained with non-affinity-purified polyclonal antisera provided the antibody titer is high enough. Examples of immunostaining with both monoclonal and non-affinity-purified polyclonal antibodies are shown in Figures 3 and 4, in which sections of embryonic day 16 (Figure 3) and adult (Figure 4) mouse kidney were double immuno-stained with rabbit polyclonal and rat monoclonal antibodies. The appropriate dilution of primary antibody must be empirically determined. In all cases, the highest dilution of antibody which gives adequate staining should be used. This is especially important when using non-affinity-purified polyclonal antisera since it helps to minimize non-specific binding of other serum proteins to the tissue sections. For dilution of primary antisera, we use PBS containing 20% normal serum and 0.2% BSA. If localization

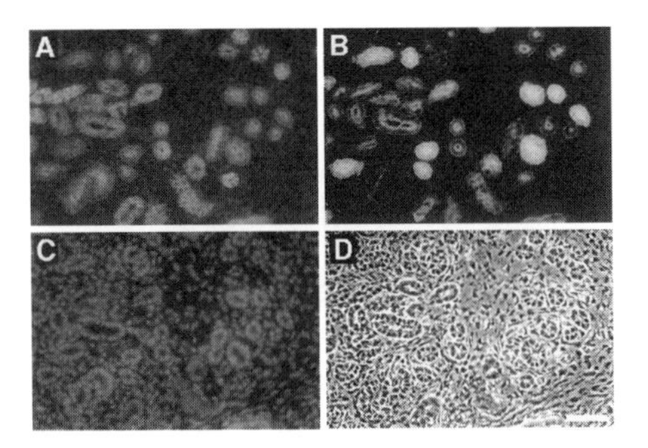

FIGURE 3

Expression of E-cadherin and cytokeratin 8 within embryonic day 16 kidney. Kidneys from embryonic day 16 mice were fixed, cryoprotected, and cryosectioned at 5 μm according to the protocol described in the text. Sections were blocked using the standard blocking solution described in the text supplemented with unlabeled goat anti-rat IgG at a final concentration of 0.4 mg/ml. Sections were then double immunostained with a rabbit polyclonal antibody to the calcium-dependent cell-cell adhesion protein E-cadherin and a rat monoclonal antibody directed against the intermediate filament protein cytokeratin 8. Primary antibodies were detected using TRITC-conjugated goat anti-rabbit and FITC-conjugated goat anti-rat secondary antibodies. After incubation with secondary antibodies, sections were stained with Hoechst 33258 as described in the text. Panel A shows (TRITC) staining for E-cadherin. Panel B shows (FITC) staining for cytokeratin 8. Panel C shows nuclei stained with Hoechst 33258. Panel D shows a phase contrast image of the section to illustrate preservation of tissue morphology. Scale bar in panel D is 50 μm.

of two different proteins within the same tissue sections is to be conducted, primary antisera to the two proteins can be diluted into the same PBS solution and applied simultaneously — provided, of course, that they were raised in different species. While incubation for 2 h at room temperature is adequate for some primary antibodies, we have found that incubation overnight at 4°C results in much stronger signal with lower background for many primary antibodies. While the reasons for this are somewhat mysterious, the phenomenon is highly reproducible.

F. Secondary Antisera/Detection Systems

Unless the primary antibody has been affinity purified and directly labeled with an enzyme marker or fluorophore, it is necessary to apply labeled secondary antibody to detect the primary antiserum. Detection systems fall into two categories: enzyme-linked and fluorogenic. While many investigators have successfully used enzyme-linked secondary antibodies to examine intercellular protein distributions within kidney,[6,8,14,15,18,30,32] diffusion of reaction products generated by enzyme-linked secondary antibodies makes this system less than ideal for examination of subcellular protein localization. Another drawback of enzyme-linked secondary antibodies is the difficulty of assessing relative protein expression levels. While immunofluorescence allows only semi-quantitative evaluation of relative protein expression levels,[1,2] the additional signal amplification generated by enzyme-linked secondary antibodies makes even this difficult. For these reasons, use of fluorescently labeled secondary antibodies is strongly recommended for subcellular localization of proteins and

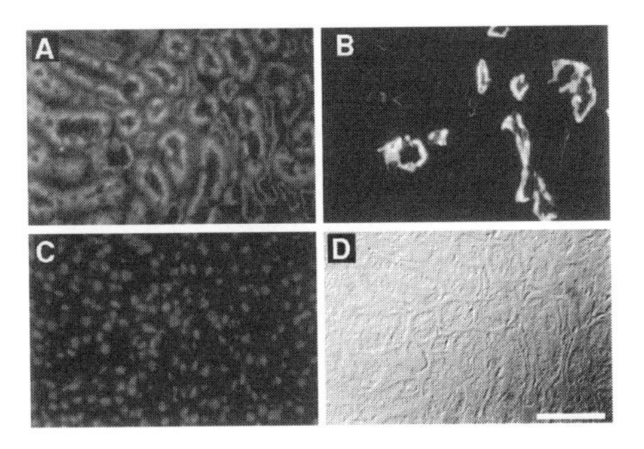

FIGURE 4

Expression of fodrin and cytokeratin 8 within adult kidney. Kidneys from adult mice were fixed, cryoprotected, and cryosectioned at 5 μm according to the protocol described in the text. Sections were blocked using the standard blocking solution described in the text supplemented with unlabeled goat anti-rat IgG at a final concentration of 0.4 mg/ml. Sections were then double immunostained with a rabbit polyclonal antibody to the submembrane cytoskeleton protein fodrin (non-erythroid spectrin) and a rat monoclonal antibody directed against the intermediate filament protein cytokeratin 8. Primary antibodies were detected using TRITC-conjugated goat anti-rabbit and FITC-conjugated goat anti-rat secondary antibodies. After incubation with secondary antibodies, sections were stained with Hoechst 33258 as described in the text. Panel A shows (TRITC) staining for fodrin. Panel B shows (FITC) staining for cytokeratin 8. Panel C shows nuclei stained with Hoechst 33258. Panel D shows a differential interference contrast image of the section to illustrate morphological preservation of the tissue. Scale bar in panel D is 50 μm.

evaluation of their *in situ* expression. Exceptions to this rule occur in samples such as *Drosophila* and *Xenopus* embryos which contain large amounts of autofluorescent yolk and in cases where generation of electron-dense enzymatic reaction products is desired for immunoelectron microscopy.[34] In most vertebrate tissues, autofluorescence is not sufficiently pronounced to pose a serious concern.

Many different fluorescent probes are available for labeling secondary antibodies, and many of these can be purchased conjugated to commonly used affinity-purified secondary antibodies. The choice of fluorophore will depend upon the type of microscope and band pass filter set available. As in the case of the primary antisera, the highest dilution of secondary antibody which provides sufficiently bright staining should be used. We generally use secondary antibodies at a final concentration of 5 μg/ml and dilute them into PBS containing 20% normal serum and 0.2% BSA as in the case of primary antisera. In contrast to primary antibodies, there appears to be no advantage to incubating slides with secondary antibody overnight at 4°C, and we instead conduct the secondary antibody incubation for 2 h at room temperature. During this incubation, the slides should be protected from light to prevent photobleaching of fluorophores. If double immunolabeling for two proteins is conducted on the same tissue sections, secondary antibodies labeled with different fluorophores and specific for each of the primary antibodies can be diluted into the same secondary antibody solution and applied simultaneously. In addition to immunostaining for proteins, it is also possible to examine nuclear localization in tissue which has been fixed, frozen sectioned, and immunofluorescently stained. Because secondary antibodies are commonly labeled with fluorescein, rhodamine, or Texas Red, it is convenient to label nuclei with dyes which can be detected using an ultraviolet filter. Two such dyes are Hoechst 33258[38] and DAPI[39,40] which can be diluted into PBS and incubated with slides after removing secondary antibody and washing.

After incubation with Hoechst 33258 or DAPI, the slides are washed once again and mounted. A number of different mounting media are available, but whatever mounting media is chosen, it should contain reagents to prevent photoquenching of the fluorophores during observation and photography. Two mounting media used successfully by our laboratory are the commercially available Vectashield (Vector Laboratories) and elvanol [PBS containing 16.7% Mowiol (Calbiochem), 33% glycerol, and 0.1% paraphenylene diamine]. One reported drawback of immunofluorescence histochemistry is that slides are not stable once they have been prepared. However, we have found that once immunofluorescently stained slides are mounted, they can be stored for up to two weeks in the dark at –80°C will no ill effects.

III. Protocol

What follows is the protocol which our laboratory uses for fixing, frozen sectioning, and immunofluorescently staining kidney from embryonic and adult mouse. With minor modification, it should be applicable for the immunofluorescence staining of most other organs and tissues as well.

A. Fixing and Sectioning Tissue

1. Remove tissue rapidly and cut it into smaller pieces if required (tissue pieces should be no more than 0.5 cm thick to allow adequate penetration of fixative).
2. Fix tissue in 10 volumes of PLP fixative (Reference 33; Section IV) at 4°C for 30 min (with rocking, if possible).
3. Wash tissue three times for 10 min with Dulbeco's phosphate buffered saline (PBS) at 4°C (with rocking, if possible). Each wash should be conducted in approximately 10 volumes of PBS.
4. Place tissue into a container of 2.5 M sucrose in PBS at 4°C for 24 h; each incubation with 2.5 M sucrose should be conducted in approximately 20 volumes of sucrose solution. Occasionally mix the sucrose solution with the tissue pieces in it by vigorously swirling the tube or inverting it. After the first 24 h incubation, transfer the tissue pieces to fresh 2.5 M sucrose solution and incubate at 4°C for another 24 h with occasional mixing. After the second incubation with 2.5 M sucrose solution, transfer the tissue pieces to another tube of 2.5 M sucrose solution and store at 4°C. The fixed tissue is stable in this condition for up to several months.
5. As required, remove tissue pieces from storage at 4°C and freeze in cryo-embedding medium (e.g., OCT, Miles). For this, we use a scintillation vial to form a cup-shaped container of aluminum foil; after freezing, the aluminum foil can be peeled away, leaving the frozen block of embedding medium with tissue embedded in it. Place the tissue pieces at the bottom of the foil cup, and then fill the cup with liquid embedding medium up to approximately 0.5 cm above the top of the tissue piece. Next, grab the

edge of the foil cup with a forceps and rapidly freeze the sample by partially lowering the foil cup into a container of liquid N_2; the liquid N_2 should not spill over into the inside of the foil cup. We find that simply immersing the foil cup with sample into the liquid N_2 often leads to fracturing of the embedding medium and sample as they freeze.

6. Once the sample is completely frozen, peel off the aluminum foil and place the frozen sample block onto a bed of dry ice. Next, place a cryostat chuck onto the bed of dry ice, face up. Place a large drop of liquid embedding medium onto the face of the chuck and then place the frozen sample block with tissue facing up onto the drop of embedding medium before it freezes; once frozen, the drop of embedding medium will cement the sample block to the cryostat chuck. To assure that the sample block is firmly attached to the chuck, a bead of liquid cyro-embedding medium should then be run around the edge of the sample block where it meets the cytostat chuck.
7. Trim the tissue block to get rid of excess embedding medium around tissue where sections will actually be cut. This is important because the smaller the cross-sectional area to be sectioned through, the easier it will be to obtain consistent sections. Approximately 2 mm of embedding medium should be left on all sides of the tissue piece, and the sides should be trimmed down to approximately 1 mm below the bottom of the tissue piece so that the sample is essentially embedded in a pedestal of frozen embedding medium which will allow one to section completely through the sample. Use a new razor blade for trimming; scrape horizontally to trim rather than trying to cut vertically; this will eliminate the possibility of shattering the block which can be brittle when cold enough. Do not allow the block to melt during the trimming process. If you remove the sample block from the bed of dry ice to trim it, and it starts to get soft or melt, immediately place it back onto the bed of dry ice and allow it to re-freeze completely before trimming further.
8. Once the sample block on the cryostat chuck is trimmed, place it into the cryostat and allow it to equilibrate to the desired temperature. If the cryostat and cryostat knife are already at the desired temperature, this should take approximately 15 min. For tissue cyro-protected with 2.5 M sucrose, cryo-sectioning should be carried out at –40°C to –45°C. If the refrigeration unit on your cryostat is unable to cool to this temperature range, it may be necessary to place dry ice and/or small quantities of liquid N_2 into the cryostat chamber for further cooling.
9. Once the cryostat and tissue block have equilibrated to the desired temperature, begin sectioning. For subcellular localization of proteins, sections should not be more than 10 μm thick and should preferably be 3 to 5 μm thick. The exact protocol to be followed here will vary depending upon the type of cryostat used and the arrangement of its controls. Successful cryo-sectioning usually involves some degree of practice, minor adjustments of temperature, and the angle of the knife blade with respect to the tissue block.
10. After a section has been cut, it must be transferred to a glass slide which has been "subbed" (see Section IV) or chemically treated to cause the section to adhere tightly to the slide ("SuperFrost Plus" slides from Fisher work well). The use of "subbed" or otherwise treated slides is necessary to prevent sections from detaching off the slides during the staining procedure. We find that a rapid and convenient way to transfer frozen sections to glass slides is as follows. In most cryostats, a plastic guard of some sort is used to prevent frozen sections from curling as they are cut. After a section has been cut, lift the plastic guard, and place one end of a room-temperature slide on the knife below the section (use one thumb to hold this end of the slide on the knife blade and apply slight pressure) while holding the other end of the slide up off the knife

blade surface over the section (use thumb and forefinger of other hand). Next, allow the raised end of the slide to "fall" onto the section; this should immediately flatten the section onto the slide. Pull the previously raised end of the slide off of the knife blade; because the knife is cold and the slide is warm, the section will stick to the slide and come off with it. It is important to conduct this procedure quickly because once the plastic guard is pulled away from the knife blade, the section will start to curl up on itself. Place the slides with sections onto a bed of dry ice to keep them frozen until you are ready to process them for immunofluorescence (slides with sections can also be stored at –80°C for up to several weeks).

B. Immunofluorescent Staining of Tissue Sections

1. Thaw tissue sections on slides at room temperature until all moisture has evaporated from slides.
2. Extract slides with CSK buffer + 1 mM pefabloc (protease inhibitor, Boehringer Mannheim) for 15 sec at 4°C. For 10 or fewer slides, a vertical staining jar filled with 30 to 40 ml CSK + pefabloc can be used. For larger numbers of slides, a horizontal staining dish into which an insert can be placed is preferable; this should be filled with approximately 250 ml CSK + pefabloc.
3. After extraction, wash slides twice with PBS at room temperature in staining dishes. Each wash should be carried out for 5 min. During this and all subsequent washing steps, be relatively gentle when placing or removing inserts with slides into staining dishes so that sections do not come off the slides.
4. Incubate slides with blocking solution for 2 h at room temperature in a humidified chamber. For this, a tightly sealed plastic box with wet paper towels in the bottom works well. Use 90 µl of blocking solution per slide. The blocking solution is kept on top of the section by means of a concave 1.5 cm × 1.5 cm piece of parafilm as follows:
   - place concave piece of parafilm on bench top (make parafilm concave by slightly bending it between your fingers)
   - place 90 µl drop of solution onto center of parafilm
   - remove one slide at a time from the last wash and invert slide with section
   - center section over drop of solution
   - carefully lower inverted slide so that solution makes contact with slide surface and solution spreads out between parafilm and slide with no air bubbles trapped where section is located
   - re-invert slide so that it is right side up
5. Carefully remove parafilm pieces from slides with forceps and tilt slides to remove blocking solution. Wash slides twice with PBS + 0.2% BSA at room temperature in staining dishes; each wash should be carried out for 5 min.
6. Incubate slides with primary antibodies overnight at 4°C in humidified chambers. Use 90 µl of primary antibody solution per slide, kept on top of sections by means of parafilm pieces. Primary antibodies are diluted in antibody buffer.
7. Carefully remove parafilm pieces from slides with forceps and tilt slides to remove primary antibody solution. Wash slides twice with PBS + 0.2% BSA at room temperature in staining dishes; each wash should be carried out for 5 min.

8. Incubate slides with secondary antibodies for 2 h at room temperature in humidified chambers in the dark. Use 90 μl of secondary antibody solution per slide, kept on top of sections by means of parafilm pieces. Secondary antibodies are diluted in antibody buffer.
9. Carefully remove parafilm pieces from slides with forceps and tilt slides to remove secondary antibody solution. Wash slides twice with PBS + 0.2% BSA at room temperature in staining dishes; each wash should be carried out for 5 min.
10. Optional: If nuclei within tissue sections are also to be visualized, tissue sections can be stained at this point with DAPI or Hoechst 33258. Incubate slides for 10 min in staining dishes filled with PBS + 25 ng/ml Hoechst 33258 or 100 ng/ml DAPI. After this incubation, wash slides twice with PBS at room temperature in staining dishes; each wash should be carried out for 5 min.
11. Remove slides from the last wash one at a time and mount in vectashield or elvanol [PBS containing 16.7% Mowiol (Calbiochem), 33% glycerol, and 0.1% paraphenylene diamine]. Once mounted, slides may be viewed immediately or stored at –80°C in a slide box for up to 2 weeks.

Examples of the types of results which can be obtained using this protocol to examine protein expression and localization in developing and adult mouse kidney are shown in Figures 3 and 4, respectively.

IV. Preparation of Buffers

A. CSK Buffer

50 mM NaCl
300 mM Sucrose
10 mM PIPES, pH 6.8
3 mM $MgCl_2$
0.5% (v/v) Triton X-100

Note: Once slides have been extracted, it is extremely important not to let them dry out. Move individual slides or slide-filled staining dish inserts one at a time from one fluid-filled dish to another. In addition, when going from a wash to an incubation solution, remove one slide at a time from the last wash and set it up for the incubation before removing the next slide from the washing solution. It is better to leave some slides sitting in the wash for slightly longer periods of time than to risk having some of the slides dry out before you get to them.

B. Blocking Solution

PBS
50 mM NH_4Cl
25 mM Lysine
25 mM Glycine
0.2% BSA
20% Normal Serum (from the species in which secondary antibody was raised)

Note: If primary antibodies from mouse or rat are to be used for staining mouse tissue, the blocking solution should also contain unlabeled anti-mouse or anti-rat secondary antibody at a final concentration of 0.4 mg/ml. This is required to prevent non-specific binding of these secondary antibodies to mouse tissue (Section II and Figure 2).

C. Antibody Buffer

PBS
0.2% BSA
20% Normal Serum (from the species in which the secondary antibody was raised)

D. Preparation of Paraformaldehyde-Lysine-Peroxidate (PLP) Fixative (from McClean and Nakane, 1974)

Final Concentrations: 0.075 M Lysine
0.0375 M $NaPO_4$
0.01 M $NaIO_4$
2% Paraformaldehyde

Solution A: dissolve 1.827 g lysine in 50 ml ddH_2O
bring to 100 ml with 0.1 M $NaPO_4$ buffer, pH 7.4 (19 ml NaH_2PO_4 + 81 ml Na_2HPO_4)
osmolarity = 300 mOsm, pH 7.4

Solution B: mix 8 g paraformaldehyde in 100 ml ddH_2O
heat to 60°C while stirring
add 1 to 3 drops of 1 N NaOH to dissolve
filter solution through a No. 4 Whatman paper to remove any undissolved material
store at 4°C for up to 24 h or aliquot, freeze, and store at –20°C or –80°C. Paraformaldehyde is not stable in solution and should not be used if it has been stored at 4°C for more than 24 h.

Just prior to use: mix 3 parts A with 1 part B
21.4 mg $NaIO_4$ per 10 ml of final solution and mix until dissolved
final pH = 6.2

E. Protocol for Preparing 'Subbed' Slides

Dissolve 1 g of gelatin in 1 liter of hot distilled water. Cool and add 0.1 g chromium potassium sulfate. Store at 4°C. Dip glass slides several times in the solution. Drain and dry in a vertical position. Store in a dust-free box.

V. Concluding Remarks

One important approach to understanding protein function is immuno-localization. Immunofluorescent staining techniques enable investigators to examine protein expression and localization within complex tissues and organs. Such studies test the physiological relevance of hypotheses developed through studies *in vitro* and with cultured cells, and may suggest novel protein functions heretofore unsuspected. We have provided one protocol which our laboratory has used successfully to examine protein localization within developing and adult kidney, and have discussed parameters which are important to consider when modifying this protocol or evaluating others. While this protocol was specifically developed for examining protein expression within kidney, it should be easily modified to examine intra- and intercellular protein distributions within other tissues and organs as well. It is our hope that these techniques will provide other investigators with improved tools for understanding protein function.

References

1. Piepenhagen, P. A. and Nelson, W. J., *Am. J. Physiol.*, 269, C1433, 1995.
2. Piepenhagen, P. A., Peters, L. L., Lux, S. E., and Nelson, W. J., *Am. J. Physiol.*, 269, C1417, 1995.
3. Rocco, M. V., Neilson, E. G., Hoyer, J. R., and Ziyadeh, F. N., *Am. J. Physiol.*, 262, F679, 1992.
4. Hoyer, J. R., *Kid. Int.*, 17, 284, 1980.
5. Hoyer, J. R., Resnick, J. S., Michael, A. F., and Vernier, R. L., *Lab. Invest.*, 30, 757, 1974.
6. Laitinen, L., Virtanen, I., and Saxen, L., *J. Histochem. Cytochem.*, 35, 55, 1987.
7. Durbeej, M., Soderstrom, S., Ebendal, T., Birchmeier, C., and Ekblom, P., *Development*, 119, 977, 1993.
8. Garrod, D. R. and Fleming, S., *Development*, 108, 313, 1990.
9. Jacobberger, J. W., *Methods*, 2, 207 1991.
10. Sternberger, L. A., *Immunocytochemistry*, Wiley, New York. 1986.
11. Doctor, R. B., Bennett, V., and Mandel, L. J., *Am. J. Physiol.*, 264, C1003, 1993.
12. Doctor, B. R., Chen, J., Peters, L. L., Lux, S. E., and Mandel, L. J., *Am. J. Physiol.*, 274, F129, 1998.
13. Torres, M., Gomez-Pardo, E., Dressler, G. R., and Gruss, P., *Development*, 121, 4057, 1995.
14. Sweeney, W. E. and Avner, E. D., *J. Tissue Culture Methods*, 13, 163, 1991.
15. Avner, E. D., Sweeney, W. E., and Nelson, W. J., *Proc. Natl. Acad. Sci. USA*, 89, 7447, 1992.
16. Minuth, W. W., Lauer, G., Bachmann, S., and Kriz, W., *Histochemistry*, 80, 171, 1984.
17. Minuth, W. W., Gross, P., Gilbert, P., and Kashgarian, M., *Kid. Int.*, 31, 1104, 1987.
18. Ekblom, P., et al., *Dev. Biol.*, 84, 88, 1981.

19. Sainte-Marie, G., *J. Histochem. Cytochem.*, 10, 250, 1961.
20. Ekblom, P., Alitalo, K., Vaheri, A., Timpl, R., and Saxen, L., *Proc. Natl. Acad. Sci. USA,* 77, 485, 1980.
21. Sainio, K., Hellstedt, P., Kreidberg, J. A., Saxen, L., and Sariola, H., *Development,* 124, 1293, 1997.
22. Sainio, K., et al., *Int. J. Dev. Biol.*, 38, 77, 1994.
23. Sainio, K., Saarma, M., Nonclercq, D., Paulin, L., and Sariola, H., *Cell. Mol. Neurobiol.*, 14, 439, 1994.
24. Sariola, H., Holm, K., and Henke-Fahle, S., *Development*, 104, 589, 1988.
25. Dressler, G. R. and Douglas, E. C., *Proc. Natl. Acad. Sci. USA,* 89, 1179, 1992.
26. Vestweber, D., Kemler, R., and Ekblom, P., *Dev. Biol.*, 112, 213, 1985.
27. Vestweber, D. and Kemler, R., *Exp. Cell Res.*, 152, 169, 1984.
28. Klein, G., Langegger, M., Timpl, R., and Ekblom, P., *Cell,* 55, 331, 1988.
29. Sorokin, L., Sonnenberg, A., Aumailley, M., Timpl, R., and Ekblom, P., *J. Cell Biol.*, 111, 1265, 1990.
30. Nouwen, E. J., Dauwe, S., Van Der Biest, I., and De Broe, M. E., *Kid. Int.*, 44, 147, 1993.
31. Humason, G. L., *Animal Tissue Techniques*, W. H. Freeman and Co., San Francisco. 1979.
32. Peters, L. L., et al., *J. Cell Biol.*, 130, 313, 1995.
33. McClean, I. W. and Nakane, P. K., *J. Histochem. Cytochem.*, 22, 1077, 1974.
34. Kashgarian, M., Biemesderfer, D., Caplan, M., and Forbush, B.,*Kid. Int.*, 28, 899, 1985.
35. Curan, R. C. and Greggory, J., *J. Clin. Pathol.*, 31, 974, 1978.
36. DeArmond, S. J. and Eng, L. F., *Prog. Exp. Tumor Res.*, 27, 92, 1984.
37. Amieva, M. R., Wilgenbus, K. K., and Furthmayr, H., *Exp. Cell Res.*, 210, 140, 1994.
38. Hilwig, I. and Gropp, A., *Exp. Cell Res.*, 75, 122, 1972.
39. Williamson, D. H. and Fennell, D. J., *Methods in Cell Biology,* Prescott, D.M., Ed., Academic Press, New York, 1975.
40. Russell, W. C., Newman, C., and Williamson, D. H., *Nature*, 253, 461, 1975.

Chapter

Optical Microscopy Studies of $[Ca^{2+}]_i$ Signaling

Lynda M. Pierini and Frederick R. Maxfield

Contents

I. Introduction 280
II. Is Optical Microscopy the Right Choice for Your Studies? 280
III. Designing Single-Cell $[Ca^{2+}]_i$ Experiments 282
  A. Wide-Field or Confocal Microscopy? 282
  B. Radiometric or Single-Wavelength Measurements? 286
  C. Choosing a $[Ca^{2+}]_i$ Indicator 286
  D. Dye Loading and Calibration 291
  E. Instrumentation 292
IV. Protocols 294
  A. Sample Chambers 294
  B. Loading Cells With Fura-2/AM 295
  C. Instrumentation 295
  D. Image Acquisition 296
  E. Calibration 296
  F. Image Analysis 297
V. Conclusion 299
References 299

0-8493-3385-7/99/$0.00+$.50

I. Introduction

Optical microscopy provides one of the most powerful techniques for studying intracellular second messengers in living cells. In contrast to flow cytometry and spectrofluorometry, optical microscopy can be used to make measurements on individual cells, providing subcellular spatial information in addition to temporal and quantitative information. Unfortunately, there are currently only a modest number of fluorescent indicators of physiologically important intracellular regulators (e.g., Ca^{2+}, H^+, Na^+, and cAMP) that can be used for microscopy studies.[1] Nonetheless, the advent of even this small group of indicators has been instrumental in broadening our understanding of intracellular signaling.

Studies in our laboratory have utilized various fluorescent indicators of intracellular Ca^{2+} ($[Ca^{2+}]_i$) to investigate the role of this ubiquitous second messenger in cell migration. Specifically, we have used optical microscopy to determine the spatial and temporal characteristics of changes in $[Ca^{2+}]_i$ in motile polymorphonuclear leukocytes (neutrophils),[2,3] to correlate these characteristics with motility parameters such as speed and persistence,[4] and to illustrate a possible mechanism for the regulated adhesion necessary to support motility.[5] In this chapter, we will present a practical guide to designing and setting up optical microscopy experiments for studying $[Ca^{2+}]_i$ signaling during cell migration. We will discuss available $[Ca^{2+}]_i$ indicators, methodological approaches to single-cell measurements, imaging instrumentation, and quantification using digital image analysis.

II. Is Optical Microscopy the Right Choice for Your Studies?

Optical microscopy experiments can be expensive and time consuming, so it is worthwhile to consider carefully whether this is the best method to use. Methods such as flow cytometry and spectrofluorometry are relatively easy to use to monitor changes in $[Ca^{2+}]_i$. For certain experiments, flow cytometry has several advantages over fluorometry. First, in flow cytometry, measurements are made on suspended cells, and only cell-associated fluorescence and the fluorescence of a very small volume of buffer contribute to the measurement. For this reason, the need for extensive, time-consuming washing of the sample is eliminated. This feature of the flow cytometer is important for experiments in which cells are loaded with an indicator dye, as for intracellular calcium ($[Ca^{2+}]_i$) or pH measurements. The contribution to fluorescence from indicator dyes that may leak or be actively transported out of the cell can be excluded during flow cytometry, but not fluorometry measurements in a cuvette. Second, since cells are measured individually with a flow cytometer, anomalous cells (e.g., dead cells that may be intensely fluorescent) can be removed from the analysis. Finally, flow cytometry studies give information about the distribution of fluorescence intensities within a cell population, while fluorometry measurements yield averages of the whole population. One drawback to flow cytometry is that measurements are limited to cells in suspension, whereas fluorometric measurements can be made on either suspended or adherent cells.[6]

Flow cytometry and spectrofluorometry are far better than microscopy for obtaining measurements on large numbers of cells since both of these instruments can measure thousands of cells in just a few minutes. However, quantitative fluorescence microscopy experiments can be much more informative. Like flow cytometry, microscopy gives one the ability to selectively disregard certain cells (e.g., dead or untransfected cells), and it provides information about the heterogeneity of a cell population. However, unlike other approaches, microscopy allows one to make repeated measurements on a single cell. Even though flow cytometers measure cells individually, each cell is measured at only one time point. Thus, the time-course information from these measurements is actually an average of all of the cells that pass the interrogation point within a unit of time. Likewise, fluorometric time-course measurements give the average response of all of the cells in the beam path at each time point. Because cells in a sample are usually not synchronized, temporal resolution is lost when measuring a population of cells with either of these methods. Furthermore, changes in $[Ca^{2+}]_i$ in response to stimulus vary dramatically from cell to cell. Single cells within one sample can respond with transient, oscillatory, and asynchronous increases in $[Ca^{2+}]_i$[7] that, when averaged, would appear as a smaller, more sustained rise in $[Ca^{2+}]_i$. Accordingly, optical microscopy can extract information about intracellular signaling that otherwise would be masked by averaging of cell populations. These measurements can be achieved on whole cells or on regions of a single cell using either an imaging system or a photometer mounted onto the microscope (microspectrofluorometry). Microspectrofluorometry can yield single-cell and subcellular measurements with higher temporal resolution than imaging microscopy, but only imaging microscopy can furnish subcellular spatial information.

Microscopy can provide more detailed information about cellular signaling responses, but this information comes at a cost. Microscopy experiments are typically more difficult than others. During microscopy-based, time-course measurements of living cells, cell viability can be compromised by photodamage, nutrient depletion, or drying of the sample. Drift of the stage or objective can result in defocusing and loss of data, and vibrations can cause blurring of images. Moreover, many fields of cells must be analyzed to obtain statistically significant results, and the analysis itself can be tedious despite recent advances in software that allow automation of some of the analyses. For these reasons, it is important to evaluate the biological question and then decide if the desired information warrants the use of microscopy as opposed to other techniques. Table 1 lists several possible goals of a calcium experiment and the methods that can be used to attain those goals. Commonly, the aim of calcium experiments is to determine if a treatment (e.g., addition of a ligand) stimulates a calcium response. Note that in most cases this can be achieved by techniques other than microscopy.

TABLE 1
Methodological Approaches to Biological Questions

Experimental Aim	Cell Type	Method
Detect changes in $[Ca^{2+}]_i$ in response to a stimulus.	suspended: pure population	flow cytometry, fluorometry
	mixed population	flow cytometry
	adherent: pure population	fluorometry, microscopy (microspectrofluorometry or imaging)
	mixed population	microscopy (microspectrofluorometry or imaging)
Detect subcellular changes in $[Ca^{2+}]_i$.	suspended	n.a.[a]
	adherent	microscopy (microspectrofluorometry or imaging)
Determine the $[Ca^{2+}]_i$ response of a single cell with high temporal resolution.	suspended	n.a.
	adherent	microscopy (microspectrofluorometry or imaging)
Detect heterogeneous $[Ca^{2+}]_i$ responses within a cell population.	suspended	n.a.
	adherent	imaging microscopy
Detect subcellular gradients of $[Ca^{2+}]_i$ or obtain other spatial information.	suspended	n.a.
	adherent	imaging microscopy

[a] n.a.: not applicable

III. Designing Single-Cell $[Ca^{2+}]_i$ Experiments

A. Wide-Field or Confocal Microscopy?

If single-cell imaging microscopy is the best approach to your biological question, should you use a wide-field or a confocal microscope? The advantages and disadvantages associated with conventional wide-field and confocal microscopes are summarized in Table 2. What follows is a discussion of these advantages and disadvantages, as well as suggestions on how some limitations can be overcome. The most notable feature of confocal microscopes is the enhancement in vertical resolution conferred by the confocal pinhole. The pinhole rejects a large portion of out-of-focus fluorescence, thereby eliminating the contribution of fluorescence from objects in different focal planes. This feature makes the confocal microscope the clear choice for certain applications, such as determining the extent of colocalization of two intracellular fluorophores. Additionally, monitoring the $[Ca^{2+}]_i$ response of cells deep within a tissue sample demands the use of a confocal system. But, for most $[Ca^{2+}]_i$ measurements, vertical resolution is not critical. Only in the case of samples with a very high background fluorescence, such as thick samples, does the elimination of out-of-focus fluorescence become an important consideration.

In the case of emission ratio measurements, a confocal microscope is preferable to a wide-field one for practical reasons. For emission ratio measurements the

TABLE 2
Methodological Approaches to Imaging Microscopy

Method	Advantages	Disadvantages
Wide-Field Microscopy	Illumination source is typically an arc lamp that provides flexibility in the choice of excitation wavelengths. Dynamic processes can be imaged with high temporal resolution using sensitive 2-D detectors that image the entire field at once. White-light illumination sources and detector speed make these microscopes well-suited for excitation ratio measurements.	Cannot make measurements on thick samples. A pair of images for ratio measurements are typically collected sequentially. This limits temporal resolution and can lead to artifacts. Emission ratio measurements require either two detectors or other modifications to the microscope. In either case, image misregistration can be a problem.
Confocal Microscopy	Vertical resolution enables measurements on relatively thick samples. Detectors are typically a series of PMTs[a] that allow simultaneous detection of multiple wavelengths, a desirable feature for emission ratio measurements. Because the detectors are PMTs, there are no spatial inhomogeneities as seen with 2-D detectors, and pixel misregistration is not an issue when making ratio measurements.	Lasers used as an illumination source restrict one's choice of fluorophores and can photodamage cells. Excitation ratio measurements are difficult because of the limited laser lines. Laser scanning of the sample is slow, so very rapid events may be missed.

[a] PMT: photomultiplier tube

fluorophore is excited at one wavelength and the emission is monitored at two wavelengths (e.g., $[Ca^{2+}]_i$ measurements with indo-1). The excitation source for a confocal microscope is typically a laser that is scanned over the sample. The resulting emitted light is detected by a series of photomultiplier tubes (Figure 1B), allowing the simultaneous collection of multiple wavelengths. This setup in effect maps the photons from a particular position in the sample to a specific pixel in multiple images. The same type of measurement using a wide-field microscope (Figure 1A) requires two cameras, one to detect each wavelength. Any differences in the geometry or response characteristics of the cameras would lead to misregistration of pixels from each image or photometric errors and, consequently, errors in ratio calculations.

It is possible to make emission ratio measurements on a wide-field microscope with one camera either by using a beam splitter or by switching emission filters. A beam splitter divides the emitted light from the sample and directs each beam through different emission filters. Light of each wavelength is then mapped to each side of the camera chip. Although this technique reduces concerns about differences in camera response characteristics, there is still potential for pixel misregistration, and there is a loss in spatial resolution since only half the chip is being used for each image. On the other hand, these problems are not a consideration if emission filters are placed in a filter wheel so that filters can be switched rapidly. In this latter technique, pixel registration and spatial resolution are retained because images are

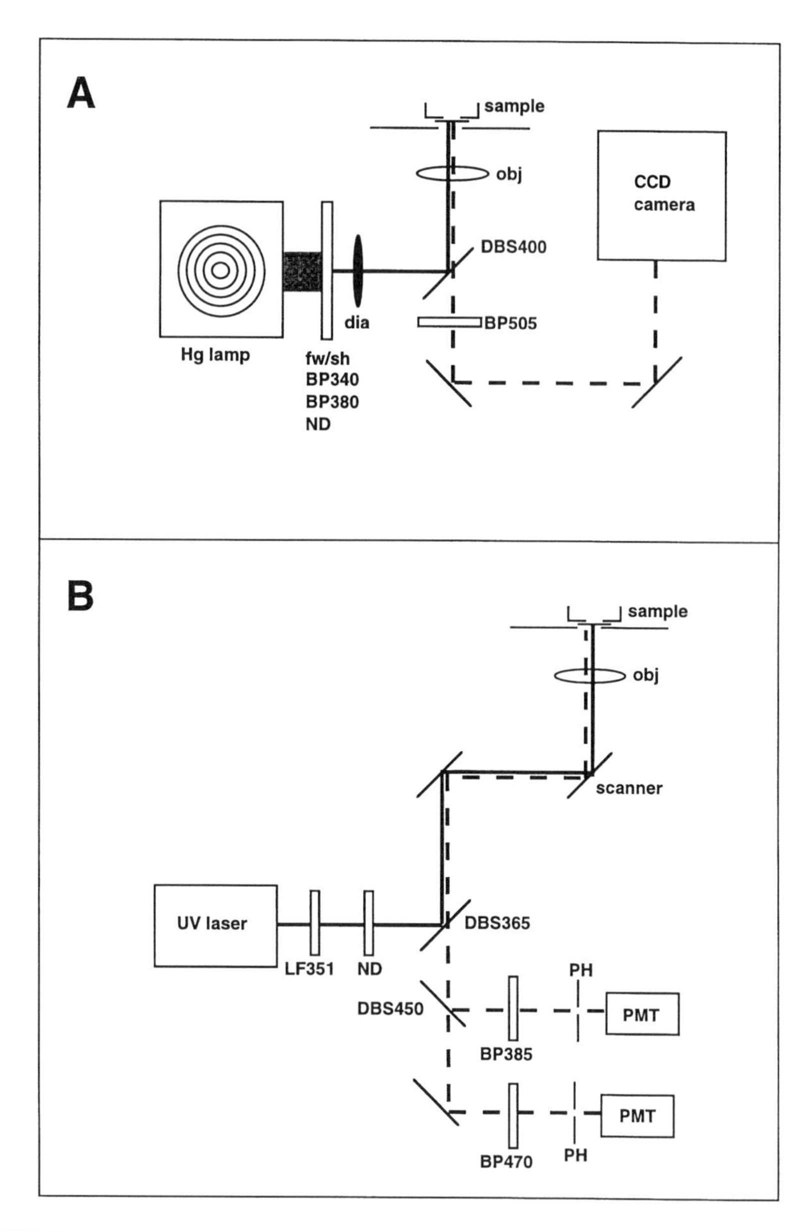

FIGURE 1

Optical paths of a wide-field microscope (A) and a laser-scanning confocal microscope (B) shown with optical components for imaging fura-2 and indo-1, respectively. DBS: dichromatic beam splitter, fw/sh: filter wheel/shutter, BP: bandpass filter, dia: adjustable diaphragm, CCD: charge coupled device, obj: objective lens, ND: neutral density filter, LF: line filter, PH: pinhole aperture, PMT: photomultiplier tube.

collected by a single camera. However, temporal resolution is compromised because images are acquired sequentially rather than simultaneously, and the time it takes to switch filters can limit the rate of image acquisition to several hundred milliseconds.

Although the vertical resolution of the confocal is beneficial for colocalization studies, it is detrimental for studies that require a large depth of field or for certain types of quantification. Because the confocal pinhole eliminates out-of-focus fluorescence, the image obtained is actually a thin "slice" of the sample. Slight shifts in the focal plane result in dramatic changes in the portion of the sample imaged and in the efficiency of collection of emitted light. For this reason, quantification and comparison of individual confocal slices from different samples is difficult to accomplish. These difficulties can be overcome by acquiring a z-series (i.e., a vertical stack of single slices) and quantifying the total fluorescence from the sample. This solution is unsatisfactory for some applications because the acquisition time per sample becomes protracted. Acquisition times of several seconds are too long to observe some dynamic processes in living cells. Even in the case of ratiometric $[Ca^{2+}]_i$ measurements, where a single confocal slice would suffice, acquisition time can be a limitation. The ability of the confocal to acquire multiple emissions simultaneously must be weighed against the fact that confocal imaging relies on the relatively slow process of laser-scanning the sample. Long acquisition times will of course limit temporal resolution, and may prohibit conclusions about subcellular spatial distribution of a response if the scan time across the cell is on the time scale of the response. In contrast, a pair of images can be acquired within a few hundred milliseconds using a wide-field microscope, and subcellular spatial information is retained because the entire cell is imaged at once.

It is possible to collect images rapidly with modified confocal microscopes. In a slit-scanning confocal microscope, the pinhole aperture and point illumination of a conventional point-scanning confocal microscope are replaced by a narrow slit aperture and a corresponding narrow line of illumination. In this design, the slit illumination and aperture span the width of the imaged field, so the sample requires scanning in only one direction and, consequently, collection time is drastically reduced. For a point-scanning confocal microscope, scan times are minimized by maximizing the number of photons obtained from the illuminated fluorophores. Typically, this is accomplished by increasing illumination intensity to optimize fluorophore excitation. However, increasing illumination intensity past a certain point is detrimental. Once the maximum fluorescence yield from a fluorophore is attained, any further increase in intensity results in an increase in background without an increase in signal. Intense illumination will also photobleach the fluorophore and photodamage the cells. Furthermore, because lasers provide only a limited number of excitation wavelengths, the use of a laser as an excitation source restricts the choice of fluorophores. This makes it difficult to perform experiments that require two specific excitation wavelengths as is the case for $[Ca^{2+}]_i$ measurements using fura-2. Many of the current generation of confocal microscopes come equipped with multiple lasers, thereby increasing the number of available excitation wavelengths. With some confocal microscopes it is possible to rapidly and precisely switch between two laser lines by using an AOTF (Acousto-Optic Tunable Filter). An AOTF is a crystal that acts as a tunable bandpass filter. Acoustic waves that are generated by radio frequencies are used to select single wavelengths of light from a multi-wavelength source. The wavelength and intensity of the illumination light can be adjusted by varying the radio frequency and the power of the radio frequency, respectively. In this way, each line in an image can be alternately excited at two different wavelengths. Nearly simultaneous collection of two images is achieved by coordinating illumination from each wavelength with collection by two detectors.[8]

B. Ratiometric or Single-Wavelength Measurements?

Detection of fluorescence is dependent on the properties of several of the components of the instrumentation, such as the detector and illumination source. The response of the detector across the image field is not necessarily uniform. This is particularly evident when using CCD cameras that frequently have pixels of variable responsiveness or even "dead" nonresponsive pixels. Spatial variation can also result from uneven illumination across the field of view, caused by either misalignment or fluctuations of the light source. Nonuniformity resulting from detector defects or illumination misalignment can be corrected digitally (see Section IV.F), but random fluctuations of the illumination source cannot. Quantification of fluorescence is further affected by properties of the sample and probe. The photobleaching characteristics of the fluorophore, uneven loading of the probe within a cell or between cells in a field, and deviations in sample thickness will generate variable fluorescence measurements between different fields, between cells in one field, or within a single cell.

Many of the above factors confound interpretations and may prohibit conclusions about subcellular spatial distribution. One way to overcome such problems is to collect two fluorescence images and ratio them. If the fluorescence images are generated from the same probe at either two different emission or two different excitation wavelengths, the resulting ratio image is relatively independent of many of the characteristics of the instrument, sample, and probe. Two single-wavelength probes can be used to generate ratios, but this is less reliable since each probe may have a different photobleaching profile or may have differential accessibility to intracellular compartments.

Although ratiometric measurements have clear advantages over single-wavelength ones, they are not a panacea. Ratiometric image acquisition and analysis are complex and can give rise to artifacts if caution is not used when interpreting data (see Section IV.F). In many instances, useful $[Ca^{2+}]_i$ measurements can be made with a single-wavelength indicator. Single-wavelength measurements are suitable for qualitative estimates of the global changes in $[Ca^{2+}]_i$ over time. When quantitative information is desired, such as relative intercellular $[Ca^{2+}]_i$ or absolute $[Ca^{2+}]_i$, it is necessary to carry out either complex calibration procedures or ratiometric measurements. Subcellular information can only be obtained by using ratiometric indicators.

C. Choosing a $[Ca^{2+}]_i$ Indicator

There are three commonly used $[Ca^{2+}]_i$ indicators that signal changes in Ca^{2+} by changing their spectroscopic properties: fura-2,[9] indo-1,[9] and fluo-3.[10] These indicators belong to a family of fluorescent Ca^{2+}-sensitive dyes that were developed by adding fluorophores to aromatic analogs of EGTA (Figure 2). The EGTA core of these dyes instills them with some common properties. All three dyes are negatively charged hydrophilic molecules that are inherently membrane impermeant. In their free acid form, the dyes can only be loaded into cells by mechanically disrupting the plasma membrane via microinjection, electroporation, or scraping. These invasive loading techniques are impractical for many cell types. Conveniently, all of the dyes are available in neutral, membrane-permeant ester derivatives. In these compounds, the negative charges, including the ones that make up the Ca^{2+}-binding site, have been esterified by acetoxy methyl (AM) groups that transform the dyes into hydrophobic, Ca^{2+}-insensitive molecules. The AM ester dyes diffuse freely across the plasma membrane and are hydrolyzed by intracellular esterases. Ester hydrolysis regenerates the membrane-impermeant, Ca^{2+}-sensitive free acid form of the dye that is now trapped inside the cell. The byproducts of hydrolysis are formaldehyde, acetate, and protons, and their buildup within cells could affect cellular responses.[11] Nevertheless, membrane-permeant hydrolyzable ester dyes provide a simple, noninvasive technique for introducing Ca^{2+} indicators into cells, and any drawbacks of these dyes can usually be overcome by careful loading (see Section III.D).

The optical properties of the Ca^{2+}-indicators are quite distinct. Fura-2 and indo-1 are UV-excitable, ratiometric indicators. That is, they exhibit a shift in either their excitation (fura-2) or emission (indo-1) wavelengths upon binding Ca^{2+} (Figure 2B and C). In contrast, fluo-3 is excited by visible light and does not undergo any shift in wavelengths upon Ca^{2+}-binding, but rather, exhibits an increase in fluorescence intensity (Figure 2A). While the single-wavelength nature of fluo-3 precludes its use for subcellular localization and makes quantification of $[Ca^{2+}]_i$ challenging (see Section III.B), it also makes it one of the easiest dyes to use for basic $[Ca^{2+}]_i$ measurements. The spectroscopic characteristics of fluo-3 are similar to fluorescein, obviating the need for specialized filter sets. Furthermore, a potential source of vibrations is abrogated and instrumentation requirements are alleviated because there is no need for filter switching. The inadequacies of fluo-3 can be overcome by loading cells simultaneously with fluo-3 and a Ca^{2+}-insensitive dye (e.g., tetramethyl rhodamine-dextran[12]) or a Ca^{2+}-sensitive dye that exhibits a decrease in fluorescence intensity upon binding Ca^{2+} (e.g., fura-red[13,14]). In this way, fluo-3 can be used for ratiometric measurements. However, the use of multiple fluorophores for ratiometric measurements can be complicated by the differences in the properties of each of the fluorophores. Ratios can be affected by differences in the susceptibility to photobleaching and accessibility to intracellular compartments of each fluorophore. These problems are not insurmountable, but they add a level of complexity to the experiment.

Ratiometric measurements are best accomplished with a single fluorophore that changes its spectroscopic characteristics in response to the parameter of interest. For emission ratio measurements of $[Ca^{2+}]_i$, indo-1 is frequently used. Indo-1 exhibits a shift in emission from ~475 nm to ~400 nm when Ca^{2+} is bound. Indo-1 and fluo-3 photobleach rapidly, and measurements with these dyes can be challenging. The minimization of acquisition times and the attenuation of illumination intensity necessary to avert photobleaching often results in a fluorescence signal that pushes the limits of detection. For practical reasons, the use of indo-1 for emission ratio

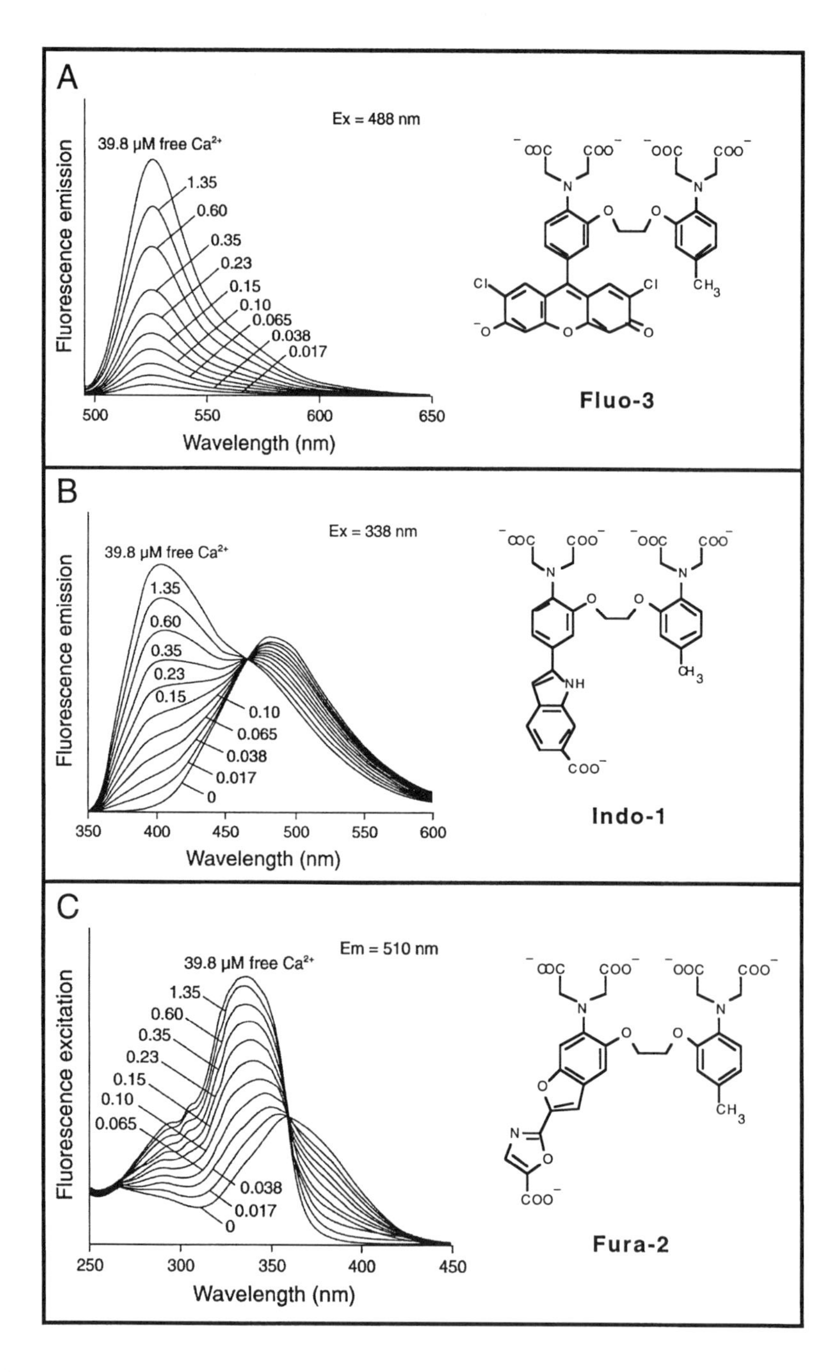

FIGURE 2

Structures and fluorescence emission (A, B) or excitation (C) spectra for fluo-3 (A), indo-1 (B), and fura-2 (C) in solutions containing a range of Ca^{2+} concentrations. Reprinted with kind permission from Molecular Probes, Inc. (Eugene, OR).

measurements is usually restricted to flow cytometry and confocal microscopy experiments. Detection of two emission wavelengths is more easily achieved with these instruments than with a conventional microscope (see Section III.A).

For excitation ratio experiments, fura-2 is the most popular dye. Fura-2 undergoes an absorption shift from 380 nm in the Ca^{2+}-free state to 340 nm in the Ca^{2+}-bound state. Fura-2 is preferable to its predecessor quin2 (Reference 15) because of its greater photostability, increased selectivity for Ca^{2+} vs. other divalent cations, and increased quantum yield. Quin2 has the lowest quantum yield of indo-1, fura-2, and fluo-3, and thus relatively high concentrations of dye must be loaded into cells to obtain a detectable signal. Quin2 at concentrations necessary to get good signal-to-noise often buffers $[Ca^{2+}]_i$ transients, making it impractical to use for most experiments. In fact, the calcium-buffering nature of quin2 has been exploited; this indicator has been used to show that calcium transients are essential for certain cellular functions, such as motility on specific substrates.[3] The practical aspects of measuring changes in $[Ca^{2+}]_i$ using quin2 have been reviewed in depth.[16]

Although fura-2, indo-1, and fluo-3 are among the best indicators currently available for calcium measurements, there are some difficulties associated with their use. The relatively high affinities and slow kinetics of these dyes make them unsuitable for accurate detection of very large spikes or very rapid changes in $[Ca^{2+}]_i$. These dyes are only useful for monitoring $[Ca^{2+}]_i$ in the range of 0-1 μM; alternative dyes with lower affinities (e.g., Calcium-green-5N) must be employed to detect $[Ca^{2+}]_i$ outside of this range. Furthermore, the quantum yields of the indicators dictate that they be loaded into the cells at relatively high concentrations compared to $[Ca^{2+}]_i$. It is important to bear in mind that these dyes are Ca^{2+} chelators, and their prescence in the cell may affect cellular processes.

As discussed below (Section III.D), the free acid form of the dyes is often extruded from the cells and/or redistributed into intracellular compartments. In some cell types, current methods for combating dye leakage or redistribution may be ineffective. One way to overcome this difficulty is to use conjugates to high molecular weight dextrans that will keep the dye trapped in the cytosol. Of course, the dextran conjugates, like the free acids, must be loaded by techniques that breach the plasma membrane, so their usefulness is dependent on the cell type. New calcium indicators are being developed that may overcome some of the problems of this current generation of dyes. A zwitterionic calcium indicator with the spectroscopic characteristics of fura-2 has been reported to be retained in the cytosol better than the anionic dyes.[17] Also, the development of dyes that can be specifically targeted to intracellular compartments may provide more detailed information about subcellular calcium responses. For example, dyes conjugated to lipophilic moieties[13] or derived from amphipathic molecules[17] allow measurement of near-membrane changes in $[Ca^{2+}]_i$ that can differ drastically from cytosolic changes in $[Ca^{2+}]_i$.[18] Table 3 summarizes the properties of fura-2, indo-1, and fluo-3 and includes descriptions of the optical systems for each indicator. Obviously, this table is not comprehensive. In fact, it comprises only a small subset of the available calcium indicators.[13] Ultimately, the decision as to which dye to use for specific $[Ca^{2+}]_i$ experiments will be governed by the biological question, the available equipment, and the properties of the cells, in addition to the relative advantages and disadvantages of available indicators.

TABLE 3
$[Ca^{2+}]_i$ Indicators

	Fluo-3	Indo-1	Fura-2
Indicator Type	Single-wavelength	Emission ratiometric	Excitation ratiometric
K_d[a] (nM)	390	250	224
Optical System[b]			
Excitation filter[c] (nm)	490 (10)	351	340 (10) 380 (10)
Dichromatic beam splitter[d]	510	365	400
Emission filter (nm)	515 (longpass)	385 (longpass) 450 (dichromatic beamsplitter)[d]	505 (bandpass)
Application	Detect global changes in $[Ca^{2+}]_i$ in a single cell or in a population of cells. Quantification is difficult. Not suitable for subcellular measurements.	Detect global changes in $[Ca^{2+}]_i$ in a single cell or in a population of cells. Compare the relative $[Ca^{2+}]_i$ responses between cells. Calculate absolute $[Ca^{2+}]_i$. Detect subcellular $[Ca^{2+}]_i$ gradients.	As for indo-1.
Comments	Suitable for use in any instrument with fluorescein optics. Photobleaches rapidly.	Most often used for flow cytometry and confocal microscopy studies. Photobleaches more rapidly than fura-2, but not as rapidly as fluo-3. Compartmentalizes readily.	Most often used for wide-field microscopy studies. Not suitable for use in microscopes with glass optical components. Glass will not pass 340 nm light. Compartmentalizes readily. Resistant to photobleaching.

[a] K_d for fluo-3 determined by Molecular Probes. K_ds for fura-2 and indo-1 from Grynkiewicz, et al.[9]

[b] Optical systems for fura-2 and fluo-3 are given for use on a wide-field microscope, whereas the optical system for indo-1 is for use on a laser scanning confocal microscope.

[c] Center transmission wavelengths with bandwidths at 50% transmission in parentheses.

[d] Light with shorter wavelength than the indicated wavelength is reflected and light with longer wavelength than that wavelength is transmitted.

D. Dye Loading and Calibration

The protocol presented in Section IV should be used merely as a starting point for developing a dye loading procedure that works well for any specific cell type. It is essential to vary parameters such as the AM-dye concentration, loading time, and temperature to optimize loading conditions. While underloading the cells makes it difficult to obtain an adequate signal over background, overloading the cells has several serious ramifications. First, the byproducts of AM ester hydrolysis can be toxic to the cells. Second, inhomogeneous dye distribution due to compartmentalization into organelles can obscure the contribution of changes in cytosolic $[Ca^{2+}]$. Third, cells that are overloaded or even cells with intrinsically low intracellular esterase activity will likely have a fraction of the dye molecules with incompletely hydrolyzed esters. These dye molecules can contribute to the fluorescence but are Ca^{2+}-insensitive, resulting in an underestimation of $[Ca^{2+}]_i$. If insufficient time is allowed for hydrolysis, a moderate rise in $[Ca^{2+}]_i$ over time might be observed as the dye molecules become deesterified and available to bind Ca^{2+}.

At least some of the pitfalls encountered when loading cells with indicator dye derive from the poor solubility of the AM dyes in aqueous solutions. If precautions are not taken when preparing the dye loading solution, a likely result is "particle loading." That is, undissolved particles of dye will become attached to and possibly endocytosed by the cells, leading to inefficient loading and uneven dye distribution. To avoid particles, it is critical that the components of the loading solution are added in the order given and that the solution is vortexed at each step. Often, detergent and/or carrier protein must be added to facilitate dye dispersion. For these instances, the non-ionic detergent, Pluronic F-127, and/or 1 to 5% heat-inactivated calf serum or BSA are typically used. Ideally, the dye loading solution should be prepared in a buffered saline solution that is free of serum or amino acids, because esterases in the serum and amines of amino acids can cleave the AM ester and result in poor loading.[13] The prospect of inefficient loading must be balanced by the need to obtain a particle-free solution and may be unavoidable in the case of cells that are affected by Pluronic F-127. Whichever formulation is used, it is prudent to look at a slide of the dye solution under the fluorescence microscope before adding it to the cells, to ensure that it is free of particles. Most dye particles can be removed by centrifuging the loading solution for 5 min at $10{,}000 \times g$.

Even if the cells are initially loaded efficiently and homogeneously with dye, the free acid form of the dye can be rapidly redistributed by active transport across membranes.[19,20] Cytosolic dye concentrations may rapidly decrease as the free acids are transported either across the plasma membrane and out of the cells, or across intracellular membranes and into organelles. For this reason, it is crucial that the distribution of dye in the cells be checked at the start and throughout the duration of every experiment. The distribution of the dye can be verified by several methods. At a minimum, dye distribution should by checked by observing the cells under a fluorescence microscope. A more sensitive method for checking for compartmentalization is to measure the fluorescence of the cells before and after the addition of a low concentration of digitonin. Low concentrations of digitonin (~0.005%) will preferentially release dye from the cytosol as opposed to intracellular organelles, so digitonin-resistant fluorescence confirms dye compartmentalization.[21] Considering that in some cell types digitonin permeabilizes endocytic vesicles in addition to the plasma membrane,[22] a better test of dye compartmentalization involves the detection of fluorescence following homogenization and subcellular fractionation (see Section IV.B).[23] In many cell types, secretion and sequestration of indicator dyes can be blocked by inhibitors of organic anion transporters, such as sulfinpyrazone and probenecid.[24] However, these inhibitors can affect cellular responses and they may be only partially effective at preventing dye compartmentalization.

After successfully loading the cells with indicator dye and monitoring $[Ca^{2+}]_i$ responses, it is occasionally valuable to correlate the fluorescence intensities reported by the indicator dyes with absolute values of $[Ca^{2+}]$. Calibrating $[Ca^{2+}]_i$ can be accomplished by either an *in vitro* or an *in situ* method. *In vitro* calibration curves are generated by measuring the fluorescence ratio of the free acid form of the indicator dye in buffers containing known concentrations of Ca^{2+}.[9,15,25,26] Because the emission spectra of many fluorophores are sensitive to their environment, it is unwise to assume that indicator dyes respond similarly both inside and outside the cellular milieu. Dyes inside the cell may be affected by adsorption to proteins, viscosity, and other ions.[27-29] Calibration measurements that are made with indicator dye loaded into cells (*in situ* calibrations) may circumvent errors due to environmental differences. In theory, $[Ca^{2+}]_i$ can be clamped to known values by making the cell membrane permeable to Ca^{2+} ions via treatment with ionomycin, a Ca^{2+} ionophore. The introduction of ionophore in Ca^{2+}-replete medium gives the maximum ratio that results from dye with saturated Ca^{2+}-binding sites, whereas the introduction of ionophore in Ca^{2+}-free medium gives the minimum ratio that results from dye with unoccupied Ca^{2+}-binding sites. In practice, it can be difficult to clamp $[Ca^{2+}]_i$ to intermediate values,[30] and it is possible to get fluorescence ratio values from indicator dyes in intact cells that are outside the range of ratio values that can be achieved in calibration buffers, especially when using indo-1.[31] It is common practice to assume that the Kd and spectroscopic properties of the indicators in cells are similar to values in calibration buffers. This may lead to systematic errors in $[Ca^{2+}]_i$ that could include systematic differences between indicator responses in the nucleus vs. the cytoplasm. Consequently, most measurements often yield at best an approximation of absolute values of $[Ca^{2+}]_i$.

E. Instrumentation

For comparison, schematic diagrams of the light paths of both a wide-field and confocal microscope are presented in Figure 1. The Protocols (Section IV) of this chapter focuses on ratiometric imaging using fura-2. These experiments are typically performed with a wide-field microscope, thus we will limit our discussion of imaging instrumentation to this system. Wide-field microscopes are usually equipped with one of two illumination sources, a xenon or a mercury arc lamp. Xenon lamps provide relatively uniform illumination across an almost continuous range of wavelengths from 200 to 700 nm. In contrast, the output of mercury lamps has several discrete "lines" or strong peaks in both the UV and visible spectra, including lines that correspond to or are near the excitation wavelengths of fluo-3 and fura-2. For ratio imaging with fura-2 the continuous output of a xenon lamp is preferable, but a mercury lamp may be used.

Specific excitation wavelengths are selected with bandpass filters that transmit only a narrow range of wavelengths. For epifluorescence microscopes, in which the objective acts as the condenser, excitation filters can be installed either in a filter wheel located just after the source or in a filter cube at the base of the objective. Control of illumination intensity is achieved with neutral density filters. These filters absorb light evenly across the spectrum so that, as opposed to bandpass filters, they reduce illumination intensity rather than eliminate illumination wavelengths. Ideally, an adjustable diaphragm should be placed at an optical image plane between the sample and the light source. The diaphragm confines the illumination to the object(s) of interest, reduces stray light, and protects surrounding areas from photobleaching. For all fluorescence studies, there should be a controllable shutter between the illumination source and the sample so that the cells will be illuminated only during image acquisition, thereby minimizing the exposure of the cells to intense light and heat. Judicious use of neutral density filters, diaphragms, and shutters will optimize signal over background while keeping photobleaching and photodamage to a minimum.

Light of the appropriate wavelength then proceeds to a dichromatic beam splitter, which diverts the excitation light along the axis of the microscope through the objective to the sample. As mentioned above, in an epifluorescence microscope the objective acts as the condenser, so fluorescence emitted by the sample is collected by the objective. The collection efficiency of the objective is proportional to the square of its numerical aperture, so for low-light fluorescence a high numerical aperture objective should be used. Fluorescence collected by the objective passes through the dichromatic beam splitter and a barrier filter, and is ultimately deflected to a detector. A barrier (or emission) filter, which selectively passes the fluorescence emitted by the specimen and blocks the excitation wavelengths, is usually installed in the filter cube just after the dichromatic beam splitter. If rapid switching between emission wavelengths is needed, it is also possible to house barrier filters in a filter wheel as for the excitation filters. It is important to make sure that all of the components of the light path pass light of the wavelengths needed for a specific experiment. In the past, most microscopes were not equipped with optical systems that would pass 340 nm light, so older microscopes may require modifications if they are to be used to make fura-2 measurements. A conventional microscope can be outfitted easily for digital imaging by purchasing the peripheral components as a package. Several manufacturers sell integrated systems that include image acquisition and analysis software in conjunction with controllers for shutters, filters, and cameras.

Speed and dynamic range are the critical features of any detector that is to be used for $[Ca^{2+}]_i$ experiments. A camera must be capable of acquiring images fast enough to capture the characteristics of the Ca^{2+} response; acquisition times of 1 image every second are usually sufficient to satisfy this criterion, but faster acquisition may be needed in some cases. More stringent requirements are put on the camera for ratiometric measurements where a ratio pair is supposed to represent one point in time. Acquisition times for each image in a pair need to be long enough to give a good signal over background noise, but short enough to permit acquisition of a pair of images within a second. Although there are several types of cameras that can meet this standard, only CCD cameras offer the dynamic range often needed to encompass the range of $[Ca^{2+}]_i$ exhibited by cells. The dynamic range of the camera refers to the number of gray values that the camera can capture and is dictated by the digitization of the fluorescence signal. For example, a camera that digitizes a fluorescence signal into an 8-bit image has a potential dynamic range of 0 to 255 (or 256 levels of gray). A camera with at least 12 bits (4096 gray levels) is recommended to ensure that the full range of $[Ca^{2+}]_i$ responses of all the cells in one field and of all the samples from one experiment are encompassed. More complete discussions of ratio imaging instrumentation can be found in several of the references.[32,33]

IV. Protocols

As mentioned above, one of the most widely used indicators for single-cell $[Ca^{2+}]_i$ measurements is fura-2. Even though we will focus our description of methods to the use of this dye, many of the techniques can be applied to each of the dyes discussed above, as well as to other ratiometric ion indicators.

A. Sample Chambers

Continuous observation of live specimens via optical microscopy may require specialized sample chambers to maintain cell viability. For long-term observation, where evaporation and nutrient depletion may present a problem, a perfusion chamber is often used. A perfusion chamber allows one to continuously replenish medium and to rapidly exchange one medium for another. A discussion of perfusion chamber design can be found elsewhere.[34]

For the observation of live cells, our laboratory uses sample chambers designed for use on an inverted microscope. These chambers are constructed from 35 mm tissue culture dishes (Corning Glass Works, Corning, NY) with a 1.2 cm hole punched in the center of each dish bottom. Hole punch (model #XX) was purchased from Roper Whitney (Rockford, IL). A No. 1 glass coverslip (VWR Scientific Products, So. Plainfield, NJ) is attached beneath the hole using a 3:1 mixture of paraffin (Tissue-Tek, Fisher Scientific Co., Pittsburgh, PA) and petroleum jelly (Vaseline) as an adhesive. For migration studies on purified substrates, coverslips are washed with Nochromix (Godax Laboratories Inc., New York, NY) prior to mounting onto the bottom of the tissue culture dishes and coating with the substrate.[35] Chemotaxis chambers can be constructed from these sample chambers by covalently attaching opsonized erythrocytes to the coverslip surface.[36] Prefabricated sample chambers with poly-D-lysine-coated coverslips are available from MatTek Corporation (Ashland, MA). These experimental chambers are open to the air, so bicarbonate-buffered media cannot be used.

B. Loading Cells with Fura-2/AM

Loading solution:

3 μL 5 mM fura-2/AM (Molecular Probes, Eugene, OR) in anhydrous DMSO
10 μL 10% (w/v) Pluronic-F127 (Molecular Probes) in H_2O
60 μL heat-inactivated fetal calf serum
2.9 mL incubation buffer (150 mM NaCl, 5 mM KCl, 1 mM $MgCl_2$, 10 mM glucose, 20 mM HEPES, pH 7.4)

Prepare a 5 mM stock solution of fura-2/AM in anhydrous DMSO. This stock solution can be aliquoted and stored desiccated at –20°C. While vortexing, add 3 μl of the 5 mM fura-2/AM stock solution to 10 μl of a 10% (w/v) solution of Pluronic-F127 in H_2O. Continuing to vortex, add 60 μl of heat-inactivated fetal calf serum to this mixture, then dilute to 3.0 ml with incubation buffer. This will yield 3.0 mL of a 5 μM fura-2/AM loading solution. Centrifuge the loading solution at 10,000 × *g* for 5 min to remove any undissolved particles of dye. Resuspend 2 to 5 × 10^6 cells in loading solution and incubate at rt for 30 to 60 min. Following incubation, wash the cells twice, resuspend in incubation buffer, then incubate at rt for an additional 10 min to ensure complete hydrolysis of the AM esters. If the $[Ca^{2+}]_i$ response will not be measured immediately, cells should be kept on ice to slow the leakage and compartmentalization of dye. Cells can usually be kept on ice for several hours. Many cell types actively redistribute and extrude the indicator dye through ion transporters, necessitating the addition of inhibitors such as probenecid (2.5 mM) or sulfinpyrazone (0.25 mM).[24] If inhibitors are called for, they should be present throughout the loading and for the remainder of the experiment. Just prior to imaging, rinse the cells into fresh incubation medium to remove extruded dye that will contribute to background fluorescence. In general, background fluorescence can be minimized by choosing medium that is free of phenol red, amino acids, and other autofluorescent components.

Cytosolic distribution of dye should be verified by first observing the cells under the fluorescence microscope. Localization of dye in mitochondria and/or endocytic vesicles is usually quite apparent. However, even in cells with a seemingly diffuse cytosolic distribution, dye could be sequestered in subcellular organelles. Cytosolic distribution should be confirmed by permeabilizing the cells with 0.005% digitonin; compartmentalized dye will be detergent-resistant. For a more sensitive test of dye distribution, homogenize cells then separate cytosol- from organelle-containing fractions by centrifuging the homogenate through 0.4 M sucrose at 120,000 × *g* for 15 min at 4°C.[23] Measure the fluorescence of the supernatant and the pellet with a spectrofluorometer. To determine if fura-2 is sequestered in organelles (i.e., Ca^{2+}-insensitive), monitor the effect on fluorescence of sequential additions of Ca^{2+} (3 mM) and EGTA (5 mM).

C. Instrumentation

For fura-2 measurements, our laboratory is currently using a Leica DMIRB inverted microscope (Heidelberg, Germany) equipped with several high numerical aperture, oil-immersion objectives. Excitation is achieved with a Hg arc 100W lamp together with 340 nm and 380 nm bandpass filters in the filter wheel. The 340 nm and 380 nm bandpass filters have built-in neutral density filters that reduce transmittance by 80%. Excitation filters for fura-2 measurements are installed in a filter wheel so that they can be rapidly switched during the experiment. For measurements with fluo-3, the excitation filter is installed in either the filter wheel or the filter cube. Excitation light, alternately 340 nm and 380 nm, is deflected by a 400 nm dichromatic beam splitter to the objective and the sample. Emitted light from the sample that is longer than 400 nm passes back through the dichromatic beam splitter to a 505 nm bandpass emission filter. Emitted light is detected with a frame-transfer CCD camera with a 512 × 512 back-thinned EEV chip (Princeton Instruments, Trenton, NJ). The filter wheel and shutters are controlled by a LUDL box (LUDL Electronic Products, Ltd., Hawthorne, NY) interfaced with a Pentium II 300 imaging workstation purchased as an integrated system from Universal Imaging Corporation (West Chester, PA). Images are acquired and analyzed using Metafluor software (Universal Imaging Corporation). The stage is maintained at 37°C by a continuous flow of warm air from an air curtain heater (Sage Instruments, Orion Research Inc., Cambridge, MA). The microscope and camera reside on a vibration isolation table (Technical Manufacturing Corporation, Woburn, MA).

D. Image Acquisition

Set acquisition times and adjust illumination intensity so that the fluorescence intensities from the cells in both the 340 nm and 380 nm images are at least 2 to 3 times the background level and the ratio of the intensities is approximately 1. Because the intensity of fura-2 at 380 nm will decrease while the intensity at 340 nm will increase upon Ca^{2+}-binding, the starting intensities of each image must be adjusted to be higher than background, but less than the maximum range of the camera. That is, the dynamic range of the indicator should be situated within the dynamic range of the camera. Use neutral density filters in conjunction with short acquisition times to achieve the appropriate starting intensities and to lessen photodamage to the cells. Acquisition times of 100 to 500 msec permit the acquisition of pairs of images every 1 to 2 sec, a time scale that should be adequate to characterize the dynamics of most Ca^{2+} responses.

E. Calibration

Calibration measurements should be performed for each sample after experimental measurements have been completed and they should be performed on the same field of cells on which the experimental measurements were made. At the end of the experiment, add 10 μM ionomycin (Sigma) in Ca^{2+}-replete medium ($[Ca^{2+}]$~3 mM) to obtain the maximum ratio value; then, after the signal reaches a plateau (~5 sec), add 10 μM ionomycin in Ca^{2+}-free medium ([EGTA]~5 mM) to obtain the minimum ratio value. The following formula can be used to calculate the $[Ca^{2+}]_i$ for each cell in the field and for each time point prior to the calibration measurements:[9]

$$[Ca^{2+}]_i = K_d \times (R-R_{min})/(R_{max}-R) \times S_{380\text{-free}}/S_{380\text{-bound}}$$

where K_d is the effective dissociation constant of Ca^{2+} from fura-2 at 37°C (~224 nM), $S_{380\text{-free}}$ is the intensity of fura-2 in the Ca^{2+}-free state at 380 nm, and $S_{380\text{-bound}}$ is the intensity of fura-2 in the Ca^{2+}-bound state at 380 nm. R, R_{min}, and R_{max} are the measured ratios of a single cell at one time point, the same cell after ionomycin treatment in the presence of EGTA and the same cell after ionomycin treatment in the presence of Ca^{2+}, respectively. Determining R_{min} and R_{max} for each cell independently avoids errors that may arise from cell to cell variation.

F. Image Analysis

As for most quantification of fluorescence microscopy images, it is necessary to remove from the images contributions from background fluorescence prior to further manipulations. Background correction can be achieved in several ways. In one method, an image of cells that have not been loaded with indicator dye is acquired under identical conditions as the experimental sample. The fluorescence signal of this image comprises contributions from detector noise, stray light, and autofluorescence from the cells and medium. If you are working with a confluent monolayer of cells, it is sufficient to subtract the average fluorescence intensity per pixel of the background image from each pixel of the sample image. However, if you are working on widely-dispersed and/or motile cells, this method will underestimate the contribution of cell-associated background since much of the field from the background sample will be cell-free. In this case, the average fluorescence intensity per pixel for pixels within the boundaries of cells in the background image should be calculated and subtracted from the cells in the sample image. A simpler, yet less rigorous method for background correction of a field of widely dispersed cells involves calculating the average fluorescence per pixel of a cell-free region of the field. This value is then subtracted from each pixel of the object of interest; contributions from cell autofluorescence are not eliminated. Regardless of the method you choose for background correction, correct each image from each wavelength using identical criteria before calculating ratios. The ratio will reduce the effects of inhomogeneities in the detector and field illumination (Section III.B). In the case of single-wavelength experiments, these inhomogeneities can be corrected for by dividing the sample images by an image of a thin film of fluorescent dye.

Obtaining useful information from ratio images is dependent on the pixel-for-pixel registration of each of the source images. Misregistration can occur due to differences in the light path for each wavelength, stage drift, or movement of the cells between acquiring each ratio image. Systematic misregistration due to optics manifests itself as a shift in images that is easily identifiable because of its uniformity. That is, when a systematic misregistration occurs between two images in a ratio pair, every pixel from one image will be shifted by the same amount and in the same direction as compared to its corresponding pixel in the other image. Consistent shifts such as these can be corrected by image processing using simple geometric transformations. Random shifts resulting from stage or sample drift cannot easily be corrected and should be avoided.

Ratio imaging of motile cells presents additional challenges, as movement of cells during image acquisition is expected. Effects of misregistration of sequential ratio images are often most conspicuous at the edges of the cells. Thus, if a polarized motile cell locomotes between ratio image acquisitions, the ratios at the leading and trailing edges of the cell will not be valid and may manifest as an apparent gradient along the axis of the cell. This type of artifact can be minimized by acquiring images in rapid succession. Acquire images at equal intervals, then calculate the 340/380 ratio using, in turn, the 380 images preceding then following each 340 image. If significant movement has occurred on the time scale of the image acquisition, the 340/380 ratio at the edges of motile cells will be alternately high and low. It is important to use this technique to avoid specious conclusions about the subcellular spatial distribution of the $[Ca^{2+}]_i$ response in motile cells.

Rapid image acquisition, which is often necessary for the reasons discussed above, can result in images with poor signal to noise. Intensity spikes due to random noise can emerge as fluctuations in ratio values that give the ratio image a granular appearance. Inhomogeneities resulting from noise can be diminished by applying a Gaussian convolution filter to each image before generating the ratio image.[2] A Gaussian filter calculates a new intensity value for each pixel in an image taking the values of neighboring pixels into account. As the neighborhood of pixels used in the calculation is increased (e.g., from 8×8 to 16×16 pixels), the image will be further smoothed, but it will also become blurred. The size of the Gaussian filter should be chosen such that the reduction in noise outweighs the loss of spatial resolution.

Detecting local changes in $[Ca^{2+}]_i$ may require additional digital image processing. Regions of local high or low $[Ca^{2+}]_i$ can be made more apparent by representing ratio values as pseudocolors. Even after the application of a Gaussian filter, noise may be a significant contribution to the ratio values, especially in regions of the image with low signal such as at the edges of the cell. To de-emphasize the contribution of pixels with low signal, the pseudocolored image can be intensity weighted.[2,37] That is, the pseudocolored image can be multiplied by the background-corrected sum of the Gaussian-filtered 340 and 380 images. In the final image, different colors represent different ratio values, whereas different brightnesses represent different signal intensities from the original images. Brightness, therefore, gives an estimate of dye concentration and cell thickness. When analyzing images, it should be kept in mind that noise contributions are more pronounced in less bright regions of cells, so caution should be used when making conclusions about local $[Ca^{2+}]_i$ in these regions. Even for bright regions of cells, the significance of local variations in colors (i.e., $[Ca^{2+}]_i$) must be tested by comparing the image to that obtained of a homogeneous solution of indicator dye.[2]

V. Conclusion

Microscopy-based, single-cell $[Ca^{2+}]_i$ measurements provide a powerful method for elucidating features of $[Ca^{2+}]_i$ responses that are either masked or unobtainable when using other techniques. However, many unpredictable difficulties may be encountered in the design and interpretation of experiments to measure $[Ca^{2+}]_i$. Therefore, this chapter provides merely a framework for the design of $[Ca^{2+}]_i$ experiments, comprising discussions about selecting an indicator dye and instrument, developing a dye loading protocol, and quantifying images. Considering that every cell system is different and that it is not possible to foresee every problem, no general protocol can be presented that will work well in every situation. Consequently, these experiments call for an alert and critical investigator. Despite the numerous caveats associated with single-cell $[Ca^{2+}]_i$ experiments, many groups have successfully accomplished these studies and gained valuable information as a result. Continued advances in instrumentation and development of new indicator dyes hopefully will make single-cell $[Ca^{2+}]_i$ studies easier and even more insightful.

References

1. Tsien, R. Y., Intracellular signal transduction in four dimensions: from molecular design to physiology, *Am. J. Phys.*, 263, 723, 1992.
2. Marks, P. W. and Maxfield, F. R., Local and global changes in cytosolic free calcium in neutrophils during chemotaxis and phagocytosis, *Cell Calcium*, 11, 181, 1990.
3. Marks, P. W. and Maxfield, F. R., Transient increases in cytosolic free calcium appear to be required for the migration of adherent human neutrophils, *J. Cell Biol.*, 110, 43, 1990.
4. Mandeville, J. T. H., Ghosh, R. N., and Maxfield, F. R., Intracellular calcium levels correlate with speed and persistent forward motion in migrating neutrophils, *Biophys. J.*, 68, 1207, 1995.
5. Lawson, M. A. and Maxfield, F. R., Ca^{2+}- and calcineurin-dependent recycling of an integrin to the front of migrating neutrophils, *Nature*, 377, 75, 1995.
6. Salzman, N. H. and Maxfield, F. R., Fusion accessibility of endocytic compartments along the recycling and lysosomal endocytic pathways in intact cells, *J. Cell Biol.*, 109, 2097, 1989.
7. Kruskal, B. A. and Maxfield, F. R., Cytosolic free calcium increases before and oscillates during frustrated phagocytosis in macrophages, *J. Cell Biol.*, 105, 2685, 1987.
8. Carlsson, K., Aslund, N., Mossberg, K., and Philip, J., Simultaneous confocal recording of multiple fluorescent labels with improved channel separation, *J. Microsc.*, 176, 287, 1994.
9. Grynkiewicz, G., Poenie, M., and Tsien, R. Y., A new generation of Ca^{2+} indicators with greatly improved fluorescence properties, *J. Biol. Chem.*, 260, 3440, 1985.
10. Minta, A., Kao, J. P., and Tsien, R. Y., Fluorescent indicators for cytosolic calcium based on rhodamine and fluorescein chromophores, *J. Biol. Chem.*, 264, 8171, 1989.
11. Tsien, R. Y., A non-disruptive technique for loading calcium buffers and indicators into cells, *Nature*, 290, 527, 1981.

12. Gillot, I. and Whitaker, M., Calcium signals in and around the nucleus in sea urchin eggs, *Cell Calcium*, 16, 269, 1994.
13. Haugland, R. P., *Handbook of Fluorescent Probes and Research Chemicals*, Molecular Probes, Inc., Eugene, 1996.
14. Schild, D., Jung, A., and Schultens, H. A., Localization of calcium entry through calcium channels in olfactory receptor neurones using a laser scanning microscope and the calcium indicator dyes fluo-3 and fura-red, *Cell Calcium*, 15, 341, 1994.
15. Tsien, R. Y., Pozzan, T., and Rink, T. J., Calcium homeostasis in intact lymphocytes: cytoplasmic free calcium monitored with a new intracellularly trapped fluorescent indicator, *J. Cell Biol.*, 94, 325, 1982.
16. Tsien, R. Y. and Pozzan, T., Measurement of cytosolic free Ca^{2+} with quin2: practical aspects, *Methods Enzymol.*, 172, 230, 1989.
17. Vorndran, C., Minta, A., and Poenie, M., New fluorescent calcium indicators designed for cytosolic retention or measuring calcium near membranes, *Biophys. J.*, 69, 2112, 1995.
18. Etter, E. F., Minta, A., Poenie, M., and Fay, F. S., Near-membrane [Ca^{2+}] transients resolved using the Ca^{2+} indicator FFP18, *Proc. Natl. Acad. Sci. USA*, 93, 5368, 1996.
19. Di Virgilio, F., Fasolato, C., and Steinberg, T. H., Inhibitors of membrane transport system for organic anions block fura-2 excretion from PC-12 and N2A cells, *Biochem. J.*, 256, 959, 1988.
20. Di Virgilio, F., Steinberg, T. H., and Swanson, J. A., Fura-2 secretion and sequestration in macrophages. A blocker of organic anion transport reveals that these processes occur via a membrane transport system for organic anions, *J. Immunol.*, 140, 1988.
21. Kao, J. P., Practical aspects of measuring [Ca^{2+}] with fluorescent indicators, *Methods Cell Biol.*, 40, 155, 1994.
22. Yamashiro, D. J., Fluss, S. R., and Maxfield, F. R., Acidification of endocytic vesicles by an ATP-dependent proton pump, *J. Cell Biol.*, 97, 929, 1983.
23. Ratan, R. R., Shelanski, M. L., and Maxfield, F. R., Transition from metaphase to anaphase is accompanied by local changes in cytoplasmic free calcium in PtK2 kidney epithelial cells, *Proc. Natl. Acad. Sci. USA*, 83, 5136, 1986.
24. Di Virgilio, F., Steinberg, T. H., and Silverstein, S. C., Inhibition of fura-2 sequestration and secretion with organic anion transport blockers, *Cell Calcium*, 11, 57, 1990.
25. Marks, P. W. and Maxfield, F. R., Preparation of solutions with free calcium concentration in the nanomolar range using 1,2-bis(o-aminophenoxy)ethane-N,N,N′, N′-tetraacetic acid, *Anal. Biochem. (United States)*, 193, 61, 1991.
26. Kruskal, B. A., Keith, C. H., and Maxfield, F. R., Thyrotropin-releasing hormone-induced changes in intracellular [Ca^{2+}] measured by microspectrofluorometry on individual quin2-loaded cells, *J. Cell Biol.*, 99, 1167, 1984.
27. Uto, A., Arai, H., and Ogawa, Y., Reassessment of fura-2 and the ratio method for determination of intracellular Ca^{2+} concentrations, *Cell Calcium*, 12, 29, 1991.
28. Williams, D. A. and Fay, F. S., Intracellular calibration of the fluorescent Ca^{2+} indicator fura-2, *Cell Calcium*, 11, 75, 1990.
29. Hove-Madsen, L. and Bers, D. M., Indo-1 binding to protein in permeabilized ventricular myocytes alters its spectral and Ca^{2+} binding properties, *Biophys. J.*, 63, 89, 1992.

30. Malgaroli, A., Milani, D., Meldolesi, J., and Pozzan, T., Fura-2 measurement of cytosolic free Ca^{2+} in monolayers and suspensions of various types of animal cells, *J. Cell Biol.*, 105, 2145, 1988.
31. Wahl, M., Lucherini, M. J., and Gruenstein, E., Intracellular Ca^{2+} measurement with indo-1 in substrate-attached cells: Advantages and special considerations, *Cell Calcium*, 11, 487, 1990.
32. Dunn, K. and Maxfield, F. R., Ratio imaging instrumentation, *Methods Cell Biol.*, 56, 217, 1998.
33. Inoue, S. and Spring, K. R., *Video Microscopy, The Fundamentals*, Plenum Press, New York, 1997.
34. Rieder, C. L. and Cole, R. W., Perfusion chambers for high resolution video light microscopic studies of vertebrate cell monolayers: Some considerations and design, *Methods Cell Biol.*, 56, 253, 1998.
35. Marks, P. W., Hendey, B., and Maxfield, F. R., Attachment to fibronectin or vitronectin makes human neutrophil migration sensitive to alterations in cytosolic free calcium concentration, *J. Cell Biol.*, 112, 149, 1991.
36. Pytowski, B., Maxfield, F. R., and Michl, J., Fc and C3bi receptors and the differentiation antigen BH2-Ag are randomly distributed in the plasma membrane of locomoting neutrophils, *J. Cell Biol.*, 110, 661, 1990.
37. Tsien, R. Y. and Poenie, M., Fluorescence ratio imaging: a new window into intracellular ionic signaling, *Trends Biochem. Sci.*, 11, 450, 1986.

Chapter 20

Wnt Signaling in *Xenopus* Embryos

François Fagotto

Contents

I. Introduction 304
A. Generalities 304
B. A Summary of Early Patterning During *Xenopus* Development 305
C. Molecular Nature of the Inducers 306
D. The Wnt Pathway 309
II. Experimental Manipulations of Axis Formation 318
III. Testing Dorsalizing and Ventralizing Activities by Injection of mRNA 321
IV. Interfering With Endogenous Components: An Overview 327
V. Preparation of Embryo Extracts 330
VI. Subcellular Distribution: Microscopic and Biochemical Methods 331
VII. Detecting Molecular Interactions 339
VIII. Detecting GSK Activity 340
IX. Analyzing β-Catenin Stability 341
X. Detection of Gene Expression 342
XI. Concluding Remarks 345
A. mRNA Concentrations, Expression Levels, Activity Levels 345
B. This is Not a "Clean" System! Further Considerations on Expression/Overexpression Experiment 346
C. Taking Advantage of a Complex System: "Playing" on Different Backgrounds 347
Acknowledgments 347
References 348

0-8493-3385-7/99/$0.00+$.50

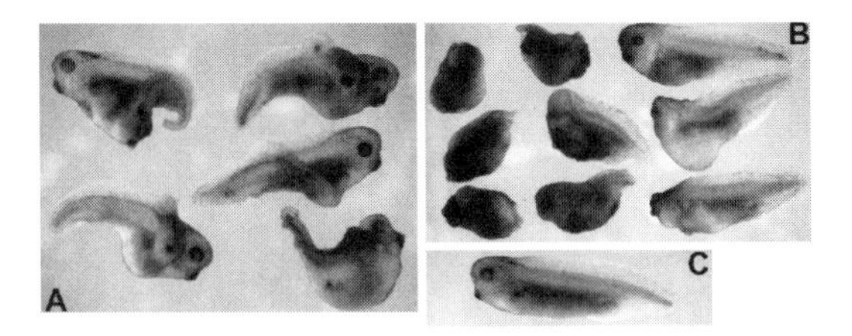

FIGURE 1

The Wnt pathway regulates axis formation in *Xenopus* embryos. A. Ectopic activation of the pathway (here by expression of an Axin dominant negative construct) in the ventral side of early cleaving embryos results in induction of a secondary axis. B. Inhibition of the endogenous pathway (here by overexpression of wild type Axin) in the dorsal side leads to loss of axial structures. A range of axial defects has been obtained in this experiment: most embryos were completely ventralized (no visible axis) (left), or almost completely ventralized (some trunk structures still present) (center). Some embryos had weaker phenotypes, with defects in head structures but a normal trunk (lower right), and a few were normal (upper right). A control, uninjected tadpole is shown in panel C.

I. Introduction

A. Generalities

The very same conserved signal transduction pathways active in adult differentiated cells also regulate embryonic development. Inhibition of a pathway, or its ectopic activation, will often perturb embryonic patterning and cause morphological abnormalities. Because such phenotypes can be readily observed and are highly reproducible, they constitute an exquisite read-out for the activity of the underlying signaling pathways.

One spectacular phenotype in *Xenopus* embryos is obtained by modulating the maternal Wnt pathway, involved in early specification of the dorsoventral axis: inhibition of the pathway leads to loss of body axis, while its ectopic activation causes formation of a complete second body axis (Figure 1). Several laboratories have exploited this system extensively to dissect the molecular mechanism of the Wnt pathway. *Xenopus* embryos are particularly well fitted for this purpose: straightforward mRNA injections allow rapid screens of mutants and coexpression of various molecules in various doses and combinations. Only a few of these large embryos are sufficient for analysis by standard biochemical/cell biological techniques, and molecular data can be directly related to functional analysis (phenotype read-out). Because the stage and localization of signaling activities is known, it is possible to test signaling molecules in various contexts (e.g., upstream signal “on” or “off”), and determine possible changes in molecular interactions, cellular distribution, post-translational modifications, and activity. Finally, despite the lack of genetics, and thus of real “knock-outs” in *Xenopus,* several methods are now available to deplete or inhibit specific molecules.

This chapter contains a short update on early induction in *Xenopus* and on the Wnt pathway, provides guidelines and strategies to characterize and analyze components of the Wnt pathway, and presents a selection of embryological, cell biological and molecular techniques useful for this purpose. However, the field is too large to be satisfactorily covered within this chapter, and readers will therefore be referred to previously published works both for basic protocols, and for some specialized techniques.

A wide spectrum of approaches is now available for the study of *Xenopus* embryos. However, some techniques are rather challenging when applied to embryos. Thus, this chapter will also discuss the problems and limitations of this model, and the various strategies that have been devised to circumvent them. To conclude with the obvious, there is no such thing as an ideal experimental system, but the most efficient progresses arise from combined approaches in various models. The recent history of unraveling the Wnt pathway is an excellent demonstration of this axiom.

B. A Summary of Early Patterning During *Xenopus* Development

1. Establishment of dorsoventral asymmetry: cortical rotation

The unfertilized *Xenopus* egg has a single axis of symmetry, running from the top of the pigmented animal hemisphere to the non-pigmented, yolk-rich vegetal hemisphere. The symmetry of this radial egg is disrupted shortly after fertilization, by a 30° rotation of the egg cortex relative to the inner cytoplasm. This rotation, which relies on microtubules, causes relocalization of a yet unidentified dorsal determinant.[1,2] Initially located at the vegetal pole, this dorsal determinant is shifted upward and spreads over a rather wide area on one side of the embryo, with a maximal density in the subequatorial region, about 60° from the vegetal pole (see Figure 4).

2. Mesoderm induction

As the embryo proceeds through successive cell divisions (cleavage) and becomes a hollow ball called the blastula, a signal(s) from the vegetal hemisphere induces the equatorial region to become mesoderm.[3,4] The top of the embryo, the "animal cap," is apparently out of reach of mesoderm-inducing signals, and develop into ectoderm. Mesoderm induction, like dorsoventral induction, relies on maternal factors. When zygotic transcription starts, at a stage called mid-blastula transition (MBT), the embryo is already clearly regionalized, with three layers of ectoderm, mesoderm, and endoderm, and a dorsoventral polarity. This regionalization can be detected at the molecular level by the localized expression of early region-specific genes, such as the mesodermal marker Xbra, or the dorsal-specific genes, Siamois, Xnr3 and Twin.

3. Dorsal induction

In the classical view of early patterning, a maternal signal from the dorsal vegetal region, called the "Nieuwkoop Center," induces the overlying mesoderm to become the Spemann Organizer.[3] Recent data have brought some nuances to this model: 1) the dorsal signal is not only present in the vegetal hemisphere, but in a wider area, which also includes the presumptive dorsal mesoderm and even some ectoderm.[2,5,6,7] Thus, the early "dorsal field" and the later Spemann Organizer largely overlap topographically. 2) Immediate targets of the maternal dorsal determinant, Siamois and Xnr3, are initially expressed over the whole dorsal field.[6] 3) Activation of these early "generic" dorsal genes is followed by subsequent regional inductions and formation of at least three functional domains: anterior endomesoderm (expressing Cerberus), head organizer (Goosecoid) and trunk/tail organizer (Xnot).[7,8] This regionalization of the dorsal side is likely to result from the synergistic action of the dorsalizing activity and of determinants/pathways present at various latitudes along the animal-vegetal axis.[6,9]

4. Dorsoventral patterning

The ventral fate had always been considered as a rather passive state, because most of the spectacular events, such as gastrulation, axial patterning and neurulation happen on the dorsal side. In fact, the ventral side expresses specific genes (BMPs, Wnt8, Xvents) and plays a crucial role in patterning during gastrulation.[7,10–12] Indeed, the antagonizing activities of ventralizing factors produced in the ventral side, and of their inhibitors secreted by the Spemann Organizer further refines dorsoventral patterning. Within the mesoderm, the Spemann Organizer itself will develop into head mesoderm and axial mesoderm (notochord). The ventro-lateral mesoderm will be divided into paraxial mesoderm (muscle), lateral mesoderm (pronephros), and ventral mesoderm (giving rise in particular to the heart).

C. Molecular Nature of the Inducers

Until recently, the role of various signaling pathways in early patterning in *Xenopus* has been very controversial, largely because overexpression of several different growth factors produced similar or overlapping phenotypes. Some confusion also arose from the use of some dominant-negative receptor constructs, which later turned out to display imperfect specificity. Although the nature of the endogenous secreted factors is still unknown, the role of the various pathways has now been quite satisfactorily clarified. Here is a brief synthesis of early molecular patterning as viewed by the author (Figure 2).

1. Mesoderm competence: FGF

The FGF pathway is clearly required for mesoderm formation.[13] However, FGF does not appear to act as the long-sought signal emanating from the vegetal hemisphere, but rather as a competence modifier, active in the animal/equatorial region.[14] Thus, mesoderm probably forms in the equatorial region through the combined influences of FGF and of the vegetal signal.[15,16]

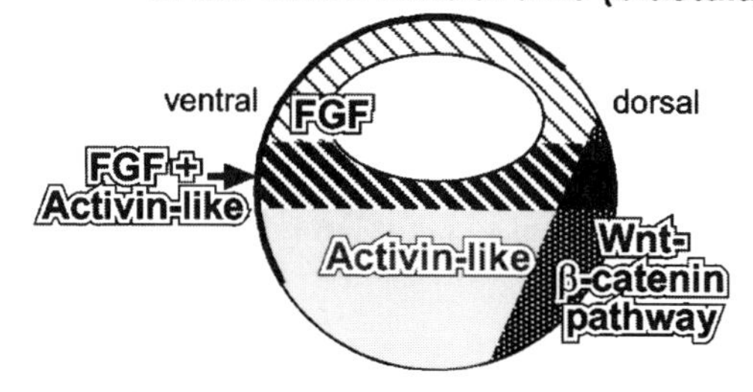

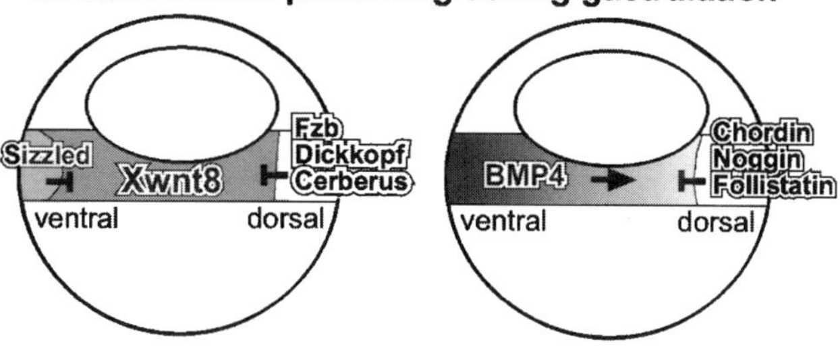

FIGURE 2

Early signaling in *Xenopus* embryos. A. At blastula stage, mesoderm is induced in the equatorial region, probably by a combination of FGF and TGFβ-like (Activin-like) signaling. The Wnt-β-catenin pathway is activated in the dorsal side, where it will synergize with mesoderm inducers to form the Spemann Organizer. B. During gastrulation, the mesoderm is further patterned through the antagonistic activities of two ventral factors, BMP4 and Xwnt8, and their inhibitors, secreted by the Spemann Organizer. The influence of Xwnt8 is also limited in the ventral side by a Wnt inhibitor, Sizzled. Not shown: neural induction and neural anterior-posterior patterning (also during gastrulation).

2. Mesoderm induction: Activin/Vg1

This vegetal factor is most certainly an Activin-like member of the TGFβ family.[17] Vg1 would be a good candidate:[18] Vg1 mRNA is maternally localized to the vegetal cortex, and released into the vegetal cytoplasm during oocyte maturation. Ectopic secreted Vg1 (produced as a BMP-Vg1 or Activin-Vg1 chimera) has strong mesoderm-inducing activity. However, processing of the endogenous Vg1 precursor in an active form has not been detected so far. It is also quite possible that not one, but a combination of two or more Activin-like factors is involved. Activin/Vg1 also appear to play a role in determination of the endoderm.[19]

3. Dorsal specification: the Wnt pathway

All available evidence clearly points to the Wnt pathway as responsible for dorso-ventral axis specification (reviewed in Reference 20): 1) Activators of the pathway show strong axis-inducing activity. 2) Inhibitors of the pathway and dominant negative constructs of activators block axis formation. 3) Maternal β-catenin is required for axis induction.[21] 4) Nuclear β-catenin accumulates in the region of dorsal activity during late blastula stages.[5] 5) Early zygotic dorsal genes (Siamois, Twin, Xnr3) are direct and specific targets of the Wnt pathway. 6) Siamois is required for axis formation.[22,23] 7) Siamois and Xnr3 are activated by the vegetal cytoplasm containing the dorsal determinant.[6] Note that the Wnt pathway dorsalizes mesoderm, but cannot induce mesoderm. Formation of the Spemann Organizer thus requires a synergistic action of mesoderm inducers and of the dorsal Wnt pathway.[6,9] Attempts to identify the endogenous Wnt ligand have failed so far. It has been proposed that the pathway might in fact be activated intracellularly, in the absence of Wnt ligand.[24]

4. A role for Activin/Vg1 in dorsal specification?

In alternative models, it had been initially proposed that a combination of ventral (FGF) and dorsal (Activin/Vg1) mesoderm inducers patterned the early embryo.[25,26] Activin/Vg1 would then act both as mesoderm inducer and dorsal determinant: indeed, Activin and Vg1 can induce ventral mesoderm at low concentrations, and dorsal mesoderm at higher concentrations. They can also induce secondary axes. Recent data, however, do not support such models, but are more consistent with a synergy between mesoderm inducers and a dorsalizing Wnt pathway. In particular, activation of dorsal genes such as Goosecoid clearly requires both the Activin- and the Wnt-pathways.[27] Surely, high (non-physiological?) levels of Activin can bypass the requirement for Wnts. However, the endogenous Activin-like activity is distributed over the whole vegetal hemisphere, at similar levels in ventral and dorsal cells.[27]

5. Dorsoventral patterning of the mesoderm: BMPs, zygotic Wnt8, and antagonists

During gastrulation, dorsoventral patterning is further refined by the antagonistic action of ventral factors and their inhibitors (Figure 2B): BMPs are strong ventralizers, with a graded, ventro-lateral distribution. They are antagonized by soluble factors secreted by the organizer, Chordin, Noggin, and Follistatin, which directly bind BMPs.[11] Wnt8 is expressed ventrally and laterally, with a rather sharp boundary along the Organizer, from which it is excluded. Unlike BMPs, Wnt8 is not so much involved in ventralization, but rather in determination of the lateral mesoderm,[28,29] where its activity is restricted: the dorsal mesoderm secretes several Wnt-inhibitors, in particular Frzbs,[30] which are homologous to the extracellular domain of the Wnt-receptors, Frizzled. Wnt8 is also probably inhibited ventrally by Sizzled (another secreted Wnt-inhibitor).[31]

6. Later patterning processes

Description of other inducing events is beyond the scope of this summary. Note that Wnts are involved in many later aspects of development. In particular, Wnt1 has an important role in patterning of the midbrain, and Wnt3a is involved in dorso-ventral patterning of the neural tube. Later inductions are not easily accessible to experimental manipulations in *Xenopus,* but other well-established models are available; for instance, the chick for the role of Wnts in limb patterning or the mouse for the study of Wnts in kidney morphogenesis.

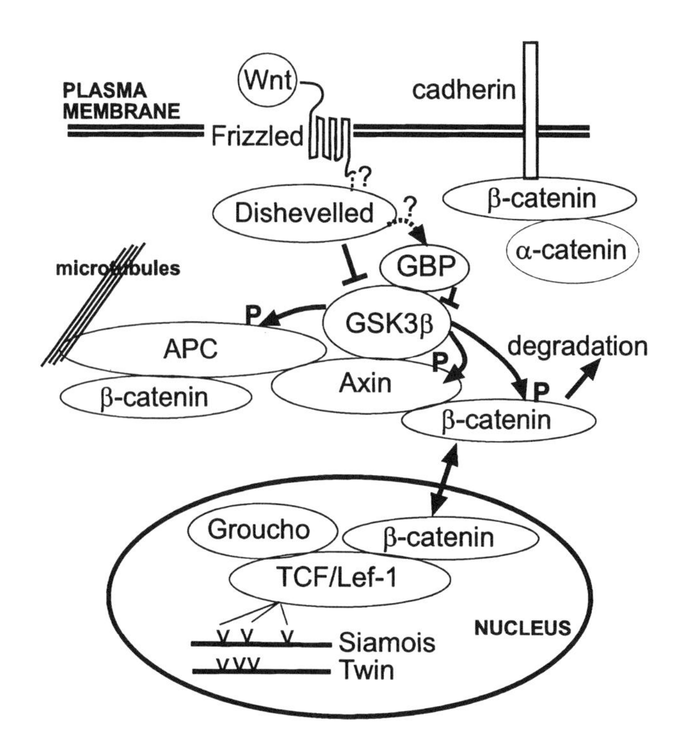

FIGURE 3
The Wnt pathway. Direct contacts between components represent direct biochemical interactions. Indirect evidence suggests a possible interaction between Frizzled and Disheveled (?). Note that all the components of the Axin-based complex (Axin, APC, GSK, β-catenin) may potentially (but not necessarily) bind simultaneously. However, β-catenin interactions with cadherins and APC are mutually exclusive, and the same is probably true for binding to Axin and TFC. (P) represents phosphorylation targets of GSK.

D. The Wnt Pathway

Figure 3 and Table 1 present the components of the Wnt pathway and summarize their interactions (see also References 20 and 24).

1. Upstream activators: Wnts, Frizzled, Dishevelled, GBP

Wnts are short-range secreted factors (they remain mostly bound to the surface of secretory cells[32]). Wnts probably bind directly to Frizzled (Fz) receptors,[33,34] which in some way activate Dishevelled. Some Fzs contain a C-terminal PDZ-binding consensus sequence that may interact directly with the PDZ domain of Dishevelled. Dishevelled negatively regulates the serine/threonine kinase Glycogen Synthase Kinase-3β (GSK3β). It may also be involved in other pathways (Notch,[35] Wnt5A pathway, see below). A novel GSK inhibitor, GBP (GSK-binding protein) is required for axis formation.[36] It is presently not known whether GBP can be activated by the upstream Wnt pathway, or whether it is part of an alternative, Wnt-independent alternative pathway. Note that some evidence suggests that in early *Xenopus*

TABLE 1
Properties of the components of the Wnt pathway

A. Activators

Activators	mRNA Doses for Axis Induction	Molecular Properties	Domains	Interactions	Post-Translational Modifications	Cellular Distribution	Tissue Distribution	Homologues	Refs.
Xwnt1, 2B, 3A, 8, 8B Xwnt11 (partial)[a]	1-20 pg	35-50kD Secreted		Frizzled (receptors) Frzb (extracellular inhibitors) ECM		Secreted, largely cell surface (ECM)-associated. Internalized (endocytosis) by receptor cells.[121]	Maternal: Xwnt8B: animal Xwnt11: vegetal Zygotic: Xwnt8: lateral mesoderm Xwnt1, 3A, 8B: neural Xwnt11: somites	Wingless (D), MOM-2, lin-44 (Ce) Many Wnts in vertebrates	122, 123, 124
Frizzled[b] (Xfz8, Rfz1)	0.25-1 ng	~60kD 7-TM receptor	CRD ⟶ PDZ-binding (X-S-X-V) ⟶	Wnts (specificity for various Wnts unclear) Dsh??			Xfz8: zygotic (Organizer)	Fz, Dfz2 (D), MOM-5, lin-17 (Ce) Many Fz in vertebrates	34, 125

Xenopus dishevelled	0.4-1.5 ng	70kD Cytoplasmic	PDZ DEP DIX	?? PDZ essential	HyperPated by wg (D)	Cytoplasmic Recruited at the membrane by Fz	Ubiquitous	Dsh (D), Dvl-1 (M)	34 68
GBP	0.25-1 ng	19kD	GSK-binding	→ GSK	Pated		Ubiquitous	Frat1 (M, H), Frat2 (H)	36
β-catenin	0.25-1 ng	92kD Cytoplasmic	12 central armadillo (42 aa) repeats	armadillo repeats: Cadherins APC TCF/Lef-1 Axin Others: α-catenin site not determined: Nucleopore, Fascin, EGF-R	GSK3β-dependent S/T-Pated (N-terminus) Wnt-independent Y-Pated	Membrane-associated (cadherins), Cytoskeleton-associated (APC, fascin), Cytosolic and Nuclear (Wnt signaling)	Ubiquitous abundant at sites of active morphogenesis (e.g., involuting mesoderm) and at embryonic boundaries	Armadillo (D), WRM-1 (Ce) Plakoglobin is a close vertebrate homologue mainly localized in desmosomes.	5 87, 105, 108, 115, 126, 127
Siamois	50-100 pg	27kD Transcription factor	Homeobox					Twin (X)	54, 128, 129

TABLE 1 (continued)
Properties of the components of the Wnt pathway

B. Negative regulators

Negative Regulators	mRNA Doses for Ventralization	Molecular Properties	Domains	Interactions	Post-Translational Modifications	Cellular Distribution	Tissue Distribution	Homologues	Refs.
XGSK-3β	2x 2-5 ng	~46kD Cytoplasmic S/T-kinase		Axin Substrates: Axin, APC, β-catenin, Glycogen Synthase, Tau	Inactivated when S-Pated (role in Wnt signaling?) Inactivated when Y-dePated ?		Ubiquitous	GSK3β (vertebrates), zeste white-3/shaggy (D), GSK-A (Di)	115, 130, 131, 132
Axin[c]	2x 1-2 ng	105/110 kD (spliced forms) Cytoplasmic	RGS-like → DIX Dimerization	APC GSK3β, β-catenin, PP2A	S/T-Pated by GSK3β	NOT cytosolic. Intracellular structures and PM-bound[c]	Ubiquitous[d]	Axin (M, R, H, C) Axin-2 (conductin, H; Axil, R)	37, 38, 39, 40, 41, 42, 43

C-cadherin[e]	2x 3-5 ng	120 kD Membrane Ca-dep. adhesion molecule	EC: 5-cadherin repeats → → Proximal IC → Distal IC →	Lateral dimers Homotypic binding p120 β-catenin plakoglobin	Basolateral in epithelia	Ubiquitous until gastrulation (maternal), then present in mesoderm (muscle)	Many cadherins in vertebrates/ invertebrates Often tissue-specific	21, 108, 114
XGrg-4	2x 1.5 ng			XTCF-3 Chromatin	Nuclear	Maternal	Groucho (D) Grg-1,3,4 (M) TLE-1 to 4 (H) XGrg-5[f] (X)	56
Frzb (Frizbees)	25-400 pg	~ 40 kD Secreted	Fz CRD	Wnts		Zygotic: Spemann Organizer		88, 89
Sizzled		Secreted				Zygotic: Ventral mesoderm		31

TABLE 1 (continued)
Properties of the components of the Wnt pathway

C. Components with unclear function

Activators/ Repressors ???	mRNA Doses for Activity	Molecular Properties	Domains	Interactions	Post-Translational Modifications	Cellular Distribution	Tissue Distribution	Homologues	Refs.
XAPC[g]	10 ng (axis induction)	310 kD cytoplasmic	Dimerization Arm repeats 15-aa-repeats → 20-aa-repeats → VAMP → basic →	 β-catenin β-catenin Axin microtubules	Pated by GSK3β Multiple forms on gel (splicing, processing, or degradation?)		Ubiquitous	APC (H) DACP (D) APR-1 (Ce)	46
XTCF3	No effect on axis Weak repressor of Siamois	70 kD	N-term → Grg-binding → HMG-box → C-term	β-catenin Xgrg DNA		nuclear	Ubiquitous	TCF1,3,4, Lef-1 (vertebrates) DTCF/ pangolin (D) POP-1 (Ce)	52, 53
Lef-1	50-500pg (axis induction)	55 kD	Similar to XTCF, but shorter C-term	β-catenin DNA			Not present maternally		133, 134

Abbreviations and symbols: CRD, Cystein-Rich extracellular Domain; EC, extracellular domain; IC, intracellular domain; ECM, extracellular matrix; Pated, phosphorylated; PM, plasma membrane; PP2A, protein phosphatase 2A; 7-TM, seven-transmembrane. Species: C, Chicken; Ce, C. elegans; D, Drosophila; Di, Dictyostelium; H, human; M, mouse; R, rat; X, Xenopus. Arrows indicate the established role for some domains in the corresponding interaction.

Notes:

[a] No axis induction by Xwnt5A, Xwnt4. Expression of Xwnt5A, 4 and 11 produce shortened, kinked embryos.[66]

[b] No maternal, axis-inducing Xenopus Frizzled has been reported so far. Xfz8 is only expressed zygotically.[125] Xfz3 and Xfz7 produce a Wnt5A phenotype.[67,69]

[c] No full length Xenopus Axin sequence published so far. All data in Xenopus obtained with heterologous (mouse) Axin.

[d] In mouse.

[e] Also called EP-cadherin. Major maternal cadherin in Xenopus. 94% identical to a minor maternal cadherin, XBU-cadherin.

[f] Long forms of Grg (Xgrg-4) act as repressors, while short forms (Xgrg-5) have the opposite effect.[56]

[g] APC is considered as negative regulator of β-catenin in mammalian cell lines.[44] In C. elegans and in Xenopus, it acts as activator of β-catenin.[45,46]

embryos, the pathway may be activated *downstream* of Dishevelled, independently of Wnts.[24]

2. Regulation of β-catenin: GSK, Axin, APC

GSK3β is constitutively active in the absence of Wnt signal, and has multiple targets, including β-catenin, APC, and Axin along the Wnt pathway. Axin is a negative regulator of the pathway. It stimulates β-catenin phosphorylation by GSK and β-catenin degradation via the ubiquitin pathway.[37-43] APC is a tumor-suppressor protein with an obscure and controversial role in this pathway: in some cell lines, it appears to downregulate β-catenin by promoting its degradation.[44] In C. elegans, it acts as a positive regulator of β-catenin, in parallel rather than along the Wnt pathway.[45] In *Xenopus*, its overexpression also activates β-catenin signaling.[46] APC binds to microtubules,[47,48] and a possible function in cell migration has been hypothesized.[49]

3. Multiple functions of β-catenin

β-catenin certainly has two, and possibly more functions:[24,50] 1) it is an essential component of the cadherin cell-cell adhesion complex, and 2) while cadherin-bound β-catenin appears to be rather stable, soluble β-catenin is rapidly degraded (through the action of GSK and Axin). Activation of the Wnt pathway induces a stabilization of soluble β-catenin, which then accumulates in the nucleus and activates downstream genes. 3) β-catenin may also have additional activities through its interactions with APC and with fascin,[51] which are associated, respectively, with microtubules and actin filaments, and could thus regulate cell architecture/motility.

4. Regulation of gene targets: β-catenin, TCF, Groucho

β-catenin can directly interact with HMG-box transcription factors of the TCF/Lef-1 family, including the maternal XTCF3.[52] Multiple XTCF3 sites have been found in each promoter of the three target genes, Siamois, Twin, and Xnr3.[53–55] In the absence of β-catenin, XTCF3 acts as a repressor.[53] This repression appears to be mediated by Xgrg-4, a member of the Groucho family of transcriptional repressors, which can bind directly to XTCF3.[56] On the contrary, the target genes are activated in the presence of β-catenin. It has been thus proposed that β-catenin is a transcriptional co-activator when associated with TCF (reviewed in Reference 57). In particular, the C-terminus of Armadillo (Drosophila β-catenin) behaves as a transactivation domain.[58] In this model, both β-catenin and TCF would be required for activation of downstream genes. Note, however, that in C. elegans, loss of POP-1, the TCF homologue, causes activation of the pathway.[59–61]

5. An alternative model for β-catenin signaling activity

Analysis of the mechanism of β-catenin nuclear import has revealed striking similarities between β-catenin and the nuclear transporters of the importin β family.[62] In fact, it appears that β-catenin is capable to mediate *export* of XTCF.[63] These data suggest an alternative model, in which TCF would act as a constitutive repressor, and Wnt-signaling would de-repress its target genes through β-catenin-mediated nuclear export of TCF. Note that this model and the co-activation model may not be mutually exclusive. More generally, it is quite possible that β-catenin might regulate either the import or the export of specific nuclear factors in a signal-dependent manner.

6. Wnt inhibitors

Frzbs and Sizzled are novel secreted molecules, which appear to directly inhibit Wnts.[30,31] Indeed, they have a region of high homology with the extracellular, cystein-rich domain of the Frizzled receptors, and can bind Wnts and prevent their interaction with their receptors. Dickkopf and Cerberus are secreted by the head mesoderm, and appear to inhibit the Wnt pathway, but the mechanism is still unknown.[10,64,65]

7. "Heterodox" Wnts and Frizzled: The "Wnt5A pathway"

Expression of Wnt5A and some other related Wnts does not induce a secondary axis, but produces instead bent embryos with shortened axes (and some head, neural tube, and tail defects).[66] Late gastrulation movements (axis elongation) are inhibited in Wnt5A-overexpressing embryos. Identical or similar phenotypes are caused by overexpression of Xfz3 and Xfz7, and by a dominant negative Xdsh (deletion of the PDZ domain).[67–69]

It is believed that Wnt5A triggers a signaling cascade completely different from the "classical" Wnt1/β-catenin pathway. This "Wnt5A pathway" appears to involve specific Fz receptors,[69,70] and to "share" Xdsh with the "classical" pathway. It has been observed that the Wnt5A pathway causes an elevation in intracellular Calcium,[70] a process mediated through G proteins. Siamois and Xnr3 expressions are not affected.[69] This pathway may well correspond to the Frizzled-1 pathway in *Drosophila,* which is wingless (Wnt1 homologue) and armadillo (β-catenin) independent, but requires Dsh as well.[71]

It remains puzzling that, although partial rescue of the dominant negative Xdsh effect by wild-type Xdsh has been reported,[68,69] Xdsh is unable to rescue the effect of Wnt5A and of the Frizzled. It will be important to firmly establish the specificity of all the aspects of the "Wnt5A" phenotype. Defects in gastrulation movements have been often considered in the past (rightly or wrongly?) as non-specific effects. Furthermore, bent axes are obtained not only with full length Xfz3, but also with severely truncated mutants, and even with high amounts of control β-galactosidase mRNA. It should also be mentioned that Wnt5A seems to behave differently in the presence of other Frizzled. It can indeed synergize with human Fz5 to activate the "classical" pathway and induce a secondary axis.[72] Obviously, the function(s) of Wnt5A remains to be clarified.

II. Experimental Manipulations of Axis Formation

Basic techniques for raising frogs, obtaining eggs and fertilizing them for *in vitro* transcription and mRNA injections are described in Kay and Peng.[73] Developmental stages are determined according to Nieuwkoop and Faber.[74]

In their original experiments, Spemann and Mangold[75] showed that a dorsal blastoporal lip grafted in the ventral side of a donor embryo induced the formation of an ectopic, secondary axis. Secondary axes can also be obtained by grafting dorsal vegetal cells,[3] by injecting cytoplasm obtained from the dorsal region of fertilized embryos,[76,77] or by transplantation of pieces of dorsal cortex.[78] Similar grafts and injections can rescue an axis in UV-irradiated, ventralized embryos. Grafts and injections of cytoplasm are rather challenging experiments. They will not be described here, but detailed protocols are available in the original publications.

Dorsoventral can also be manipulated by very simple treatments: normal axis induction can be inhibited (ventralization) by UV irradiation before the first cleavage. On the contrary, dorsal induction can be extended to the whole embryo (dorsalization) by treatment with D_2O (not described here, see Reference 79), or with Lithium ions.

1. Ventralization by UV irradation

Dorsoventral polarity is established by a microtubule-dependent rotation of the egg cortex (see Introduction). UV-irradiation is a simple and effective method to prevent cortical rotation through depolymerization of the cortical microtubules.[80]

When rotation is completely blocked, embryos develop as balls or cylinders of ventral tissues, lacking any anterior and dorsal structures. The first sign of ventralization can already be detected during gastrulation, by the absence of dorsal blastoporal, replaced by a late, weaker ventral lip appearing simultaneously all around the vegetal cells. Early dorsal markers such as Chordin (but not Siamois and Xnr3, see below), are inhibited. Incomplete ventralization is obtained with sub-optimal UV doses, or if cortical rotation has already started at the time of irradiation. A whole range of defects can then be observed, from reduction of the most anterior structures (forehead, cement gland, eyes) to progressive loss of anterior, then dorsal structures (notochord, then somites and neural tube, ...).

Note that the dorsal determinant remains intact at the vegetal pole of UV-irradiated embryos,[81] and is able to locally activate the Wnt pathway and turn on direct target genes.[5,6,82] However, it is unable to trigger further downstream events,[6] which require additional determinants/pathways active near the equator of the embryo. In fact, a normal axis can be rescued simply by tilting UV-irradiated eggs on the side, allowing the dorsal determinant(s) to be relocalized near the equator by gravity.[83]

UV irradiation

Fertilize eggs
Wait for visible activation (15 to 20 min)[a]
Immediately remove jelly coat using 2.5% cystein, pH #8
Wash rapidly, but extensively, in 0.1 × MMR
Immediately[b] place batches of embryos in an UV-transparent dish[c] above a UV source[d] and irradiate the vegetal pole
Incubate irradiated embryos in 0.1 × MMR. Discard abnormal embryos.

[a] Fertilized eggs rotate according to gravity, pigmented animal pole up.

[b] Irradiation must be completed before the onset of cortical rotation, i.e., 30 min post-fertilization at room temperature.

[c] Can use large quartz cuvettes or dishes, or even handmade dishes with a bottom made of "Saran" paper.

[d] Distance from the source, intensity and length of irradiation (typically 1 to 2 min) should be carefully adjusted for each UV source. I use an UV Stratalinker (Stratagene) placed upside-down (light bulbs at the bottom). When the dish is immediately placed on top of a bulb, an irradiation equivalent to 90 mjoules is optimal. Excessive irradiation is lethal. In particular, it inhibits cleavage. Optimal conditions should produce ventralized embryos with an average DAI < 0.5 (see below). Note, however, that, even in the best cases, a few embryos generally "escape" ventralization and develop with a more or less complete axis. UV-ventralized embryos tend to be more fragile than normal embryos, and often poorly withstand further experimental manipulations such as injections.

2. Dorsalization by LiCl treatment

When early cleaving embryos are incubated in the presence of high LiCl, the Wnt pathway becomes activated over the whole embryo,[5] and a radial Organizer is formed around the equatorial ring.[84] Mildly dorsalized embryos develop with larger heads and reduced trunk and tail. Completely dorsalized embryos develop as huge heads, radially symmetrical, with multiple eyes and cement glands, without any ventral/posterior structures. Lithium is a well-known modulator of the inositol phosphate pathway, but also a potent inhibitor of GSK3. Inhibition of GSK appears to be the major target of Lithium in axis formation.[85,86]

Lithiumn treatment

Incubate stage 32 to 64[a] cell embryos for 10 to 20 min in 0.3 M LiCl[b] in 0.1 × MMR
Rinse extensively in 0.1 × MMR.

[a] Optimal dorsalization between 16- and 64-cell stage. At later stages, the effect is much weaker. After midblastula transition (stage 8½), LiCl treatment causes partial ventralization, through activation of the zygotic Wnt8 pathway.

[b] These conditions produce maximally dorsalized embryos (DAI 8-10). Longer treatments are toxic. Milder dorsalization can be achieved by lowering LiCl concentration (0.1 to 0.2 M).

3. Quantitation of ventralized/dorsalized phenotypes: the dorsoanterior index

The external morphology of UV-irradiated and LiCl-treated embryos (analyzed at late tailbud-early tadpole stages) provides an easy, very reliable criteria to estimate the degree of dorsal induction. It tightly correlates with internal criteria (such as the presence of neural tissues, notochord, muscle), and with the dosage of early dorsal markers. Using the graded phenotypes of UV- and lithium-treated embryos as "standards," a scale for dorsal induction, the dorsoanterior index (DAI), has been established,[79,84] which is widely used by *Xenopus* embryologists. Index 0 corresponds to completely ventralized, 5 to normal, and 10 to fully dorsalized embryos. Examples of normal and ventralized embryos are presented in Figure 1.

Note that the DAI has been established for a very precise process, i.e., the variation of the amount of endogenous dorsal "Organizer" activity, and should be applied specifically in this context. For instance, when the zygotic Wnt8 pathway is ectopically activated in the dorsal side, embryos develop without head and notochord, thus resembling partially ventralized embryos (DAI ~2).[29] Nevertheless, the real phenotype is clearly different: the paraxial somitic mesoderm (a dorso-lateral, not ventral tissue) has expanded at the expenses of the notochord (see Section III), but the total balance between ventral and dorsal tissues is not significantly changed, unlike in UV-irradiated embryos. More complex phenotypes can also be obtained, for instance when the maternal and the zygotic Wnt8 pathways are both partially affected. It is therefore important to firmly establish in each case the legitimacy of using the DAI.

4. Quantitation of secondary axes

Axis duplication can be sometimes detected as early as gastrulation, by the appearance of a second blastoporal lip, and secondary axes are often clearly visible during neurulation. However, axis duplication is best scored at late neurula–early tailbud stages, when the extent of the process can be appreciated through morphological signs of differentiation along the body axis. I routinely score four categories of secondary axes using simple criteria: "complete" (cement gland), "partial" (no cement gland), "vestigial" (short posterior protrusions, or small pigment area), and "normal." "Partial" can be further subdivided in "long, almost complete" and "short" (say, less then half of the body length). However, the use of more detailed criteria (e.g., a bona fide complete axis should have well-formed eyes) is difficult at these early stages, and usually superfluous. In many instances, the frequency of axis duplication, irrespective of the completeness of the axes, is sufficient (e.g., screening expression libraries or panels of deletion mutants). Percentage of axis duplication is then calculated as the number of "complete" + "partial" axes divided by the total number of scored (surviving) embryos × 100. "Vestigial" axes can hardly be considered as real axes. I do not usually include embryos with gastrulation defects in the final score.

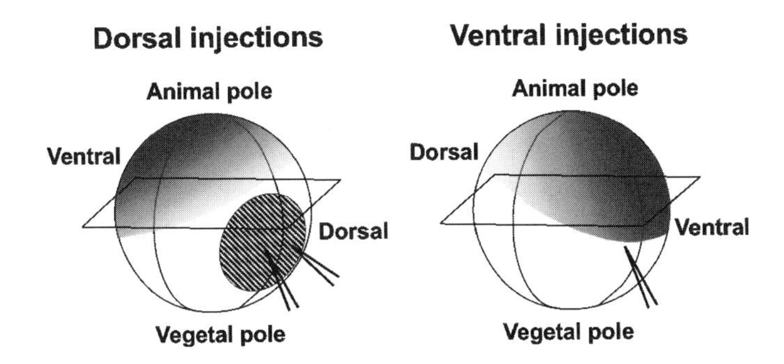

FIGURE 4
Targeting injections to the dorsal side (inhibition of the endogenous dorsalizing center) or to the ventral side (axis duplication). Dorsal and ventral sides are identifiable by the difference in pigmentation, a consequence of cortical rotation. The hatched area represents the dorsal, subequatorial region containing the endogenous Wnt-like dorsalizing activity.

III. Testing Dorsalizing and Ventralizing Activities by Injection of mRNA

Two main assays are used to study the effect of exogenous molecules on axis induction: dorsal injections in attempts to inhibit normal axis formation and ventral injections to induce an ectopic axis (Figure 4). Alternatively, dorsalizing activities can be tested for their ability to rescue an axis in UV-irradiated, ventralized embryos.*

For successful inhibition of axis formation, the dorsalizing field must be targeted as precisely and completely as possible. This is best achieved by injecting the subequatorial region of the two dorsal blastomeres of 4-cell embryos (Figure 4A). At the 8-cell stage, injections in the two dorsal-vegetal blastomeres usually result in only partial ventralization, probably because some dorsalizing activity is present in the animal blastomeres as well.

Ventral injections are usually performed subequatorially in one ventral blastomere at the 4 to 8-cell stage, which is the optimal site for induction of a secondary axis. However, the Wnt pathway and its direct target genes can be activated in both animal and vegetal cells.

This section presents guidelines to determine the effect of a putative modulator of the Wnt pathway (called factor "X"). See examples in References 37 and 87.

* Note that lower doses of activators are required for axis rescue in UV-ventralized embryos compared to induction of a secondary axis in normal embryos (I observed about a twofold difference in mRNA amounts for various factors such as Wnt8 and Noggin, unpublished). One can imagine two reasons for this difference. 1) In UV-irradiated embryos, the endogenous dorsal determinant is still active at the vegetal pole. While unable to induce an axis by itself, it might supplement the activity of an exogenous factor. 2) It is also conceivable that the dorsal pathway may become actively repressed in the ventral region of normal embryos, thus increasing the threshold required for induction of an ectopic axis.

1. Dorsalization

1a. Can "X" activate the Wnt pathway?

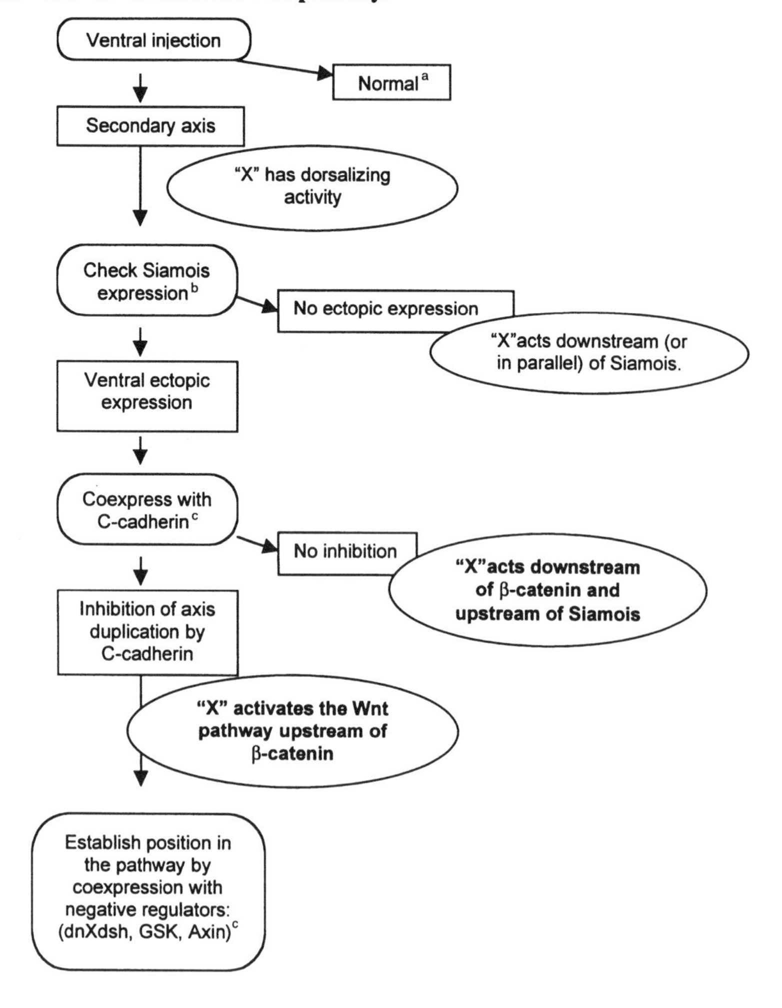

Comments:

[a] It is important to test a range of mRNA concentrations. Too high expression may also fail to induce axes in some cases (either due to toxicity, or because the process requires a precise ratio between the inducer and other endogenous components). This is, for instance, the case of a dominant negative (ΔRGS) Axin mutant (unpublished).

[b] By RT-PCR on dissected ventral halves (see Section X).

[c] A careful titration of the various mRNAs is absolutely essential for these coexpression experiments (see next paragraph)

1b. Titration of dorsalizing/ventralizing factors for epistasis experiments

1. Determine the minimal amount of "X" mRNA capable of inducing secondary axes at high frequency (typically >80% secondary axes, including a large number of complete axes).
2. Titrate similarly upstream (Xwnt8, Xdsh) and downstream (Siamois) dorsalizing components of the pathway.[a]
3. Determine the concentration of each ventralizing factor (dnXdsh, GSK, Axin, C-cadherin) sufficient to efficiently block axis duplication by an upstream component, but not by Siamois.[b] This criterion is important to verify that the observed ventralization is not a trivial consequence of toxicity (high levels of exogenous mRNA/protein). Use the determined doses for all subsequent experiments.

3b. Alternatively, determine the concentration of ventralizing factors that produces strong (DAI <2), but *specific* (i.e., rescued by Siamois) ventralization when injected in each of the two dorsal blastomeres at 4-cell stage.

4. Test factor "X" at the dose determined in (1) coexpressed with various ventralizing factors.[c]

[a] Typical mRNA doses are given in Tables 1 and 2. However, the specific activity can vary significantly for different mRNA constructs, depending on various parameters (e.g., untranslated regions, mRNA stability, epitope tags), and for different mRNA preparations (purity, ratio of capped mRNA).

[b] For instance, relatively high amounts of C-cadherin mRNA are required to inhibit axis induction by Xwnt8 and β-catenin (I use 3 to 5 ng mRNA), while 1 to 2 ng Axin mRNA are sufficient to inhibit axis induction by Wnt or Xdsh. In both cases, the same doses do not affect induction by Siamois.

[c] Because most components of the pathway interact directly with each other, it may be difficult in some cases to obtain a precise epistasis, even after careful titration. Obviously, the best confirmation of epistatic data comes from the understanding of the underlying biochemical processes.

2. Ventralization

2a. Can "X" inhibit axis formation?

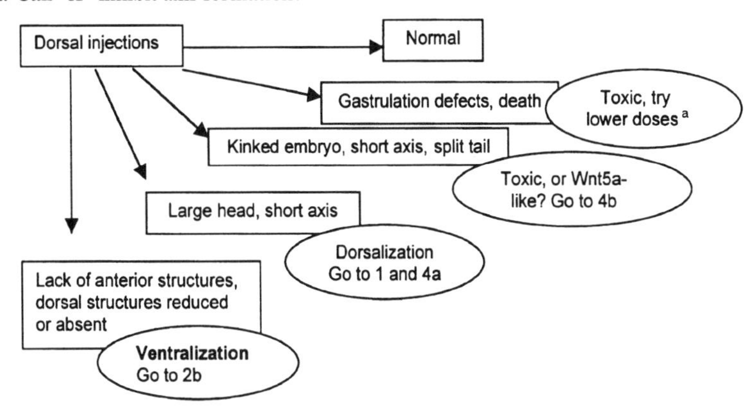

[a] 5 to 10 ng of mRNA is usually considered as the upper "safe" limit, but some constructs are toxic at lower doses.

2b. Does "X" inhibit the Wnt pathway?

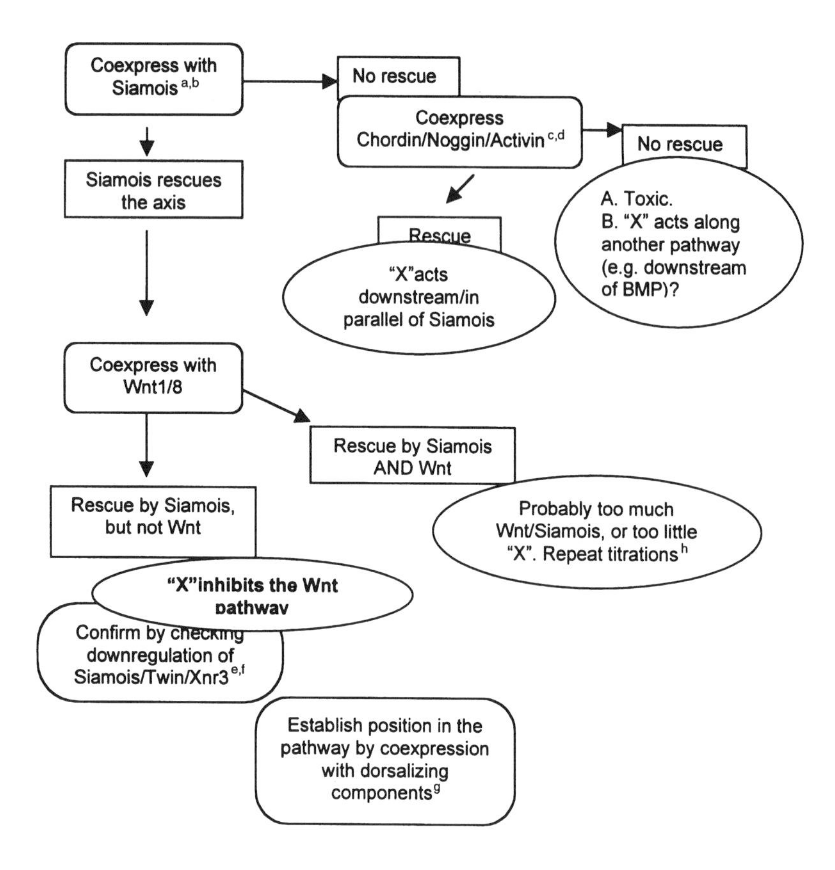

Comments:

[a] It is absolutely essential to carefully titrate the various mRNAs used for coinjections (see 1b).

[b] "X" mRNA is coexpressed with each of the various dorsalizing factors in the two dorsal blastomeres, and DAI is scored. Alternatively, injections can be performed in one ventral blastomere, and the capacity of each dorsalizing factor to induce an axis, in the presence and in the absence of "X," is determined.

[c] Activin/Vg1 can induce dorsal mesoderm directly, in the absence of Wnt signal (see above). Noggin and Chordin can dorsalize mesoderm downstream of the primary, Wnt-mediated induction.

[d] These factors do not induce complete secondary axes. Therefore, they are titrated to the minimal amount that produces partial axes at high frequency.[87] Usually, a steep threshold is observed for axis induction: for instance, I found that 50 pg of Noggin mRNA had almost no effect, 100 pg produced some axes, 200 to 500 pg >70% secondary axes. Thus, 200 pg was determined to be the optimal amount. With Activin, partial axis duplication could only be observed in a narrow range (5 to 10 pg). Indeed, Activin is a strong, highly diffusible mesoderm inducer. Above a certain concentration, the morphology of the embryo is so drastically modified that body axes cannot be recognized.

[e] By RT-PCR.

[f] All known components of the Wnt pathway appear to regulate expression of Siamois (and Twin/Xnr3) in early *Xenopus* embryos. Even if factors other than TCF/Lef-1 may exist downstream of β-catenin, which may not affect Siamois, it is not clear whether they would have an interpretable, or even detectable, phenotype in this assay.

[g] Use wild-type activators of the pathway (Xwnt8, Xdsh, β-catenin, and Siamois, see Table I) as well as dominant negative mutants of negative regulators of the pathway (kinase mutant GSK, ΔRGS-Axin, Table 2).

[h] In theory, "X" could be upstream of Wnt. However, the endogenous maternal pathway seems to start only downstream of Wnt.

3. Specificity

Is "X" specific for the Wnt pathway?

Even if "X" has fulfilled the criteria mentioned above, and has been shown to activate/inhibit the Wnt pathway, this molecule may also have a more general effect, and could influence other pathways as well. It is of course impossible to test every single pathway. However, there are a few very simple tests, which can rule out an involvement in some of the major early inducing pathways in *Xenopus*.

Activation of target genes: (RT-PCR)

Express "X" and check for activation of marker genes.

Markers: Activin and FGF pathways: *Xenopus* Brachyury (Xbra) (in animal caps)
Activin, not FGF: Goosecoid (in animal caps)
BMP pathway: Xvent-1 (in dorsal mesoderm)

Repression of target genes: (RT-PCR)

Coexpress "X" and FGF/Activin/BMP4 and check for inhibition of marker genes (as above).

Inhibition of axis duplication:

Coexpression with Activin/Vg1, and Noggin/Chordin/dominant negative BMP receptor.

4. Other Wnt pathways

4a. Is "X" modulating the zygotic Wnt8 pathway?

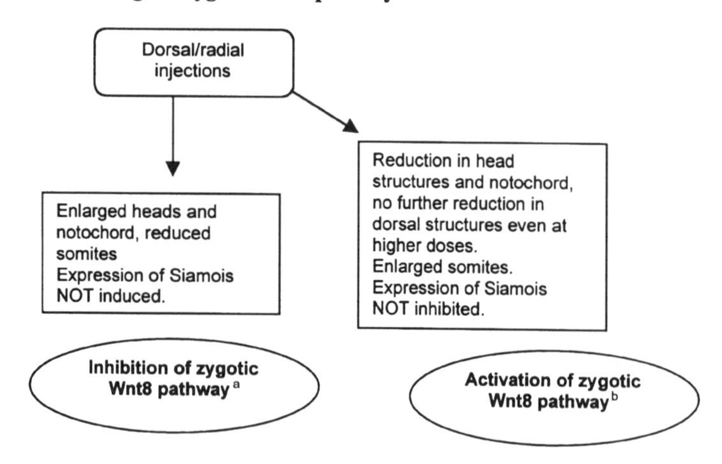

Comments:

[a] This phenotype has been obtained with Wnt inhibitors, Frzbs and Sizzled,[30,31,88,89] and with a truncated dominant negative Wnt8 construct.[28] It is also observed with Dickkopf and Cerberus,[10,64,65] two other factors which appear to inhibit the Wnt8 pathway, although probably more indirectly.

[b] This phenotype has been reported for injections of Wnt8 plasmid DNA.[29] Plasmid DNA is not transcribed before midblastula stages, and thus has no influence on the maternal Wnt pathway, unlike injected mRNA. Injection in the dorsal equatorial region of DNA coding for other downstream activators of the Wnt pathway should also cause a Wnt8 phenotype (this is indeed the case for β-catenin DNA, unpublished). It is also possible that specific activators *upstream* of Wnt may have the same effect, even when injected as mRNA. However, none has been identified in *Xenopus* so far.

4b. Is "X" involved in the Wnt5A pathway?

The "Wnt5A phenotype" has been observed for Wnt5A, Wnt4, Xfz3, Xfz7 and a dominant negative Xdsh.[66-69] It is characterized by a shortened body axis, a result of impaired elongation of axial tissues at the end of gastrulation. Furthermore, bent, "kinked" axes, head defects, neural closure defects, and sometimes short or split tail can be observed. Axial structures (notochord, somites) are present, although with some morphological defects. The phenotype is strongest when injections are performed in the dorsal upper marginal zone. The effect is zygotic, since a similar phenotype is obtained by injection of Wnt5A DNA. It is however, difficult to assess to what extent some of these abnormalities are specific, since the phenotype has not been rescued so far.

In addition to morphological criteria, two assays may be tested.

1) Animal cap assay: Wnt5A, Xfz7, and dominant negative Xdsh inhibit elongation of Activin-treated animal caps, without affecting early mesodermal markers (Xbra) or the differentiation of dorsal structures (notochord, muscle).

2) Calcium measurements: Wnt5A has been reported to induce a G-protein-dependent increase in intracellular Calcium in Zebrafish embryos.[70] Similar experiments can be performed in *Xenopus*, using dissected animal caps loaded with the Calcium-sensitive fluorescent probe Fura2.[69] However, Calcium measurements require some specialized imaging equipment (see chapter by Maxfield and Pierini).

IV. Interfering with Endogenous Components: An Overview

1. Depletion of maternal components

In the absence of genetics, depletion of maternal RNA is the closest experiment to a real knock-out of maternal components. This approach has been successfully established for *Xenopus* by Heasman and co-workers.[21,90-92] Specific mRNAs are depleted by injection of anti-sense, chemically modified oligonucleotides in full-grown oocytes, oocyte maturation is induced by progesterone-treatment, and finally mature oocytes are fertilized.

In fact, these are quite challenging experiments.

1) A panel of oligonucleotides has to be tested to establish conditions for efficient RNA depletion.

2a) *Xenopus* oocytes are not easily fertilized *in vitro*. Instead, injected oocytes are reimplanted in a female, which is then induced to lay by hormone injection. Large numbers of oocytes must be injected, because yields of injected, transplanted oocytes developing into viable embryos are rather low, about 25% in the best cases, much lower in many instances.

2b) A very useful shortcut to the transplantation technique is provided by the possibility to fertilize mature *Xenopus* oocytes by direct injection of sperm nuclei.[93] While the yields of viable embryos are certainly not higher than in the original transplantation method, the protocol is significantly alleviated.

3) Even when complete RNA depletion is achieved, a variable (and sometimes considerable) pool of pre-stored protein usually remains, and is transmitted to the developing egg. This is clearly the case for cadherins and β-catenin.[91,92] In the case of β-catenin, the cadherin-pre-bound pool is very stable, and barely affected by mRNA depletion. Soluble β-catenin, however, is very unstable in the absence of Wnt signal (see introduction), and is completely absent in mRNA-depleted embryos.[92] As a consequence, depletion of β-catenin mRNA has little or no effect on cell-cell adhesion, while β-catenin signaling and axis formation are inhibited.[21]

Despite these limitations, depletion of maternal factors is an invaluable technique. The new possibility of direct oocyte fertilization should decrease the "energy barrier" that has so far held a more widespread use of this approach. Very detailed protocols for both anti-sense depletion and sperm injection have been published elsewhere.[90,93,94]

2. Depletion of zygotic components

Synthesis of zygotic products can be inhibited by injection of anti-sense DNA or RNA. This strategy has been effective, for instance, in demonstrating the antagonistic role of BMPs and Goosecoid.[95] Note that the efficiency of this technique varies dramatically from molecule to molecule. See References 95 and 96 for protocols.

3. Dominant negative constructs

This much more commonly used approach takes advantage of the property of some mutant constructs to inhibit the activity of endogenous factors. A list of available dominant negative constructs for components of the Wnt pathway is presented in Table 2. A few other useful dominant negatives can be found in Table 3.

TABLE 2
Available dominant negative constructs along the Wnt pathway

Wild-Type Protein	Name of Dominant Negative Construct	Characteristics	mRNA Doses for Phenotype	Phenotype	Refs.
Xwnt8	dnXwnt8	C-terminal truncation	2x 300 pg	No effect on endogenous maternal induction. Inhibits axis duplication by Xwnt8. Inhibits zygotic Xwnt8 (large heads, reduced somites).	28
Xdsh	Xdd1	PDZ deleted	2x 0.5-1 ng	No effect on endogenous maternal induction. Inhibits axis duplication by Xwnt8 and Xdsh, not β-catenin. Wnt5A phenotype.	68
GSK3	kmGSK3	Kinase mutant	2 ng	Axis duplication	130, 131
Axin	deltaRGS	RGS deleted	1-2 ng	Axis duplication	37
(XTCF3)	(ΔN-XTCF3)*	N-terminal truncation	2x 250 pg	(Ventralization)*	52

* It is not clear whether this construct is really a dominant negative, since wild-type XTCF does not induce an axis. The mechanism of action of this construct has not been established. Wild-type XTCF does not appear to rescue the phenotype.

A serious problem of this approach is the difficulty to assess the specificity of the constructs. For instance, it was originally found that a dominant negative Activin receptor, which was used to block mesoderm induction,[17] was also able to cause direct neural induction.[97] In fact, it later appeared that this construct was

TABLE 3
Some other growth factors and dominant negative receptor constructs

Name	mRNA Doses	Phenotype	Refs.
eFGF bFGF	5-20 pg 500 pg	Induces ventral mesoderm in animal caps	135, 136
dnFGF-R (XFD)	1-4 ng	Inhibits mesoderm	13
Activin	1-10 pg 5-10 pg	Induces ventral and dorsal mesoderm in animal caps Partial axis duplication in ventral injections	17, 137
BMP2-Vg1	20-200 pg	Induces ventral and dorsal mesoderm in animal caps Partial axis duplication in ventral injections	18
dnActivin-R (XActRIIB)	1-2 ng	Inhibits mesoderm Neural induction in animal caps	17, 99
dnActivin-R (XALK4)	1-2 ng	Inhibits mesoderm No effect on neural induction	99
BMP4	500 pg	Inhibits neural induction Ventralizes mesoderm	138, 139
dnBMP-R	1 ng	Neural induction in animal caps Partial axis duplication	138, 140
Noggin	100-200 pg	Neural induction in animal caps Partial axis duplication	141
Chordin	200 pg	Neural induction in animal caps Partial axis duplication	142

not only inhibiting the Activin/Vg1 pathway, but also the BMP pathway.[98,99] More recently, constructs with more selective specificity have shown that blocking the Activin/Vg1 pathway inhibits mesoderm, while blocking the BMP pathway induces neural tissue.[99] An absolute rule in the use of dominant negative constructs should be the rescue of the normal phenotype by co-expression of the wild-type protein. Furthermore, the specificity should also be tested by comparing the effect on related pathways.

4. Engrailed repressor chimeras

These are a particular type of dominant negative constructs, which can be used to study the role of transcription factors. The repressor domain of Engrailed is fused to the transcription factor of interest. The overexpressed chimera will compete with the endogenous factor on the DNA, and actively repress its target genes. Similarly, an activator domain, VP16, can be used to force activation of target genes. This approach has been successfully applied to demonstrate the role of Siamois in axis formation,[22,23] and protocols can be found in the original articles (see also Reference 100).

Note that specificity is again crucial, as for any dominant negative molecule. Excess of a DNA-binding protein is likely to cause non-specific binding at irrelevant DNA sites. Again, rescue experiments are absolutely required. Additional controls for specificity can be obtained by single-point mutations within the DNA-binding domain, or comparison of closely related factors.

V. Preparation of Embryo Extracts

Because *Xenopus* embryos are loaded with large amounts of yolk proteins, direct extraction in buffers containing SDS is generally not advised. Instead, embryos are routinely extracted using a non-ionic detergent, Nonidet-40 (NP-40), under conditions where the yolk remains largely insoluble.

In embryos, most components of the Wnt pathway are fully soluble in NP-40, including cadherins and catenins (in mammalian culture cells, a fraction of the cadherin/catenin complexes appears to be resistant to extraction by non-ionic detergents). APC, however, is only partially extracted by NP-40, and higher yields can be obtained by addition of deoxycholate.[46] Thus, one of the first steps in characterizing a new molecule should be to determine its extractability in non-ionic and ionic detergents.

An alternative method for extraction after rapid separation of the yolk is also presented.

1. Standard NP-40 extraction

Collect embryos in a 1.5 ml centrifuge tube, remove as much medium as possible.

Add ice-cold NP-40 buffer (up to 25 embryos/100 µl buffer), and homogenize by pipetting through a 200 µl pipette tip.

Centrifuge for 5 min at full speed in a tabletop centrifuge, at 4°C.

Collect supernatant in a clean tube. The extract can then be mixed with SDS-PAGE loading buffer for analysis on SDS-PAGE and immunoblotting, or used for further biochemical experiments, such as immunoprecipitations (see below).

NP-40 buffer: 1% NP-40, 150 mM NaCl, 10 mM HEPES-NaOH, 2 mM EDTA, pH 7.4, with protease inhibitors (1 mM PMSF, 0.5 mM iodoacetamide, 1 µg/ml pepstatine A, 2 µg/ml leupeptin, 4 µg/ml aprotinin, 10 µg/ml antipain, 50 µg/ml benzamidine).

2. Variations of the NP-40 buffer and use of other detergents

NP-40 concentration can be lowered to 0.5%.

0.5% deoxycholate can be added to the NP-40 buffer.

Embryos can also be extracted in RIPA buffer (0.2% SDS, 0.5% deoxycholate, 1% Triton X-100, 150 mM NaCl, 20 mM HEPES-NaOH, pH 7.4). Under these conditions, however, the yolk will be solubilized. Smears will appear on the gels, which can be minimized by loading smaller amounts of extracts.

3. Alternative preparation (unpublished)

Homogenize up to 20 embryos in 100 µl of ice-cold isotonic buffer without *detergent* by pipetting through a 200 µl tip in a 1.5 ml tube.

Isotonic buffer is 100 mM NaCl, 20 mM HEPES, pH 7.4, 2 mM $MgCl_2$, 1 mM EGTA, 0.1 mM EDTA, protease inhibitors.[a]

Centrifuge immediately in a tabletop centrifuge for 20 to 30 sec at 1500 rpm (150g).[b]

Collect the supernatant in a clean tube.

The supernatant is now largely free of yolk, and can be either used for further biochemical analysis, or extracted in SDS-PAGE loading buffer.[c]

Comments:

[a] The isotonic buffer aims to minimize disruption of yolk granules and other endosomes, and release of hydrolytic enzymes.

[b] Centrifugation conditions are such that most of the yolk granules sediment, with minimal pelleting of other cellular components. Fast sedimenting components (nuclei!) will obviously be lost in this preparation!

[c] Because extraction can be completed in less than 1 min, this protocol is particularly suitable to analyze labile components (proteins with high sensitivity to proteolytic degradation, phosphorylated proteins).

VI. Subcellular Distribution: Microscopic and Biochemical Methods

Regulated subcellular localization is thought to be a crucial factor for many signaling activities. Dishevelled, for instance, can apparently be either cytoplasmic or recruited to the plasma membrane (in response to Frizzled). β-catenin is present in many different cellular pools, with very distinct properties and functions: at the plasma membrane (cadherin-bound), associated with the cytoskeleton (microtubules, via APC, and actin filaments, via fascin), cytosolic and nuclear upon activation of the Wnt pathway. Likewise, Axin distributes in two distinct pools at the cell periphery and in the cytoplasm (unpublished). Analysis of these various pools requires combined optical and biochemical approaches.

1. Immunofluorescence

Various protocols have been used in embryos.

Staining of paraffin or plastic sections is generally suboptimal, due to loss of antigens during sample preparation. Definitely not recommended!

Whole-mount staining works very well if mild fixatives (methanol-based) are used (see References 101 and 102 for detailed protocol). Whole-mount stained embryos can be cleared and observed with a confocal microscope,[103] or embedded in resin to obtain thin sections.[5,104] A drawback of whole-mount techniques is the extremely poor penetration of antibodies in aldehyde-fixed samples, even followed by extensive permeabilization. Basically, staining is then limited to the first superficial cell layer,[a] unless extended incubations are performed (2 to 3 days!).

Staining on frozen sections is faster and more versatile than whole-mount staining, and avoids the problems of antibody penetration. Originally established for methanol-DMSO-fixed embryos,[105] the protocol can be used with paraformaldehyde fixations as well. Conditions for sectioning are most crucial (under sub-optimal conditions, embryos are very brittle).

Fixation: 4% paraformaldehyde, 100 mM NaCl, 100 mM HEPES, pH 7.4, for 1 h at room temperature, with slow constant stirring. Remove aldehyde fixative and add straight cold (–20°C) Dent's fixative (20% DMSO, 80% methanol). –20°C overnight (can usually be stored for long periods at – 20°C)[b].

Embedding: Wash in 100 mM NaCl, 100 mM Tris-HCl, pH 7.4 (quenching of unreacted aldhehydes) for 30 min. Embed in 15% cold water fish gelatin, 15% sucrose overnight at room temperature,[d] then in 25% fish gelatin[c], 15% sucrose again overnight.[b] Can store at 4°C for at least a week.

Mounting: Place and orient embryos in a mold filled with 20% fish gelatin, 15% sucrose (can group up to three embryos together), freeze on dry ice. Trim sample down to a small cube (about 5 to 7 mm sides), paste to the object holder using O.C.T.[e] as glue. Angles and edges of the gelatin block must be sharp.[f] May be re-trimmed if required. Orient the surface of the block parallel to the knife, and rotate 45° compared to the edge of the knife (see diagram).[f]

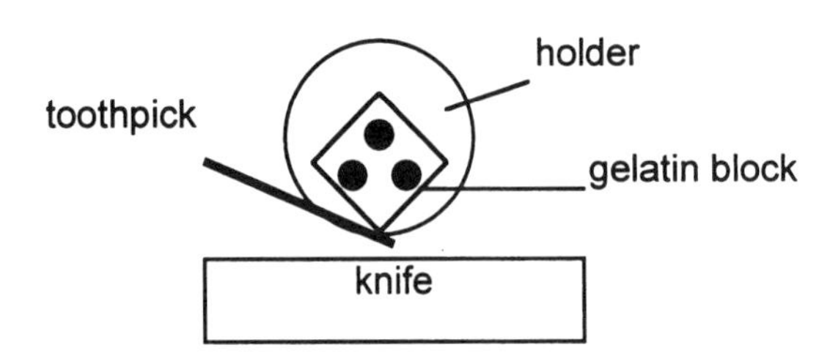

Sectioning: Equilibrate the block and the knife at –16°C to –18°C. Cut 10 μm-thick sections (can also do 5 μm). Collect section by grabbing the leading edge with a toothpick or tweezers, in order to prevent rolling of the section. Quickly transfer section to a coated or pretreated glass slide.[g] Store slides at –20°C (–80°C for long periods).

Immunofluorescence:

Air dry for at least 1 h.

1 min in acetone, air dry (2 min), rehydrate in PBS, then incubate 30 to 60 min in blocking solution (5% milk in PBS).

Incubate with primary antibody in 5% milk/PBS.

Wash three times in 1% milk/PBS.

Incubate with secondary (green) fluorescent antibody[h] in 1% milk/PBS.

Wash three times in PBS.

(Optional: Incubate with tertiary (green) fluorescent antibody[i] in 1% milk/PBS, wash three times in PBS.)

(Optional: Incubate for 10 min with 20 μg/ml DAPI or Hoechst 33342 in PBS for nuclear staining, wash two times in PBS.)

Incubate for ~10 min in 0.1% Eriochrome Black[j] in PBS to counterstain the yolk granules in red (quenching of the green autofluorescence of the yolk).

Wash in PBS.

Mount slides using anti-fade agent in glycerol.[k]

The green signal and the red-counterstained yolk can be viewed simultaneously using a blue excitation filter and a long-pass emission filter. They can be viewed separately using filter sets for standard double-staining with fluorescein and rhodamine/Texas Red.

Double-staining:

Double-staining is still problematic in *Xenopus* embryos due to the high autoflorescence of the yolk. The use of Eriochrome eliminates its green autofluorescence, but strongly increases its red fluorescence. Double-staining will require compromise, e.g., by omitting Eriochrome staining (autofluorescence will then be present in both colors). If one of the two antibodies gives a very strong signal (e.g., epitope tags), it might be possible to stain it with Cy3-conjugated antibodies and still distinguish it from Eriochrome-stained yolk granules.

Comments:

[a] This is the reason why, in whole-mount staining, nuclear accumulation of endogenous β-catenin appears much stronger at the surface of the blastula.[5] In fact, β-catenin accumulates in nuclei of deep cells as well (Figure 5, unpublished).

[b] Alternatively, embryos can be directly fixed in cold Dent's fixative, in which case embedding in 15% sucrose is sufficient, and the 25% gelatin step can be omitted. Sectioning of Dent's fixed samples is generally easier than for aldehyde-fixed embryos. However, retention of soluble components, including nuclear β-catenin, is poor.

[c] Diluted 1:2 from a 45% stock solution (Fluka, #48717).

[d] Cold water fish gelatin is liquid at room temperature.

[e] O.C.T. compound, Miles Inc., #4583.

[f] Necessary to prevent sticking of sections to the knife.

[g] SuperFrost Plus, #041300, Fisher Instruments.

[h] I use Oregon Green 488 goat anti-mouse (#O6380) and anti-rabbit (#O3681) IgGs from Molecular Probes.

[i] e.g., DFAT-F(ab′)$_2$ donkey anti-goat IgG (Dianova, #313166006).

[j] Eriochrome Black T, # E2377, Aldrich.

[k] Slow Fade (#S2828) or Slow Fade Light (#S7461), Molecular Probes.

2. Immunogold labeling

Most of what is known today about subcellular localization in *Xenopus* is limited to the low resolution achieved by immunofluorescence, without any knowledge on the underlying ultrastructure. Immunocytochemistry in frog embryos is clearly not trivial. Due to the very large size of the samples, penetration of fixatives is slow, and it is difficult to find a compromise between acceptable preservation of both ultrastructure and antigenicity. Furthermore, the concentration of endogenous antigens is usually very low, at the limit of detection, or below. The following

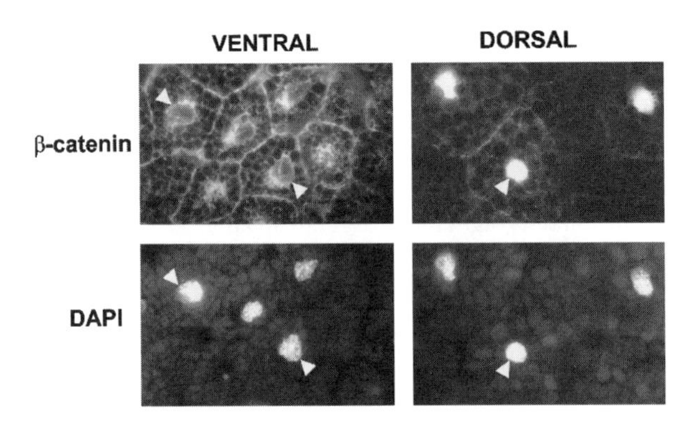

FIGURE 5

β-catenin nuclear accumulation in the dorsal side of a late blastula stage. Endogenous β-catenin was detected on frozen sections from paraformaldehyde-fixed embryos by indirect immunofluorescence (Oregon Green/DTAF staining). Nuclei (arrowheads) were stained with DAPI. Spheres appearing black in the fluorescein channel, and gray in the DAPI channel, are the yolk platelets, counterstained with Eriochrome Black. β-catenin accumulates in the nuclei of dorsal vegetal blastomeres. In the ventral side, β-catenin is excluded from the nuclei, although it is abundant at the plasma membrane and in the perinuclear region.

approach has given very promising results both with endogenous and exogenous (epitope-tagged) antigens[a]. This method is based on pre-embedding staining using nanogold, which has a good penetration through the tissues, and silver enhancement.[b, 106]

Fixation: 4% paraformaldehyde, 0.02% glutaraldehyde, 100 mM HEPES, pH 7.4, for 4 h at room temperature, then store at 4°C.

Embedding and thick sectioning: Embed in 3% Low Melting Point Agarose, cut 75 to 100 μm-thick sections with a Vibrotome.

Permeabilization and blocking: Incubate 2 h in 20% normal goat serum (NGS) and 0.1% Saponin in PBS.

1° antibody: Incubate overnight with 1° antibody in NGS/PBS/Saponin, wash several times with PBS (for at least a total of 6 h).

2° antibody: Incubate overnight with Nanogold-coupled 2° antibody,[c] wash with PBS as above.

Post-fixation: 1 h at room temperature in 2% glutaraldehyde/PBS, wash in PBS, then water.

Silver enhancement: 1 to 1.5 h in the dark.[107] The reaction can be followed by observing darkening of the cells under a dissecting microscope. Rinse in water, contrast with 1% Uranyl Acetate for 1 to 2 h, rinse in water.

Embedding and thin sectioning: Dehydrate progressively in ethanol, embed in Spurr. Form a flat block containing the Vibratome section and polymerize at 60°C. Paste the thin block on a larger Epon block as support, using a drop of Epon as glue. Cut thin sections parallel to the plane of the Vibratome section.

Comments:

[a] As an alternative method, very small pieces of embryos can be dissected, fixed, and embedded in Lowicryl for standard post-embedding staining. Only applicable for very strong signals (epitope-tags).

[b] Nanogold is too small to be useful for direct observation under the electron microscope. However, larger agglomerates are obtained by silver enhancement. These agglomerates appear by EM as high contrast, irregularly shaped (about 20 to 50 nm diameter) spots. Thicker sections can also be viewed with a light microscope, where silver staining appears dark brownish.

[c] 1 nm Gold-coupled Goat anti-mouse/anti-rabbit IgG (Nanoprobes).

3. Cell fractionation: analysis of soluble/sedimentable/membrane/nuclear fractions

a) Soluble (cytosolic)[a] and sedimentable (i.e., membranes, organelles, cytoskeleton) fractions are obtained by simple centrifugation procedures.[105]

b) Most plasma membrane proteins, including cadherin, are glycosylated. Binding to the lectin Concanavalin A (ConA) coupled to beads can very efficiently purify glycoproteins. This is a particularly straightforward assay to determine, and even quantify, membrane associated-pools of cytoplasmic proteins, such as β-catenin.

c) Various β-catenin pools can be analyzed by a combination of differential centrifugation and ConA-precipitation (Reference 108, see below).

d) Finally, the nucleus has become of increasing interest in the study of the Wnt pathway, since β-catenin accumulates in the nucleus and associates with the transcription factors TCF/Lef-1. Although purification of intact embryonic nuclei is quite a challenge, I propose here a protocol which yields nuclei in an acceptable good shape. This is clearly still a work in progress (see comments below).

[a] "Detergent (Triton)-soluble" fractions are often confused with "soluble," "cytosolic" fractions. The latter are obtained by centrifugation at high speed in the *absence* of detergent. Extracts obtained in the presence of Triton X-100 (or equivalent mild, non-ionic detergents) contain a complex mixture of cytosolic, membrane, and other proteins. Because many cytoskeletal elements are insoluble in non-ionic detergents, Triton insolubility has been sometimes used as criterion for association with the cytoskeleton. However, there are many other Triton-insoluble cellular components, including some plasma membrane lipid domains and various associated proteins, as well as chromatin.

a. Separation of soluble and sedimentable fractions: Differential centrifugation

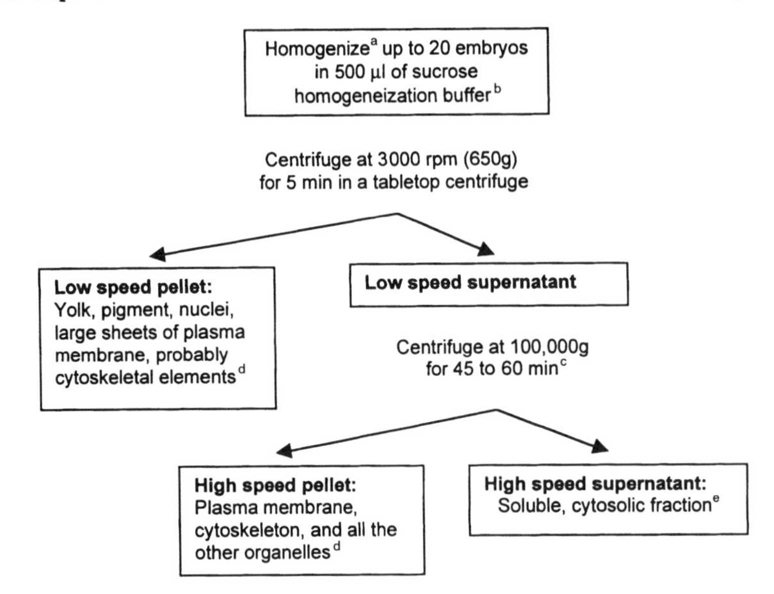

[a] Homogenization is performed in a 1.5 ml microfuge tube by pipetting embryos through a 200 μl pipette tip, about 10 to 15 times. All steps are performed at 4°C.

[b] Homogeneization buffer: 250 mM sucrose, 10 mM HEPES, 2 mM $MgCl_2$, 1 mM EGTA, 0.5 mM EDTA, pH 7.2, containing proteinase inhibitor cocktail.

[c] I use a Beckman TL100 tabletop ultracentrifuge, and either thick wall, polycarbonate 1 ml tubes (#343778) or 1.5 ml polyallomer tubes (#357448).

[d] Low- and high-speed pellets can be used as such, or membrane associated proteins can be further separated by ConA precipitation (see below).

For direct analysis by SDS-page and immunoblotting, the high-speed pellet can be directly extracted in SDS-PAGE loading buffer. However, when boiled in loading buffer, the yolk-rich low-speed pellet becomes a thick paste. Routinely, the low-speed pellet is first extracted in a small volume of NP-40 buffer (see "extractions").

For ConA precipitation, both low- and high-speed pellets must be extracted in NP-40 buffer. They are usually combined in a final volume of 500 μl.

[e] The high-speed supernatant can be concentrated by acetone precipitation (add 4 volumes of cold acetone (–20°C), precipitate at –20°C for 15 to 20 min, centrifuge for 15 min at 14,000 rpm in a table top centrifuge), or by TCA precipitation (add ice-cold TCA to 5% final concentration, centrifuge, rinse pellet with acetone, air dry, and resuspend in SDS-PAGE loading buffer).

b. Separation of glycoprotein-associated (membrane-associated) components by binding to Concanavalin A

Prepare 500 µl of a NP-40 extract, containing up to 20 embryos (or equivalent fractions) in a 1.5 ml microfuge tube, on ice.
Add 80 µl/10 embryos of a 70% slurry of ConA-sepharose beads.[a]
Incubate for 1 h at 4°C with continuous gentle mixing.
Spin down the beads, collect supernatant (**unbound fraction**), wash the beads (**bound fraction**) three times with 1 ml NP-40 buffer.
For direct analysis by SDS-page and immunoblotting, extract beads directly with boiling SDS-PAGE loading buffer. Depending on the sensitivity of the detection, the unbound fraction may require to be concentrated (see above).

[a]This is about the minimal amount of ConA beads required to completely deplete the cadherin content of an extract of early embryos. Concanavalin A - sepharose 4B (Sigma #C9017). Beads should be washed before use, since the slurry contains 20% ethanol..

c. Analysis of β-catenin cellular pools

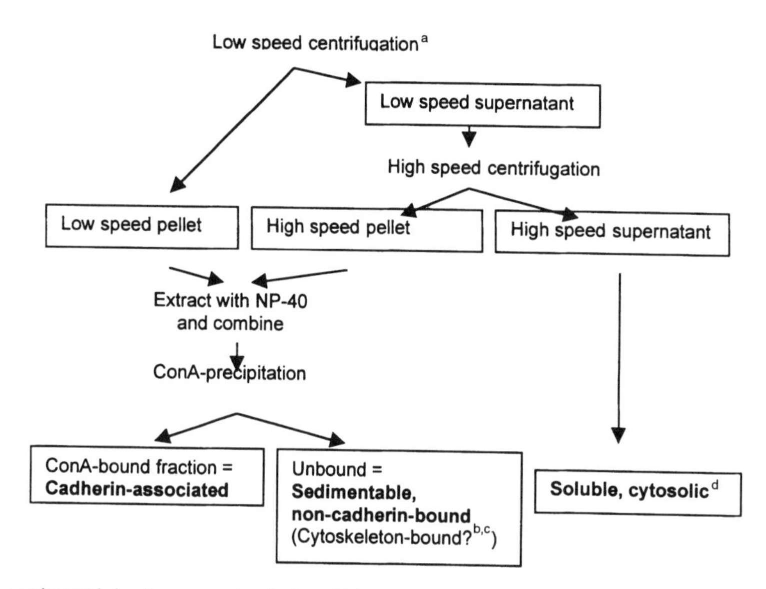

[a] For experimental details, see protocols A and B.

[b] Endogenous APC is mostly recovered in the high-speed pellet (unpublished). APC-bound β-catenin is probably found in this pool, although this has not been tested directly.

[c] There is also a pool of sedimentable Axin, which also binds β-catenin (unpublished).

[d] Nuclear β-catenin is also mostly recovered in this fraction (unpublished).

d. Isolation of nuclei

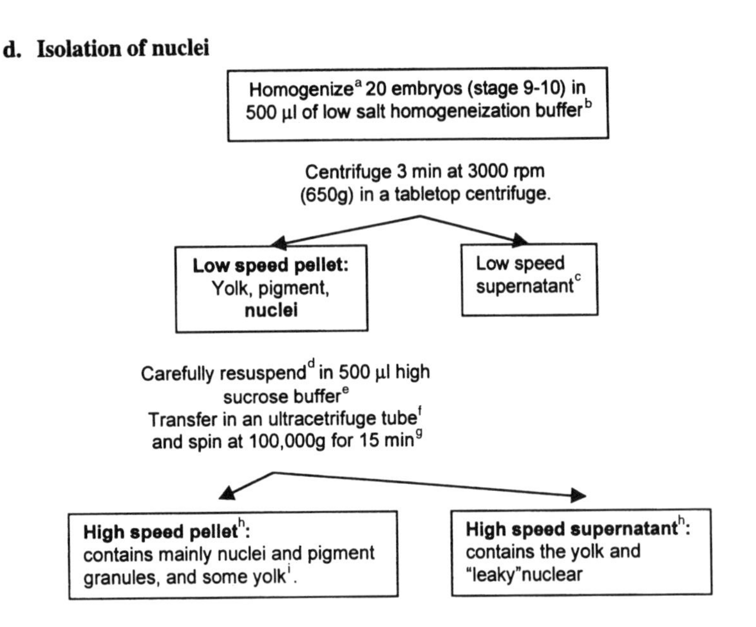

[a] Homogenize in a 1.5 ml centrifuge tube, using a 200 µl pipette tip.

[b] Homogeneization buffer: 10 mM HEPES, pH 7.4, 1 mM EGTA, 1 mM EDTA, 0.5 mM spermidine, 0.5 mM spermine, protease inhibitors.

[c] Can be subsequently fractionated into cytosolic and sedimentable fractions by high-speed centrifugation as described above.

[d] Resuspend using a clipped (1 mm aperture) 200 µl pipette tip.
The yolk is somehow dissolved during this step.

[e] High sucrose buffer: 1.8 M sucrose, 80 mM KCl, 15 mM NaCl, 15 mM HEPES, pH 7.4, 5 mM EDTA.

[f] Beckman TL 100 ultracentrifuge, thick wall polycarbonate 1 ml tubes. (#343778)

[g] Optional: Addition of a 50 µl 2.3 M sucrose cushion at the bottom of the tube improves preservation of the nuclei. However, clean separation of the pellet from the supernatant is more difficult.

[h] For direct analysis by immunoblotting, resuspend the high-speed pellet directly in SDS-PAGE loading buffer. For the supernatant, add 1/6 volume of 6X SDS-loading buffer *without sucrose/glycerol.*

[i] Some C-cadherin is observed in the pellet.

General comments:

This protocol is inspired from a variety of published and unpublished protocols for isolation of nuclei.

Note that nuclei tend to leak during purification. What is obtained, in fact, is a fraction of "nuclear ghosts," i.e., chromatin-surrounded nuclear membranes, but soluble or weakly associated components are largely lost. I have tested the retention of soluble nuclear proteins by expressing a nuclear β-galactosidase construct (β-galactosidase containing an NLS). By immunofluorescence, this protein appears strongly concentrated in the nucleus. By cell fractionation, close from 100% of it is recovered in the first supernatant! The same seems to happen with nuclear β-catenin (see previous section). On the other hand, a large fraction of exogenous, epitope-tagged XTCF is recovered in the nuclear pellet.

Another problem for purification of nuclei from early embryos is the high mitotic rate, especially before midblastula transition. Indeed, mitotic chromatin does not co-sediment with interphase nuclei. Furthermore, the size of the nuclei varies with developmental stages and tissues: nuclei from early blastomeres are huge, then decrease in size as cleavage proceeds. Nuclei from the large vegetal cells, and later endodermal cells, remain larger than average. Their sedimentation rate varies accordingly. The present method has been established for late blastula nuclei (stages 9 to 10).

VII. Detecting Molecular Interactions

1. Immunoprecipitations

Immunoprecipitations are routinely carried out using crude NP-40 extracts, or extracts from cell fractionation (see above). If the source of antibody is abundant, it is recommended to crosslink the antibody directly to protein A/protein G-beads.[a] Thus, leakage of antibody during boiling in SDS-PAGE loading buffer is completely prevented, and very clean immunoblots are obtained.

Immunoprecipitation protocol

Extract 10 to 20 embryos in 500 µl 1% NP-40 extraction buffer[b] (see V) in a 1.5 ml microfuge tube. All steps are at 4°C.

Add soluble antibody (for rabbit sera, usually 1 to 2 µl), or antibody coupled to beads (10 to 50 µl of 50% slurry) and incubate 1 h to overnight with constant stirring.[c]

In the case of soluble antibodies, add 20 µl 50% slurry of Protein A/Protein G-agarose beads and incubate for 1 h.

Spin 20 to 30 sec, collect supernatant (= unbound fraction).

Wash beads 3 to 4 times with 1 ml NP-40 buffer.

After the last wash, remove as much buffer as possible.

Add SDS-PAGE loading buffer to the beads and boil. Spin, collect supernatant, and analyze by immunoblotting.

[a] Use Protein A/Protein G coupled to beads through *non-hydrolyzable* bonds (e.g., Immobilized Protein A/Protein G, #20333, #20398, Pierce) and crosslink the antibody with dimethylpimelimidate (# 20666, Pierce) according to the manufacturer's instructions.

[b] Note that the cadherin-β-catenin complex, which is exceptionally tight, can be isolated in the presence of SDS (RIPA buffer[109]).

[c] Use pre-immune or non-immune antibodies as negative controls. When coimmunoprecipitations with an exogenous molecule are to be tested, uninjected embryos can be used as controls.

2. Localization/colocalization by immunofluorescence

In some cases, technical problems hinder the detection of protein-protein interactions by immunoprecipitation. For instance, the association may be labile in NP-40, or, on the contrary, the putative partners may not be extractable in mild detergents. In some cases, immunofluorescence may provide indirect evidence. Obviously, colocalization in itself cannot be used to postulate a molecular interaction. However, some clue can be obtained when experimental manipulations lead to consistent changes in subcellular localization of putative interacting partners. For instance, ectopically expressed epitope-tagged Dishevelled distributes intracellularly, but redistributes at the plasma membrane when Frizzled molecules are coexpressed,[34] suggesting a possible interaction. Of course, such observations cannot distinguish between direct or indirect association. In fact, they do not even demonstrate that any interaction occurs (e.g., Frizzled may indirectly induce Xdsh to associate with another membrane protein complex), an assumption which must be confirmed biochemically.

VIII. Detecting GSK Activity

Regulation of GSK activity by upstream components of the Wnt pathway is a major unsolved step in the Wnt pathway. An *in vivo* assay for GSK activity has been established using a natural substrate of GSK, the microtubule-binding protein tau.[85] Antibodies are available, which recognize either total tau protein,[110] or specifically the GSK-phosphorylated forms of tau.[111]

GSK kinase activity

Isolate and defolliculate stage 6 Xenopus oocytes.[112]

Coinject mRNA coding for the putative GSK regulator and 1 ng of GSK mRNA.

Positive control: Coinject the same amount of a control mRNA and 1 ng of GSK mRNA.

Negative control: use uninjected oocytes.

Incubate overnight at 18°C in OR2 medium.[112]

Inject 10 ng of recombinant tau protein, incubate 2 h at room temperature.

Homogenize 10 oocytes in 100 μl tau lysis buffer (1 M NaCl, 100 mM Tris, pH 6.5, 0.5 mM $MgCl_2$, 1 mM EDTA, 2 mM DTT, 50 mM NaF, 0.1 mM $NaVO_4$), centrifuge for 5 min, collect supernatant, and analyze by immunoblotting.

Blot with a mixture of anti-T14 and anti-T46 to detect total tau protein.

Blot with anti-PHF1 to detect phosphorylated tau protein.

Note that GSK-phosphorylated forms migrate slower than unphosphorylated tau.

Quantify bands by densitometry.

IX. Analyzing β-Catenin Stability

Detection of changes in cytosolic levels of β-catenin are a very useful criterion to monitor activation of the Wnt pathway, and an essential assay to study the β-catenin degradation machinery (APC/Axin/GSK3β).

1. Endogenous β-catenin

In *Xenopus,* stabilization of endogenous β-catenin is difficult to detect, mainly due to the large pool of membrane-associated β-catenin, which is stable in the absence of signal, and only weakly affected by Wnt signaling.[24,113] Even maximal activation (complete dorsalization by Lithium treatment) causes only a slight increase in total β-catenin (unpublished). Changes are best seen after fractionation of the soluble pool (see above and Reference 92), or after depletion of the cadherin-bound pool by Con-precipitation (see above and Reference 114). So far, I have been unsuccessful in attempts to detect differences in levels of endogenous soluble β-catenin between dorsal and ventral regions.

2. Exogenous β-catenin

An easier method to detect stabilization of β-catenin is based on co-injection of epitope-tagged β-catenin.[115] Even then, β-catenin mRNA must be carefully titrated: if levels are too high, exogenous β-catenin will overwhelm the regulation/degradation machinery, and will be stabilized independently of upstream signals. For unknown reasons, stabilization is not observed when levels of tagged β-catenin are too low. Thus, the assay works only within a narrow range of β-catenin mRNA (about 5 to 10 times lower than the minimal amount for axis duplication).

2a. Testing stabilization of β-catenin

Coinject 50 to 100 pg epitope-tagged β-catenin mRNA and dorsalizing factor mRNA in the ventral side of a 4 to 8-cell stage embryo, or in the animal pole at the 2-cell stage.

Use a β-catenin/Wnt8 coinjection as positive control (10 to 20 pg Wnt mRNA, sufficient for axis induction), and β-catenin alone as negative control. β-galactosidase mRNA should be added to adjust equal amounts of total mRNA in all conditions.

Extract embryos at stage 9 to 10 with NP-40 buffer and analyze levels of tagged β-catenin by immunoblotting

Unlike stabilization, destabilization of β-catenin can be detected over a much wider range of β-catenin mRNA levels (0.075 to 2 ng), at least after overexpression of Axin. However, requirements for β-catenin degradation appear to vary with the extent of β-catenin overexpression (unpublished). It is therefore advisable to perform the assay with low amounts of β-catenin mRNA. Because free β-catenin represents only a fraction of total β-catenin under these conditions, destabilization is best observed on samples depleted of cadherin-bound β-catenin by treatment with ConA.

2a. Testing destabilization of β-catenin

Coinject 50 to 100 pg epitope-tagged β-catenin mRNA with a ventralizing factor (e.g., 1 to 3 ng Axin mRNA) or equivalent amounts of β-galactosidase mRNA in the dorsal side of a 4 to 8-cell stage embryo.
Extract 6 embryos at stage 9 to 10 in 60 μ NP-40 buffer.
Add 50 μl ConA-sepharose slurry, incubate with constant stirring for one hour, spin.
Collect supernatant and analyze levels of free tagged β-catenin by Western Blot.

X. Detection of Gene Expression

1. RT-PCR

Changes in early dorsal markers (e.g., Siamois, Xnr3, Goosecoid, Chrodin) after manipulations of normal axis formation (dorsalization or ventralization) can be determined in RNA preparations from whole embryos. Induction of an ectopic dorsalizing center in the ventral side can be similarly studied in RNA preparations from dissected ventral halves. Analysis of early ventral markers (e.g., Xwnt8, Xvents) can be performed as well, although in the case of local (dorsal or ventral) injections, changes are usually less spectacular than for dorsal markers.

The following protocol is modified from Wilson and Melton.[116]

1a. RNA extraction

Collect 3 embryos,[a,b] or 3 embryo halves,[c] in a 1.5 ml microfuge tube, and freeze on dry ice.
Add 1 ml of homogenization buffer [d, e] and disrupt immediately by pipetting through a 1-ml tip and by vortexing.
5 min on ice
Add 0.2 ml chloroform, vortex 20 sec, 5 min on ice
Spin for 20 min at 14,000 rpm in a tabletop centrifuge
Transfer the upper phase (450 μl) to a clean tube
Add 1 μl glycogen[f] and 450 μl isopropanol, 10 to 15 min on ice
Spin for 15 min at 14,000 rpm, discard supernatant
Wash pellet with 1 ml 75% ethanol, spin
Wash pellet with 1 ml 100% ethanol, spin
Air-dry pellet and resuspend in 40 μl DEPC water

[a] Use stage 10 to $10^1/_2$ embryos. Expression of several dorsal marker genes peaks around this stage, including Siamois and Goosecoid. However, expression of some genes raises steadily until later stages (e.g., Chordin and MyoD).

[b] Use uninjected or β-galactosidase-injected embryos as controls. Do not use UV-irradiated embryos as negative controls, since Siamois and Xnr3 are still strongly expressed under these conditions.

[c] Dorsal halves are used as positive controls for dorsal markers, and unmanipulated/uninjected, or β-galactosidase-injected ventral halves are used as negative controls.

[d] All subsequent steps are performed at 4°C using clean RNase-free tubes, pipette tips, and reagents (DEPC water), and wearing gloves.

[e] Ultraspec™ RNA from Biotecx Laboratories Inc., Houston, Texas.

[f] Nucleic acid carrier. Glycogen for molecular biology, 20 mg/ml, Boehringer, #901393.

1b. Reverse transcription

Take an 8 μl aliquot of RNA,[g] add 2 μl random hexamer primer (50 μg/ml)
Denature for 4 min at 65°C, then cool on ice
Prepare RT mix:

For one aliquot:	For 5 aliquots:
4 μl 5X RT buffer	20 μl
0.2 μl 20 mM DTT	1 μl
0.5 μl Rnasin (40 U/μl)	2.5 μl
1 μl dNTP mix (10 mM each)	5 μl
3.3 μl H_2O (DEPC)	16.5 μl

Add 9 μl mix to the negative control (–RT), containing RNA from whole (or dorsal halves) control embryos
Add Superscript II Reverse Transcriptase (200 U/μl, Gibco BRL) to the rest of the mix (1 μl per aliquot) and distribute 10 μl to the remaining tubes
Incubate for 30 min at 42°C

[g] To avoid saturation of the reaction, use as little RNA as possible. In the case of whole embryos, can further dilute RNA 1:1.

1c. PCR

Add 2 μl template (RT reaction) to tubes. One of the templates should be –RT
Add 1 μl of upstream/downstream primers[h,i] mix (0.1 mg/ml each) and 8 μl H_2O
Add PCR mix:

For one tube:	For 20 tubes:
2.5 μl 10x Taq buffer	50
0.25 μl dNTP (10 mM each)	5
0.25 μl Taq polymerase (Promega, 1.25 units)	5
0.1 μl ^{32}P dCTP (or dATP) (10 μCi/μl)	2
12 μl H_2O	240

If possible, use thin "micro" PCR tubes. If larger, 0.5 ml tubes are used, cover with mineral oil.
Run PCR program:[j]

93°C 2.5 min; 25 cycles: 93°C 30 sec, 55°C 1 min, 72°C 30 sec
Finally, 72°C 5 min, then cool to 4°C

[h] One set of reactions should contain primers to detect a "loading control," i.e., a gene not affected by the experimental manipulation, such as EF-1.

[i] If the PCR products have clearly distinct sizes, it is possible to include two or more pairs of primers in the same reaction. Should then check if PCR reaction is still linear under these conditions.

A few primers are listed below. Many others can be found in the literature.

[j] This typical program works well with listed primers. May require adjustments (annealing temperature and number of cycles) for other primers.

1d. Primers:
Siamois:[k] forward: 5′-TTGGGAGACAGACATGA-3′;
reverse: 5′-TCCTGTTGACTGCAGACT-3′
Goosecoid: forward: 5′-ACAACTGGAAGCACTGGA-3′;
reverse: 5′-TCTTATTCCAGAGGAACC-3′
Xnr3: forward: 5′-ATGGCATTTCTGAACCTG-3′;
reverse: 5′-TCTACTGTCACACTGGTA-3′
Chordin: forward: 5′-TGCAGTGTCCCCCCATC-3′;
reverse: 5′-GCAGTGCATAACTCCGAA-3′
Noggin: forward: 5′-ATGGATCATTCCCAGTGC-3′;
reverse: 5′-TCTGTGCTTTTTGCTCTG-3′
EF-1: forward: 5′-CAGATTGGTGCTGGATATGC-3′;
reverse: 5′-ACTGCCTTGATGACTCCTAG-3′

[k] This set of primers was designed to recognize only endogenous Siamois: the forward primer corresponds to a 5′-untranslated sequence absent in our Siamois expression construct.

1e. Analysis by PAGE[l]
Prepare a 5% acrylamide (stock: 29% acrylamide/1% bisacrylamide) gel in 0.5x TAE buffer
Add 3 µl 10x DNA loading buffer to each PCR aliquot, and load 5 µl/lane
Run gel at 150 V. Stop before free radioactivity (front) leaks out of the gel.
Dry gel and expose X-ray film from 2 h to overnight at –70°C (use an intensifier screen)

[l] See basic molecular biology method books for recipes.

2. *In situ* hybridization

This technique is the nucleic acid counterpart of immunofluorescence, and is extremely useful to determine spatial patterns of gene expression, in whole mounts or on sections. Double staining is also possible. Methods have been described in detail elsewhere.[117,118] Note that detection of very early gene transcripts is difficult, especially in the vegetal cells, largely because of the low amount of transcripts and their dilution in these very large cells.

3. Reporter genes

The promoters of the three known direct target genes of the maternal Wnt pathway in *Xenopus* have been recently identified and analyzed (Siamois,[53] Twin,[54] and Xnr3[55]). They have been shown to bind XTCF3, and to require β-catenin for activation. Data are also available for other downstream promoters (Goosecoid,[27] Forkhead[119]). Finally, artificial promoters containing multiple TCF-binding sites have been used in cell lines,[52] and could be used in *Xenopus*. The activity of the promoters in *Xenopus* embryos has been tested using Luciferase as a reporter gene. Luciferase activity is detected *in vitro* by luminometry, from extracts of whole embryos or of dissected pieces. This method is very sensitive and quantitative, but obviously provides only very limited spatial resolution.

Great hopes were initially set on the green fluorescent protein (GFP) technology to detect signaling activities in *Xenopus*. However, newly synthesized GFP becomes fluorescent only after a time lag, which is a serious handicap to study transient processes such as early induction. So far, GFP has been mostly used merely as a lineage tracer. A certainly less exciting, but probably "safer" marker for "in situ" detection of promoter activity would be the classical β-galactosidase. Any reporter construct that can be detected with high sensitivity could also be used (e.g., multiple epitope-tags localized by immunofluorescence).

XI. Concluding Remarks

A. mRNA Concentrations, Expression Levels, Activity Levels

Like surgery for humans, introducing a foreign molecule in cells is always a "trauma." Let's minimize it! Any mRNA should always be titrated and used at the minimal dose giving a specific phenotype. 10 ng mRNA is about the upper limit that a blastomere can withstand, but as already mentioned, some constructs are toxic at lower doses. This is not so much a problem when a positive phenotype is observed (e.g., axis duplication), but it is a serious concern when negative effects are obtained (inhibition of axis formation). Then, rescue experiments become crucial.

An everlasting question is how close/far a given experimental manipulation is from a physiological situation. This is obviously a complex problem, but some basic information can be easily obtained. Levels of an exogenously expressed protein can be compared with endogenous levels, if suitable antibodies are available. Exogenous epitope-tagged proteins can usually be distinguished from the endogenous molecules on immunoblots, because they migrate slightly slower. Also, comparison of the cellular distribution, for instance by immunofluorescence, may also give some indication of how "normally" the exogenous protein behaves.

When different mutants of a same protein are compared, expression levels should be determined, if possible around the stage where they are expected to act. Epitope-tags (better as multiple copies) are invaluable for such purposes. Note that even two closely resembling constructs may be expressed at very different levels. It may also be useful to determine if a deletion mutant is properly localized (e.g., immunofluorescence). For instance, a mutant receptor may be trivially inactive because it is misfolded and never reaches the cell surface.

Obviously, factors of different nature (Wnt and β-catenin) cannot be compared according to their expression levels. However, if they are tested for the same phenotype, their activities can (and must) be standardized, as described in Section III.

B. This Is Not a "Clean" System! Further Considerations on Expression/Overexpression Experiment

1. Competition with the endogenous molecule

One major parameter to be considered in the interpretation of all overexpression experiments (including transgenics!) is the "omnipresence" of the endogenous components. Note that endogenous proteins can still be present after mRNA depletion in antisense experiments (Section IV). Even in genetic systems such as *Drosophila* embryos, zygotic null mutants are often still loaded in early stages with maternal components.

The potential for competition between exogenous and endogenous proteins is well illustrated in the case of β-catenin mutants. A membrane-tethered β-catenin mutant had been found to induce axis duplication,[114] which was very surprising considering the prevalent model of signaling via nuclear β-catenin (the same results were obtained with a similar plakoglobin construct[120]). However, it appeared that the effect of this mutant was indirect: indeed, its expression induced stabilization and nuclear localization of endogenous β-catenin, probably by sequestering regulatory factors, such as APC.[24] Furthermore, its axis duplication activity was completely inhibited by cadherin overexpression, even though this mutant, unlike wild-type β-catenin, could not bind directly to cadherins. These results clearly identified endogenous β-catenin as the actual signaling molecule, indirectly activated by the membrane-tethered mutant. The results were somewhat different with a similar but cytosolic β-catenin mutant, which also did not bind directly to cadherins, and showed strong axis duplication activity.[108] Unlike the membrane-tethered mutant, the activity of this cytosolic mutant was only marginally inhibited, even by high cadherin levels. While the observed partial inhibition is a sign that some competition with endogenous β-catenin occurred, the strong residual cadherin-insensitive activity argues for an independent, direct signaling activity of this mutant.

In conclusion, results from functional analysis of mutants must be interpreted with caution, especially when little is known about interacting partners. Control experiments, such as the above-mentioned inhibition by cadherins, are useful to detect possible artifacts.

2. "Forced" activation of the pathway

It is important to keep in mind that activation/inhibition of a pathway by overexpression is a very artificial situation. Except in rare cases (stabilization of β-catenin), pathways are normally not regulated by changes in levels of protein, but rather by more subtle changes in protein-protein interaction and biochemical activities. Modulation by overexpression is usually "forced" by "flooding" the system. It is, for instance, easy to imagine that excess GSK will bypass endogenous regulation of GSK activity, and destabilize β-catenin despite the presence of upstream activation of the pathway. Conversely, expression of dominant negative GSK will compete with endogenous GSK in the ventral side. Similarly, overexpressed β-catenin will escape its constitutive downregulation by overwhelming the GSK/Axin machinery. Overexpressed Frizzled may "autoactivate" by forcing recruitment of signaling components (Dishevelled?) at the membrane. Overexpression of other factors may not necessarily lead to any detectable phenotype: for example, lack of activity of overexpressed XTCF3 can be explained by the fact that specific DNA sites are probably few, and readily saturated by the maternal protein.

C. Taking Advantage of a Complex System: "Playing" on Different Backgrounds

As mentioned in the Introduction, the embryo can be viewed as a combinatorial system of several localized signals. Each particular region can be defined by a set of active and inactive pathways (Figure 2). This situation can be exploited to test *in situ* the properties of a signaling component in a particular environment. One can determine whether a given factor is affected by the activity of upstream signals, for instance by expressing epitope-tagged constructs in the ventral region (Wnt pathway OFF) or in the dorsal side (Wnt pathway ON). Molecular interactions (immunoprecipitation), subcellular localization (cell fractionation, immunofluorescence), or posttranslational modifications (e.g., stability, phosphorylation) can then be compared in the ventral and dorsal side. Note that it is then necessary to express very low amounts of the exogenous factor, as close as possible to physiological levels, in order to avoid saturation of endogenous regulatory mechanisms (see β-catenin stability, Section IX).

Localized expression can similarly be used to study the inter-relationship between Wnt signaling and the other major signaling pathway in early embryos, also taking advantage of specific inhibitors of these pathways (dominant negatives). It is thus not unrealistic to predict that an integrated description of early-inducing cascades may be achieved in the near future.

Acknowledgments

I thank the authors who have kindly communicated unpublished data, and Herbert Steinbeisser for discussions and advice. I apologize for the many relevant works that could not be cited in this chapter.

References

1. Vincent, J.-P. and Gerhart, J. C., Subcortical rotation in Xenopus eggs: an early step in embryonic axis specification, *Dev. Biol.*, 123, 526, 1986.
2. Elinson, R. P. and Holowacz, T., Specifying the dorsoanterior axis in frogs: 70 years since Spemann and Mangold, *Curr. Top. Dev. Biol.*, 30, 253, 1995.
3. Nieuwkoop, P. D., The formation of mesoderm in Urodelean amphibians. I. Induction by the endoderm, *Wilhelm Roux's Arch. EntwMech. Org.*, 162, 341, 1969.
4. Smith, J. C., Mesoderm induction and mesoderm inducing factors in early amphibian development, *Development*, 105, 665, 1989.
5. Schneider, S., Steinbeisser, H., Warga, R. M., and Hausen, P., β-catenin translocation into nuclei demarcates the dorsalizing centers in frog and fish embryos, *Mech. Dev.*, 57, 191, 1996.
6. Darras, S., Marikawa, Y., Elinson, R. P., and Lemaire, P., Animal and vegetal pole cells of early Xenopus embryos respond differently to maternal dorsal determinants: implications for the patterning of the organiser, *Development*, 124, 4275, 1997.
7. Lemaire, P. and Kodjabachian, L., The vertebrate organizer: structure and molecules, *Trends Genet.*, 12, 525, 1996.
8. Bouwmeester, T. and Leyns, L., Vertebrate head induction by anterior primitive endoderm, *Bioessays*, 19, 855, 1997.
9. Kimelman, D., Christian, J. L., and Moon, R. T., Synergistic principles of development: overlapping patterning systems in Xenopus mesoderm induction, *Development*, 116, 1, 1992.
10. Glinka, A., Wu, W., Onichtchouk, D., Blumenstock, C., and Niehrs, C., Head induction by simultaneous repression of Bmp and Wnt signaling in Xenopus, *Nature*, 389, 517, 1997.
11. Thomsen, G.H., Antagonism within and around the organizer: BMP inhibitors in vertebrate body patterning, *Trends Genet.*, 13, 209, 1997.
12. Gawantka, V., Delius, H., Hirschfeld, K., Blumenstock, C., and Niehrs, C., Antagonizing the Spemann organizer: role of the homeobox gene Xvent-1, *EMBO J.*, 14, 6268, 1995.
13. Amaya, E., Musci, T. J., and Kirschner, M. W., Expression of a dominant negative mutant of the FGF receptor disrupts mesoderm formation in *Xenopus* embryos, *Cell*, 66, 257, 1991.
14. Cornell, R. A., Musci, T. J., and Kimelman, D., FGF is a prospective competence factor for early activin-type signals in Xenopus mesoderm induction, *Development*, 121, 2429, 1995.
15. Cornell, R. A. and Kimelman, D., Activin-mediated mesoderm induction requires FGF, *Development*, 120, 453, 1994.
16. LaBonne, C. and Whitman, M., Mesoderm induction by activin requires FGF-mediated intracellular signals, *Development*, 120, 463, 1994.
17. Hemmati-Brivanlou, A. and Melton, D. A., A truncated activin receptor inhibits mesoderm induction and formation of axial structures in Xenopus embryos, *Nature*, 359, 609, 1992.
18. Thomsen, G.H. and Melton, D.A., Processed Vg1 protein is an axial mesoderm inducer in Xenopus, *Cell*, 74, 433, 1993.

19. Henry, G.L., Brivanlou, I. H., Kessler, D. S., Hemmati-Brivanlou, A., and Melton, D. A., TGF-β signals and a pattern in *Xenopus laevis* endodermal development, *Development*, 122, 1007, 1996.
20. Heasman, J., Patterning the Xenopus blastula, *Development*, 124, 4179, 1997.
21. Heasman, J., Crawford, A., Goldstone, K., Garner-Hamrick, P., Gumbiner, B., McCrea, P., Kintner, C., Noro, C. Y., and Wylie, C., Overexpression of cadherins and underexpression of β-catenin inhibit dorsal mesoderm induction in early *Xenopus* embryos, *Cell*, 79, 791, 1994.
22. Fan, M. J. and Sokol, S. Y., A role for Siamois in Spemann organizer formation, *Development*, 124, 2581, 1997.
23. Kessler, D. S., Siamois is required for formation of Spemann's organizer, *Proc. Natl. Acad. Sci. USA*, 94, 13017, 1997.
24. Miller, J. R. and Moon, R. T., Signal transduction through β-catenin and specification of cell fate during embryogenesis, *Genes Dev.*, 10, 2527, 1996.
25. Green, J. B., New, H. V., and Smith, J. C., Responses of embryonic Xenopus cells to activin and FGF are separated by multiple dose thresholds and correspond to distinct axes of the mesoderm, *Cell*, 71, 731, 1992.
26. Gurdon, J. B., Harger, P., Mitchell, A., and Lemaire, P., Activin signaling and response to a morphogen gradient, *Nature*, 371, 487, 1994.
27. Watabe, T., Kim, S., Candia, A., Rothbacher, U., Hashimoto, C., Inoue, K., and Cho, K. W., Molecular mechanisms of Spemann's organizer formation: conserved growth factor synergy between Xenopus and mouse, *Genes Dev.*, 9, 3038, 1995.
28. Hoppler, S., Brown, J. D., and Moon, R. T., Expression of a dominant-negative Wnt blocks induction of MyoD in Xenopus embryos, *Genes Dev.*, 10, 2805, 1996.
29. Christian, J. L. and Moon, R. T., Interactions between Xwnt-8 and Spemann organizer signaling pathways generate dorsoventral pattern in the embryonic mesoderm of Xenopus, *Genes Dev.*, 7, 13, 1993.
30. Zorn, A. M., Cell-cell signaling: frog frizbees, *Curr. Biol.*, 7, R501, 1997.
31. Salic, A. N., Kroll, K. L., Evans, L. M., and Kirschner, M. W., Sizzled: a secreted Xwnt8 antagonist expressed in the ventral marginal zone of Xenopus embryos, *Development*, 124, 4739, 1997.
32. Bradley, R. S. and Brown, A. M., The proto-oncogene int-1 encodes a secreted protein associated with the extracellular matrix, *EMBO. J.*, 9, 1569, 1990.
33. Bhanot, P., Brink, M., Samos, C. H., Hsieh, J. C., Wang, Y., Macke, J. P., Andrew, D., Nathans, J., and Nusse, R., A new member of the frizzled family from Drosophila functions as a Wingless receptor, *Nature*, 382, 225, 1996.
34. Yang-Snyder, J., Miller, J. R., Brown, J. D., Lai, C. J., and Moon, R. T., A frizzled homolog functions in a vertebrate Wnt signaling pathway, *Curr. Biol.*, 6, 1302, 1996.
35. Axelrod, J. D., Matsuno, K., Artavanis-Tsakonas, S., and Perrimon, N., Interaction between Wingless and Notch signaling pathways mediated by dishevelled, *Science*, 271, 1826, 1996.
36. Yost, C., Farr, G. H., IIIrd, Pierce, S. B., Ferkey, D. M., Chen, M. M., and Kimelman, D., GBP, an inhibitor of GSK-3, is implicated in Xenopus development and oncogenesis, *Cell*, 93, 1031, 1998.

37. Zeng, L., Fagotto, F., Zhang, T., Hsu, W., Vasicek, T. J., Perry, W. L., IIIrd, Lee, J. J., Tilghman, S. M., Gumbiner, B. M., and Costantini, F., The mouse Fused locus encodes Axin, an inhibitor of the Wnt signaling pathway that regulates embryonic axis formation, *Cell*, 90, 181, 1997.
38. Behrens, J., Jerchow, B. A., Wurtele, M., Grimm, J., Asbrand, C., Wirtz, R., Kuhl, M., Wedlich, D., and Birchmeier, W., Functional interaction of an axin homolog, conductin, with β-catenin, APC, and GSK3β, *Science*, 280, 596, 1998.
39. Hart, M. J., de los Santos, R., Albert, I. N., Rubinfeld, B., and Polakis, P., Downregulation of β-catenin by human Axin and its association with the APC tumor suppressor, β-catenin and GSK3β, *Curr. Biol.*, 8, 573, 1998.
40. Ikeda, S., Kishida, S., Yamamoto, H., Murai, H., Koyama, S., and Kikuchi, A., Axin, a negative regulator of the Wnt signaling pathway, forms a complex with GSK-3β and β-catenin and promotes GSK-3β-dependent phosphorylation of β-catenin, *EMBO. J.*, 17, 1371, 1998.
41. Kishida, S., Yamamoto, H., Ikeda, S., Kishida, M., Sakamoto, I., Koyama, S., and Kikuchi, A., Axin, a negative regulator of the wnt signaling pathway, directly interacts with adenomatous polyposis coli and regulates the stabilization of β-catenin, *J. Biol. Chem.*, 273, 10823, 1998.
42. Sakanaka, C., Weiss, J. B., and Williams, L. T., Bridging of β-catenin and glycogen synthase kinase-3β by axin and inhibition of β-catenin-mediated transcription, *Proc. Natl. Acad. Sci. USA*, 95, 3020, 1998.
43. Yamamoto, H., Kishida, S., Uochi, T., Ikeda, S., Koyama, S., Asashima, M., and Kikuchi, A., Axil, a member of the Axin family, interacts with both glycogen synthase kinase 3β and β-catenin and inhibits axis formation of Xenopus embryos, *Mol. Cell. Biol.*, 18, 2867, 1998.
44. Rubinfeld, B., Albert, I., Porfiri, E., Fiol, C., Munemitsu, S., and Polakis, P., Binding of GSK3β to the APC-β-catenin complex and regulation of complex assembly, *Science*, 272, 1023, 1996.
45. Rocheleau, C. E., Downs, W. D., Lin, R., Wittmann, C., Bei, Y., Cha, Y. H., Ali, M., Priess, J. R., and Mello, C. C., Wnt signaling and an APC-related gene specify endoderm in early C. elegans embryos, *Cell*, 90, 707, 1997.
46. Vleminckx, K., Wong, E., Guger, K., Rubinfeld, B., Polakis, P., and Gumbiner, B. M., Adenomatous polyposis coli tumor suppressor protein has signaling activity in Xenopus laevis embryos resulting in the induction of an ectopic dorsoanterior axis, *J. Cell Biol.*, 136, 411, 1997.
47. Smith, K. J., Levy, D. B., Maupin, P., Pollard, T. D., Vogelstein, B., and Kinzler, K. W., Wild-type but not mutant APC associates with the microtubule cytoskeleton, *Cancer Res.*, 54, 3672, 1994.
48. Munemitsu, S., Souza, B., Muller, O., Albert, I., Rubinfeld, B., and Polakis, P., The APC gene product associates with microtubules *in vivo* and promotes their assembly *in vitro*, *Cancer Res.*, 3676, 1994.
49. Nathke, I. S., Adams, C. L., Polakis, P., Sellin, J. H., and Nelson, W. J., The adenomatous polyposis coli tumor suppressor protein localizes to plasma membrane sites involved in active cell migration, *J. Cell Biol.*, 134, 165, 1996.
50. Gumbiner, B. M., Signal transduction by β-catenin, *Curr. Opin. Cell Biol.*, 7, 634, 1995.

51. Tao, Y. S., Edwards, R. A., Tubb, B., Wang, S., Bryan, J., and McCrea, P. D., β-catenin associates with the actin-bundling protein fascin in a noncadherin complex, *J. Cell Biol.*, 134, 1271, 1996.
52. Molenaar, M., van de Wetering, M., Oosterwegel, M., Peterson-Maduro, J., Godsave, S., Korinek, V., Roose, J., Destree, O., and Clevers, H., XTcf-3 transcription factor mediates β-catenin-induced axis formation in Xenopus embryos, *Cell*, 86, 391, 1996.
53. Brannon, M., Gomperts, M., Sumoy, L., Moon, R. T., and Kimelman, D., A β-catenin/XTcf-3 complex binds to the siamois promoter to regulate dorsal axis specification in Xenopus, *Genes Dev.*, 11, 2359, 1997.
54. Laurent, M. N., Blitz, I. L., Hashimoto, C., Rothbacher, U., and Cho, K. W., The Xenopus homeobox gene twin mediates Wnt induction of goosecoid in establishment of Spemann's organizer, *Development*, 124, 4905, 1997.
55. McKendry, R., Hsu, S. C., Harland, R. M., and Grosschedl, R., LEF-1/TCF proteins mediate wnt-inducible transcription from the Xenopus nodal-related 3 promoter, *Dev. Biol.*, 192, 420, 1997.
56. Roose, J., Molenaar, M., Hurenkamp, J., Peterson, J., Bratjes, P., van de Wetering, M., Destree, O., and Clevers, H., The Xenopus Wnt effector XTcf-3 interacts with Groucho-related transcritpion repressors, *Nature*, 395, 608, 1998.
57. Willert, K. and Nusse, R., β-catenin: a key mediator of Wnt signaling, *Curr. Opin. Genet. Dev.*, 8, 95, 1998.
58. van de Wetering, M., Cavallo, R., Dooijes, D., van Beest, M., van Es, J., Loureiro, J., Ypma, A., Hursh, D., Jones, T., Bejsovec, A., Peifer, M., Mortin, M., and Clevers, H., Armadillo coactivates transcription driven by the product of the Drosophila segment polarity gene dTCF, *Cell*, 88, 789, 1997.
59. Lin, R., Thompson, S., and Priess, J. R., pop-1 encodes an Hmg box protein required for the specification of a mesoderm precursor in early *C. elegans* embryos, *Cell*, 83, 599, 1995.
60. Lin, R., Hill, R. J., and Priess, J. R., POP-1 and anterior-posterior fate decisions in C. elegans embryos, *Cell*, 92, 229, 1998.
61. Thorpe, C. J., Schlesinger, A., Carter, J. C., and Bowerman, B., Wnt signaling polarizes an early C. elegans blastomere to distinguish endoderm from mesoderm, *Cell*, 90, 695, 1997.
62. Fagotto, F., Gluck, U., and Gumbiner, B. M., Nuclear localization signal-independent and importin/karyopherin-independent nuclear import of β-catenin, *Curr. Biol.*, 8, 181, 1998.
63. Fagotto, F., Kaufmann, C., and Wiechens, N., unpublished.
64. Glinka, A., Wu, W., Delius, H., Monaghan, A. P., Blumenstock, C., and Niehrs, C., Dickkopf-1 is a member of a new family of secreted proteins and functions in head induction, *Nature*, 391, 357, 1998.
65. Bouwmeester, T., Kim, S., Sasai, Y., Lu, B., and De Robertis, E. M., Cerberus is a head-inducing secreted factor expressed in the anterior endoderm of Spemann's organizer, *Nature*, 382, 595, 1996.
66. Moon, R. T., Campbell, R. M., Christian, J. L., McGrew, L. L., Shih, J., and Fraser, S., Xwnt-5A: a maternal Wnt that affects morphogenetic movements after overexpression in embryos of *Xenopus laevis, Development*, 119, 97, 1993.

67. Shi, D. L., Goisset, C., and Boucaut, J. C., Expression of Xfz3, a Xenopus Frizzled family member, is restricted to the early nervous system, *Mech. Dev.*, 70, 35, 1998.
68. Sokol, S. Y., Analysis of Dishevelled signaling pathways during Xenopus development, *Curr. Biol.*, 6, 1456, 1996.
69. Medina, A. and Steinbeisser, H., personal communication.
70. Slusarski, D. C., Corces, V. G., and Moon, R. T., Interaction of Wnt and a Frizzled homologue triggers G-protein-linked phosphatidylinositol signaling, *Nature*, 390, 410, 1997.
71. Krasnow, R. E., Wong, L. L., and Adler, P. N., Dishevelled is a component of the frizzled signaling pathway in Drosophila, *Development*, 121, 4095, 1995.
72. He, X., Saint-Jeannet, J. P., Wang, Y., Nathans, J., Dawid, I., and Varmus, H., A member of the Frizzled protein family mediating axis induction by Wnt- 5A, *Science*, 275, 1652, 1997.
73. Kay, B. K. and Peng, H. B., Eds., *Xenopus laevis*: Practical Uses in Cell and Molecular Biology, in *Methods in Cell Biology*, Vol. 36, Academic Press, Inc., New York, 1991.
74. Nieuwkoop, P. and Faber, P., *Normal Table of Xenopus laevis*, North-Holland Publ., Amsterdam, 1967.
75. Spemann, H. and Mangold, H., Uber Induktion von Embryonenanlagen durch Implantation artfremder Organizatoren., *Wilhelm Roux's Arch. EntwMech. Org.*, 100, 599, 1924.
76. Yuge, M., Kobayakawa, Y., Fujisue, M., and Yamana, K., A cytoplasmic determinant for dorsal axis formation in an early embryo of Xenopus laevis, *Development*, 110, 1051, 1990.
77. Holowacz, T. and Elinson, R. P., Properties of the dorsal activity found in the vegetal cortical cytoplasm of Xenopus eggs, *Development*, 121, 2789, 1995.
78. Kageura, H., Activation of dorsal development by contact between the cortical dorsal determinant and the equatorial core cytoplasm in eggs of Xenopus laevis, *Development*, 124, 1543, 1997.
79. Scharf, S. R. and Gerhart, J. C., Axis determination in eggs of Xenopus laevis: a critical period before first cleavage, identified by the common effects of cold, pressure and ultraviolet irradiation, *Dev. Biol.*, 99, 75, 1983.
80. Elinson, R. P. and Rowning, B., A transient array of parallel microtubules in frog eggs: potential tracks for a cytoplasmic rotation that specifies the dorso-ventral axis, *Dev. Biol.*, 128, 185, 1988.
81. Holowacz, T. and Elinson, R. P., Cortical cytoplasm, which induces dorsal axis formation in Xenopus, is inactivated by UV irradiation of the oocyte, *Development*, 119, 277, 1993.
82. Brannon, M. and Kimelman, D., Activation of Siamois by the Wnt pathway, *Dev. Biol.*, 180, 344, 1996.
83. Scharf, S. R. and Gerhart, J. C., Determination of the dorsal-ventral axis in eggs of *Xenopus laevis:* complete rescue of UV-impaired eggs by oblique orientation before first cleavage, *Dev. Biol.*, 79, 181, 1980.
84. Kao, K. R. and Elinson, R. P., The entire mesodermal mantle behaves as Spemann's organizer in dorsoanterior enhanced *Xenopus laevis* embryos., *Dev. Biol.*, 127, 64, 1988.
85. Hedgepeth, C. M., Conrad, L. J., Zhang, J., Huang, H. C., Lee, V. M., and Klein, P. S., Activation of the Wnt signaling pathway: a molecular mechanism for lithium action, *Dev. Biol.*, 185, 82, 1997.

86. Stambolic, V., Ruel, L., and Woodgett, J. R., Lithium inhibits glycogen synthase kinase-3 activity and mimics wingless signaling in intact cells, *Curr. Biol.*, 6, 1664, 1996.
87. Fagotto, F., Guger, K., Gumbiner, B.M., Brannon, M., and Kimelman, D., Induction of the primary dorsalizing center in Xenopus by the Wnt/GSK/β-catenin signaling pathway, but not by Vg1, Activin or Noggin, *Development*, 124, 453, 1997.
88. Wang, S., Krinks, M., and Moos, M., Jr., Frzb-1, an antagonist of Wnt-1 and Wnt-8, does not block signaling by Wnts -3A, -5A, or -11, *Biochem. Biophys. Res. Commun.*, 236, 502, 1997.
89. Leyns, L., Bouwmeester, T., Kim, S. H., Piccolo, S., and De Robertis, E. M., Frzb-1 is a secreted antagonist of Wnt signaling expressed in the Spemann organizer, *Cell*, 88, 747, 1997.
90. Heasman, J., Holwill, S., and Wylie, C. C., Fertilization of cultured Xenopus oocytes and use in studies of maternally inherited molecules, in *Xenopus: Practical Uses in Cell and Molecular Biology*, Vol. 36, Kay, B. K. and Peng, H. B., Eds., San Diego, Academic Press, 1991, 214.
91. Heasman, J., Ginsberg, D., Geiger, B., Goldstone, K., Pratt, T., Yoshida-Noro, C., and Wylie, C., A functional test for maternally inherited cadherin in Xenopus shows its importance in cell adhesion at the blastula stage, *Development*, 120, 49, 1994.
92. Kofron, M., Spagnuolo, A., Klymkowsky, M., Wylie, C., and Heasman, J., The roles of maternal α-catenin and plakoglobin in the early Xenopus embryo, *Development*, 124, 1553, 1997.
93. Amaya, E. and Kroll, K. L., A method for generating transgenic frog embryos, in *Early Development of Xenopus laevis. Course Manual*, Sive, H. L., Grainger, R. M., and Harland, R. M., Eds, Cold Spring Harbor, 1996, 63.
94. Prives, C. and Foukal, D., Use of oligonucleotides for antisense experiments in Xenopus laevis oocytes, in *Xenopus: Practical Uses in Cell and Molecular Biology*, Vol. 36, Kay, B. K. and Peng, H. B., Eds., San Diego, Academic Press, 1991, 185.
95. Steinbeisser, H., Fainsod, A., Niehrs, C., Sasai, Y., and De Robertis, E. M., The role of gsc and BMP-4 in dorsal-ventral patterning of the marginal zone in Xenopus: a loss-of-function study using antisense RNA, *EMBO J.*, 14, 5230, 1995.
96. Lombardo, A. and Slack, J. M., Inhibition of eFGF expression in Xenopus embryos by antisense mRNA, *Dev. Dyn.*, 208, 162, 1997.
97. Hemmati-Brivanlou, A. and Melton, D. A., Inhibition of activin receptor signaling promotes neuralization in Xenopus, *Cell*, 77, 273, 1994.
98. Yamashita, H., ten Dijke, P., Huylebroeck, D., Sampath, T. K., Andries, M., Smith, J. C., Heldin, C. H., and Miyazono, K., Osteogenic protein-1 binds to activin type II receptors and induces certain activin-like effects, *J. Cell Biol.*, 130, 217, 1995.
99. Chang, C., Wilson, P. A., Mathews, L. S., and Hemmati-Brivanlou, A., A Xenopus type I activin receptor mediates mesodermal but not neural specification during embryogenesis, *Development*, 124, 827, 1997.
100. Lemaire, P., Darras, S., Caillol, D., and Kodjabachian, L., A role for the vegetally expressed Xenopus gene Mix.1 in endoderm formation and in the restriction of mesoderm to the marginal zone, *Development*, 125, 2371, 1998.
101. Cary, R. B. and Klymkowsky, M. W., Differential organization of desmin and vimentin in muscle is due to differences in their head domains, *J. Cell Biol.*, 126, 445, 1994.

102. Klymkowsky, M. W. and Hanken, J., Whole-mount staining of Xenopus and other vertebrates, *Methods Cell Biol.*, 36, 419, 1991.
103. Karnovsky, A. and Klymkowsky, M. W., Anterior axis duplication in *Xenopus* induced by the over-expression of the cadherin-binding protein plakoglobin, *Proc. Natl. Acad. Sci. USA,* 92, 4522, 1995.
104. Schneider, S., Herrenknecht, K., Butz, S., Kemler, R., and Hausen, P., Catenins in *Xenopus* embryogenesis and their relation to the cadherin-mediated cell-cell adhesion system, *Development*, 118, 629, 1993.
105. Fagotto, F. and Gumbiner, B. M., β-catenin localization during *Xenopus* embryogenesis: accumulation at tissue and somite boundaries, *Development*, 120, 3667, 1994.
106. Kurth, T., Licht- und electronenmikrosopische Untersuchungen zur Lokalisation von Zelladhäsionsmolekülen und deren Wechselwirkungen in adulten und embryonalen Geweben von Xenopus laevis (Daudin, 1802), PhD thesis., Tübingen, 1997.
107. Danscher, G., Localization of gold in biological tissues, *Histochemistry*, 71, 1, 1981.
108. Fagotto, F., Funayama, N., Gluck, U., and Gumbiner, B. M., Binding to cadherins antagonizes the signaling activity of β-catenin during axis formation in Xenopus, *J. Cell Biol.*, 132, 1105, 1996.
109. McCrea, P. and Gumbiner, B., Purification of a 92-kDa cytoplasmic protein tightly associated with the cell-cell adhesion molecule E-cadherin (uvomorulin): Characterization and extractability of the protein complex from the cell cytostructure, *J. Biol. Chem.*, 266, 4514, 1991.
110. Hong, M. and Lee, V. M., Insulin and insulin-like growth factor-1 regulate tau phosphorylation in cultured human neurons, *J. Biol. Chem.*, 272, 19547, 1997.
111. Greenberg, S. G., Davies, P., Schein, J. D., and Binder, L. I., Hydrofluoric acid-treated tau PHF proteins display the same biochemical properties as normal tau, *J. Biol. Chem.*, 267, 564, 1992.
112. Opresko, L. K., Vitellogenin and *in vitro* culture of oocytes, in *Xenopus: Practical Uses in Cell and Molecular Biology*, Vol. 36, Kay, B. K. and Peng, H. B., Eds., San Diego, Academic Press, 1991, 117.
113. Peifer, M., Sweeton, D., Casey, M., and Wieschaus, E., Wingless signal and Zeste-white 3 kinase trigger opposing changes in the intracellular distribution of armadillo, *Development*, 120, 369, 1994.
114. Miller, J. R. and Moon, R. T., Analysis of the signaling activities of localization mutants of β-catenin during axis specification in Xenopus, *J. Cell Biol.*, 139, 229, 1997.
115. Yost, C., Torres, M., Miller, J. R., Huang, E., Kimelman, D., and Moon, R. T., The axis-inducing activity, stability, and subcellular distribution of β-catenin is regulated in Xenopus embryos by glycogen synthase kinase 3, *Genes Dev.*, 10, 1443, 1996.
116. Wilson, P. A. and Melton, D. A., Mesodermal patterning by an inducer gradient depends on secondary cell-cell communication, *Curr. Biol.*, 4, 676, 1994.
117. Harland, R. M., *In situ* hybridization: an improved whole mount method for Xenopus embryos, in *Xenopus: Practical Uses in Cell and Molecular Biology*, Vol. 36, Kay, B. K. and Peng, H. B., Eds., San Diego, Academic Press, 1991, 685.
118. Lemaire, P. and Gurdon, J. B., A role for cytoplasmic determinants in mesoderm patterning: cell-autonomous activation of the goosecoid and Xwnt-8 genes along the dorsoventral axis of early Xenopus embryos, *Development*, 120, 1191, 1994.

119. Kaufmann, E., Paul, H., Friedle, H., Metz, A., Scheucher, M., Clement, J. H., and Knochel, W., Antagonistic actions of activin A and BMP-2/4 control dorsal lip-specific activation of the early response gene XFD-1′ in Xenopus laevis embryos, *EMBO J.*, 15, 6739, 1996.
120. Merriam, J. M., Rubenstein, A. B., Klymkowsky, M. W., Tao, Y. S., Edwards, R. A., Tubb, B., Wang, S., Bryan, J., and McCrea, P. D., Cytoplasmically anchored plakoglobin induces a WNT-like phenotype in Xenopus, *Dev. Biol.*, 185, 67, 1997.
121. van den Heuvel, M., Nusse, R., Jonhston, P., and Lawrence, P. A., Distribution of the *wingless* gene product in Drosophila embryos: a protein involved in cell-cell communication, *Cell*, 59, 739, 1989.
122. Moon, R. T., In pursuit of the functions of the Wnt family of developmental regulators: insights from *Xenopus laevis, Bioessays*, 15, 91, 1993.
123. Cui, Y., Brown, J. D., Moon, R. T., and Christian, J. L., Xwnt-8b: a maternally expressed Xenopus Wnt gene with a potential role in establishing the dorsoventral axis, *Development*, 121, 2177, 1995.
124. Ku, M. and Melton, D. A., Xwnt-11: a maternally expressed Xenopus wnt gene, *Development*, 119, 1161, 1993.
125. Deardorff, M. A., Tan, C., Conrad, L. J., and Klein, P. S., Frizzled-8 is expressed in the Spemann organizer and plays a role in early morphogenesis, *Development*, 125(Pt 14), 2687, 1998.
126. Funayama, N., Fagotto, F., McCrea, P., and Gumbiner, B. M., Embryonic axis induction by the armadillo repeat domain of β-catenin: evidence for intracellular signaling, *J. Cell Biol.*, 128, 959, 1995.
127. Guger, K. A. and Gumbiner, B. M., β-catenin has Wnt-like activity and mimics the Nieuwkoop signaling center in Xenopus dorsal-ventral patterning, *Dev. Biol.*, 172, 115, 1995.
128. Lemaire, P., Garrett, N., and Gurdon, J. B., Expression cloning of Siamois, a Xenopus homeobox gene expressed in dorsal-vegetal cells of blastulae and able to induce a complete secondary axis, *Cell*, 81, 85, 1995.
129. Carnac, G., Kodjabachian, L., Gurdon, J. B., and Lemaire, P., The homeobox gene Siamois is a target of the Wnt dorsalisation pathway and triggers organiser activity in the absence of mesoderm, *Development*, 122, 3055, 1996.
130. Dominguez, I., Itoh, K., and Sokol, S. Y., Role of glycogen synthase kinase 3 β as a negative regulator of dorsoventral axis formation in Xenopus embryos, *Proc. Natl. Acad. Sci. USA*, 92, 8498, 1995.
131. Pierce, S. B. and Kimelman, D., Regulation of Spemann organizer formation by the intracellular kinase Xgsk-3, *Development*, 121, 755, 1995.
132. He, X., Saint-Jeannet, J. P., Woodgett, J. R., Varmus, H. E., and Dawid, I. B., Glycogen synthase kinase-3 and dorsoventral patterning in Xenopus embryos, *Nature*, 374, 617, 1995.
133. Behrens, J., von Kries, J. P., Kuhl, M., Bruhn, L., Wedlich, D., Grosschedl, R., and Birchmeier, W., Functional interaction of β-catenin with the transcription factor LEF-1, *Nature*, 382, 638, 1996.
134. Huber, O., Korn, R., McLaughlin, J., Ohsugi, M., Herrmann, B. G., and Kemler, R., Nuclear localization of β-catenin by interaction with transcription factor LEF-1, *Mech. Dev.*, 59, 3, 1996.

135. Isaacs, H. V., Pownall, M. E., and Slack, J. M., eFgf regulates Xbra expression during Xenopus gastrulation, *EMBO J.*, 13, 4469, 1994.
136. Isaacs, H. V., Tannahill, D., and Slack, J. M., Expression of a novel FGF in the Xenopus embryo. A new candidate inducing factor for mesoderm formation and anteroposterior specification, *Development*, 114, 711, 1992.
137. Thomsen, G., Woolf, T., Whitman, M., Sokol, S., Vaughan, J., Vale, W., and Melton, D. A., Activins are expressed early in Xenopus embryogenesis and can induce axial mesoderm and anterior structures, *Cell*, 63, 485, 1990.
138. Schmidt, J. E., Suzuki, A., Ueno, N., and Kimelman, D., Localized BMP-4 mediates dorsal/ventral patterning in the early Xenopus embryo, *Dev. Biol.*, 169, 37, 1995.
139. Wilson, P. A. and Hemmati-Brivanlou, A., Induction of epidermis and inhibition of neural fate by Bmp-4, *Nature*, 376, 331, 1995.
140. Graff, J. M., Thies, R. S., Song, J. J., Celeste, A. J., and Melton, D. A., Studies with a Xenopus BMP receptor suggest that ventral mesoderm-inducing signals override dorsal signals *in vivo*, *Cell*, 79, 169, 1994.
141. Smith, W. C. and Harland, R. M., Expression cloning of noggin, a new dorsalizing factor localized to the Spemann organizer in Xenopus embryos, *Cell*, 70, 829, 1992.
142. Sasai, Y., Lu, B., Steinbeisser, H., Geissert, D., Gont, L. K., and De Robertis, E. M., Xenopus chordin: a novel dorsalizing factor activated by organizer-specific homeobox genes, *Cell*, 79, 779, 1994.

Index

A

Acousto-Optic Tunable Filter (AOTF), 285
Actin filaments, *see* Stress fibers
α-Actinin, 21, 25, 142, 150, 246
Activin/Vg1, 307, 308, 329
Adhesion-dependent cell cycle progression
anchorage-dependent growth assays, 139–140
cell culture protocols
collection and extraction, 132–133
cycle progression induction, 131–132
Erk phosphorylation analysis considerations, 133
maintenance, 130–131
synchronization, 131
cyclin-cdk complex
cdk inhibitor analysis, 138
in vitro kinase activity analysis, 137
recombinant cyclin, 138–139
role of, 130
electrophoresis protocols, 133–135
fluorescent staining
actin stress fiber, 137
focal contact, 137
immunofluorescence, 135–136
mediation of, 130
recombinant protein preparation, 138–139
Adhesion, tracking integrin-mediated
experiment overview, 219
materials and buffers, 226, 228–229
protocols, 219–226
results and applications, 229–231
signaling pathway analysis approaches, 218–219
T cell activation stimuli examples, 218
Affinity chromatography, synthetic peptides study
cytoplasmic binding partner role, 20–21
protocols, 21–22
relevance of interaction, 22
Affinity modulation
CHO cell protocols, 205–206
definition and function, 204
flow cytometry used, 205–206
materials, 206–207
use in $\alpha_{IIb}\beta_3$ protocols, 204–207
α5 and α6-transfectants
α5β1 integrin role in fibronectin deposition, 185–186
differential activation of FAK and paxillin, 152
differentiation of, 155
expression vectors, 145–146, 147
mitogenic pathway role, 151–152
possible transmembrane signaling role, 146
α-actinin, 21, 25
$\alpha_{IIb}\beta_3$, *see* Platelet $\alpha_{IIb}\beta_3$
Amino acids, *see also* Proteins
leucine, 35, 37
lysine, 265
peptides, synthetic, *see* Synthetic peptides
primary paxillin sequence, 72
proline, 50–51
serine/threonine, *see* Serine/threonine kinase
tyrosine, *see* Tyrosine phosphorylation
Anchorage-dependent growth
assays for, 139–140
signals for pathway modulation, 146
Anchorage-independent cell growth, 187
Ankyrin-like repeats, 44
Anoikis
definition of, 161
stable gene expression assay, 162–163

transient assays for gene effects
β-galactosidase/DAPI double-staining method, 163, 164
keratin 18-cleavage assay, 163, 164–165
materials, 165–166
Antibodies
anti-fibronectin, 191
commercially available, 135
D57, 209
FAK specific, 52–54
fibrinogen/ligand-mimetic monoclonal, 204, 237, 238
ligation effects, β cytoplasmic domains, 52–54
non-affinity-purified, 269
phosphospecific, in pathway analysis, 119–120
primary, in protein localization protocol, 269–270
secondary, fluorescently labeled, 270–271
in Shc signaling assay, 103
Antigenic myoblast marker L4, 142, 144
Antisera, *see* Antibodies
AOTF (Acousto-Optic Tunable Filter), 285
Apoptosis, 26, 50
ATPase, 246, 250
Attachment, cell, *see* Cell attachment inhibition
Autoactivation protocol, bait testing, 38–39
Autofluorescence, 271
Avidity modulation
confocal microscopy used, 208
definition and function, 207
flow cytometry used, 207–208
tyrosine phosphorylation analysis, 210–212
use in $\alpha_{IIb}\beta_3$ protocols, 207–208

B

Bait fusion proteins
cDNA library screens, 40–42
fusion protein expression, 35
plasmid preparation
bait construct, 35
protocols, 35–37
transformation methods, 37–38
testing
autoactivation protocol, 38–39
expression protocol, 39–40
BDM, 250
β1, *see* β Cytoplasmic domains
Binding affinity
actin filament, 246
calculating, 25–26
DNA binding domain, 35
identifying with peptides, 20–21
kinase and calmodulin binding, 247
mapping, 29–30
paxillin and GST-fusion protein, 72–74
protein binding sites, 64
Biological functions of integrin signaling
anoikis
stable expression experiments, 162–163
transient assays for gene effects, 163–165
cell cycle progression analysis
anchorage-dependent growth assays, 139–140
cell culture protocols, 130–133
cyclin-cdk complex, 130, 137, 138
electrophoresis protocols, 133–135
fluorescent staining, 135–137
recombinant protein preparation, 138–139
cell migration
chemotaxis chamber study, 171–175
CHO cell stable functional analysis, 169–171
FAK biochemical pathway analysis, 175–179
FAK role in, 50, 168
fibronectin matrix regulation
biochemical analysis, 192–196
definition and function, 184, 196
fluorescence microscopy analysis, 189–192
signaling mechanisms, 185–188

MAP kinase cascades
coimmunoprecipitation of Raf and Ras/Rap, 123–124
GTP loaded Ras measurement assay, 122–123
immunoprecipitation of Raf kinase, 120–122
measurement of, 118–119
morphology and signaling examination, 124–126
pathway description and analysis, 118–120
mitogenic pathways
downstream effectors, 151–154
muscle differentiation mechanism, 142–143, 156–157
primary quail myoblast study protocols, 144–146
signal modulation, 146–150
terminal differentiation, 154–156

C

C3 exotransferase, 250
[Ca^{2+}], *see* Intracellular [Ca^{2+}]
Ca^{2+}/calmodulin binding, 247
Cancer cells
FAK overexpression and, 50
fibronectin matrix studies and, 184
oncogenes, 256
tumor malignancy and anoikis, 161
Cas (Crk-associated substrate)
cell motility and FAK association, 168
-crk signaling complex studies
crk role, 89–90
JNK pathway activation detection, 90–91
materials and buffers, 95–96
protocols, 91–95
docking module role, 82
protein-protein interactions
materials and buffers, 89
mediation by tyrosine phosphorylation, 86
protocols, 86–89
SH2-domain relationship, 82, 86, 89
tyrosine phosphorylation relationship
materials and buffers, 84–85
overview, 82–83, 86
protocols, 83–84
β-Catenin Wnt/*Xenopus* experiments, 327
exo- vs. endogenous proteins, 346–347
functions, 316
gene target regulation, 316
signaling activity model, 316–317
stability analysis, 341–342
subcellular localization, 331
Caveolin-1
associated c-Fyn tyrosine kinase, 102, 110–111
coimmunoprecipitation
of integrins, 104–106
of Shc, 109
CCD cameras, 286, 294
CD2+ HL-60 cells, *see* Green fluorescent protein
cDNA
encoding mutations for transfection, 64
library parameter, 237
library screen, 35, 37, 40–42, 236
Cell attachment inhibition
chimeric receptors expression assay, 9–12
β cytoplasmic domains and, 9
flow cytometry used, 9–11, 12
Cell cycle progression, adhesion-dependent
anchorage-dependent growth assays, 139–140
cell culture protocols
collection and extraction, 132–133
cycle progression induction, 131–132
Erk phosphorylation analysis considerations, 133
maintenance, 130–131
synchronization, 131
cyclin-cdk complex
cdk inhibitor analysis, 138
in vitro kinase activity analysis, 137
recombinant cyclin, 138–139
role of, 130
electrophoresis protocols, 133–135

fluorescent staining
actin stress fiber, 137
focal contact, 137
immunofluorescence, 135–136
mediation of, 130
recombinant protein preparation, 138–139
Cell differentiation studies
cell cycle transition regulation, 146–147
integrin regulation, 147
proliferating vs. differentiating, 143, 150
protocol for, 149–150
Cell migration
chemotaxis chamber study, 171–175
CHO cell functional analysis, 169–171
discussion of FAK role in, 50, 168
FAK biochemical pathway analysis
general considerations, 175
in vitro kinase assay, 176–177
in vivo /Src, /Cas association in CHO cells protocols, 177–178
materials and equipment, 178–179
tyrosine phosphorylation protocol, 176
fibronectin matrix assembly, 168
intracellular [Ca^{2+}] relationship, 280
Cell spreading inhibition
chimeric receptors and, 12–14
possible FAK signaling role, 50
Chelators, 289
Chemical inducement of dimerization, 207
Chemotaxis chamber, 171–175, 294
Chimeric receptors signaling studies
cell attachment inhibition
β cytoplasmic domains mediation, 9
expression level comparisons, 10–11
materials, 12
protocols, 11–12
cell spreading inhibition
β cytoplasmic domains regulating role, 12–13
materials and buffers, 14
protocols, 13–14
dominant negative inhibitor function, 4, 14–15
tyrosine phosphorylation activation
increase in FAK observed, 5
materials and buffers, 8–9
protocols, 5–8
Chinese hamster ovary (CHO) cells
in affinity modulation assay, 205–206
$\alpha_{IIb}\beta_3$ signaling
inside-out, 204–209
outside-in, 209–213
chemotaxis chamber motility study, 173–175
fibronectin matrix relationship, 185
in fluorescence microscopy assay, 189–191
functional analysis, 168
materials and equipment, 170–171
protocols, 169–170
stable overexpression pros and cons, 169
in vivo FAK/Src, FAK/Cas association, 177–178
stable overexpression of FAK, 50, 168
Chromatography, *see* Affinity chromatography
c-Jun kinases, *see* JNK pathway
Clone analysis, 241
Coimmunoprecipitation
with caveolin-1, 104–106, 109
caveolin-1 associated c-Fyn tyrosine kinase assay, 110–111
Fyn with Shc, 111
of Raf and Ras/Rap, 123–124
Color development on glucose, 42
Confocal microscopy
in avidity modulation assay, 208
resolution considerations, 285
schematic diagrams, 292–294
vs. wide-field, 282–286
Constitutively active constructs, 218–219
Contractility regulation by RhoA
and focal adhesion assembly, 249–250
inhibitors, 248
mechanism for decreased, 247
microinjected cell analysis, 250–255
in nonmuscle cells, 246–250
role of, 256
stress fiber attachment mediation, 246
Crk, *see* Cas (Crk-associated substrate)

Cryoprotected tissue, 267–268
Cyclin-cdk complex
cdk inhibitor analysis, 138
kinase activity analysis, 137
recombinant cyclin, 138–139
role of, 130
β Cytoplasmic domains
adhesion process, importance in, 4
antibody ligation effects, 146
in bait plasmid screen preparation, 35–36
binding partners identification, 20–21
cell attachment role, 9
cell phenotype determination and, 147
cell spreading role, 12–13
chimeric receptors used to define, 14–15
chimeric receptors used to study, 4–9
enhanced integrin signaling effects, 156–157
FAK activation by, 5
FAK interaction with talin, 102
fibronectin matrix regulation, 186
integrin signaling role, 12–13
myoblast proliferation and, 147
in protein isolation screen, 236
synthetic peptides studies, 20–21
Cytoplasmic tyrosine kinase, *see* Focal adhesion kinase
Cytoskeleton
adaptor protein, *see* Phosphorylation
integrin signaling role, 124
synthetic peptide interactions with integrins, 20–22, 28

D

DAI (dorsoanterior index), 320
Desmin, 142
Dimerization, chemical inducement of, 207
Dishevelled, 309, 331, 340, 347
DNA-binding domain, 35
Docking proteins, 82
Dominant-negative constructs
chimeric receptor studies, 4, 9, 14–15
engrailed repressor chimeras, 329
overexpression techniques, 218–219
Wnt pathway, 328–329
Dorsal studies
dorsalization by LiCl treatment, 319
dorsalization by mRNA injection, 322–323
dorsoventral asymmetry, 305
dorsoventral patterning, 306, 308
induction, 305
Wnt pathway specification, 307–308
Dorsoanterior index (DAI), 320
Dyes, *see also* Green fluorescent protein
Ca^{2+} indicator selection, 286–290
fluorescent staining, 135–137, 189, 274–275
loading and calibration, 291–292
properties and uses, 286

E

EGTA, 287
Electrophoresis protocols, *see also* SDS-polyacrylamide gel
bait plasmid transformation, 38–39
cell cycle progression, 133–135
cell spreading study, 13
chimeric receptor studies, 7
FAK immunoprecipitation, 54
fibronectin matrix assembly assay, 193
paxillin phosphorylation assay, 76
in yeast two-hybrid screen, 36–37
ELISA analysis, 193–194
Embedding/sectioning techniques, 266–268, 272–274
Embryogenesis, 184
Emission ratio measurements, 282–284
Endothelial cells, 103, 256
Enzymes
ATPase, 246, 250
C3 exotransferase, 250
FAK, activation by β cytoplasmic domains, 5
β-Galactosidase
/DAPI double-staining method, 163, 164
reporter gene, 35

GTPases
activating proteins, 118
regulation of fibronectin matrix assembly, 187–188
RhoA regulation of focal adhesion and actin filaments, 246
myosin light chain kinase, 247
Epithelial cells
anoikis of, 161
MDCK cells, 162
protected by FAK activation, 26
renal, 264
in RhoA study, 256
in Shc signaling assay, 103
Erks (extracellular-regulated kinases), 90
phosphorylation analysis considerations, 133
Ras, 120, 152
anti-active antibody, 119
MAP kinase activation, 118, 120
Raf-1 activation and, 118, 187
Shc activation of, 102–103
Excitation wavelengths, 293
Expression cloning
cDNA library parameters, 237
experimental considerations, 237–238
flow cytometry techniques, 237–238
inside-out signaling for protein identification, 235–236
outside-in signaling, 235
protein isolation screens, 236
protocols
cell sort, 238–240
clone analysis, 241
Hirt supernatants, 240
materials and equipment, 241–242
Extracellular-regulated kinases (Erks), *see* Erks

F

FACSing (fluorescence-activated cell sorting), *see* Flow cytometry
FAK, *see* Focal adhesion kinase
Far-Western analysis
^{32}P-labeled GST generation, 56–57
with GST-fusion proteins, 88
protein denaturation/renaturation, 57
FGF (fibroblast growth factor), 155, 306, 308
Fibrinogen, 203, 204, 205, 209, 237, 238
Fibroblasts, 26, 103, 155, 306, 308
Fibronectin matrix assembly
biochemical analyses
materials, 195–196
protocols, 193–194
technical comments, 194–195
cell control of pericellular, 184, 187
cell migration and, 168
definition and function, 184, 196
fluorescence microscopy analysis
assembly process considerations, 190–191
materials, 191–192
protocols, 189–190
signaling mechanisms
cytoplasmic domain, regulation by, 186
GTPases, regulation by, 187–188
initiation due to cell surface integrins, 185
intracelluar protein kinases, regulation by, 186–187
mediation by specific integrins, 185–186
Flow cytometry, 237
in adhesion assay, 222–226
adhesion study overview, 219
in affinity modulation assay, 205–206
in avidity modulation assay, 207–208
in cell attachment assay, 9–11, 12
experimental results and applications, 229–231
expression cloning, 237–238
vs. fluorometry, 280–281
in mitogenic pathway study, 148–149
quantitation of cell number, 222–226
Fluo-3, 286, 287, 289
Fluorescein-phalloidin, 137
Fluorescence-activated cell sorting (FACSing), *see* Flow cytometry

Fluorescence microscopy
fibronectin matrix assembly, 189–192
image analysis, [Ca^{2+}], 297–298
Fluorescent staining of cells
actin stress fiber, 137
focal contact, 137
immunofluorescence, 136, 189, 274–275
Fluorometry vs. flow cytometry, 280–281
Fluorophores, *see* Dyes
Focal adhesion kinase (FAK)
activation by β cytoplasmic domains, 5
antibodies, specific, 52–54
apoptosis inhibition role, 26
biochemical pathway analysis
general considerations, 175
in vitro kinase assay, 176–177
in vivo /Src, /Cas association in CHO cells, 177–178
materials and equipment, 178–179
tyrosine phosphorylation, 176
cell migration role, 50, 168
cellular adhesion role, 49–50
chimeric receptor study, 5, 7
contractility and, 249–250
differential activation by α5 and α6, 152
integrin ligation effect on, 151
overexpression in CHO cells, 168
phosphorylated sites, 64
protein-protein interactions
cell lysates preparation, 51–52
in vitro protein association, 55–57
in vivo protein association, 52–55
signal transducing complex, 49–51
subcellular localization, 57–58
response to integrin activation, 168
Src-family kinase combination effects, 102, 168
stress fibers and, 249
Free acid form, 289, 291
Frizzled, 308, 309, 317, 331, 340
Fura-2
cell loading with, 295
properties of, 286, 287, 289
uses of, 292, 294
Fusion proteins
expression, bait testing, 35, 38, 39–40
green fluorescent protein
adhesion study overview, 219
experimental results and applications, 229–231
flow cytometry protocol, 222–226
limitations to use with *Xenopus*, 345
protocols using, 219–226
in RhoA GTPases study, 124
for subcellular localization, 58
GST
association of Shc, with tyrosine phosphorylated β4, 113
-C3 exotransferase expressing cells, 251–252
Far Western blotting, 88
generation of ^{32}P-labeled, 56–57
paxillin binding role, 72–74
protocols, 86
purification of, 55–56, 86
recombinant protein purification, 250
SH2-domains *in vitro* association experiments, 87–88
Fyn
caveolin-1 associated, tyrosine kinase, 110–111
coimmunoprecipitation with caveolin-1
protocols, 104–105
reagents, 105–106
coimmunoprecipitation with Shc, 111
transition after mediated activation, 102

G

G0 phase, G1, 131–132, 138
G1 phase, cell cycle, 103, 130
β-Galactosidase
/DAPI double-staining method, 163, 164
reporter gene, 35
GFP, *see* Green fluorescent protein
Glutathione-S-transferase (GST), *see* GST-fusion proteins
Glycogen Synthase Kinase-3β (GSK3β), 309, 340
Grb-2 recruitment, 112

Green fluorescent protein (GFP)
adhesion study overview, 219
experimental results and applications, 229–231
flow cytometry protocol, 222–226
limitations to use with *Xenopus*, 345
protocols using, 219–226
in RhoA GTPases study, 124
for subcellular localization, 58
Growth factors and terminal differentiation, 154–155
GSK3β (Glycogen Synthase Kinase-3β), 309, 340
GST-fusion proteins
association of Shc, with tyrosine phosphorylated β4, 113
-C3 exotransferase expressing cells, 251–252
Far Western blotting, 88
generation of ^{32}P-labeled, 56–57
paxillin binding role, 72–74
protocols, 86
purification of, 55–56, 86
recombinant protein purification, 250
SH2-domains *in vitro* association experiments, 87–88
GTPases
activating proteins, 118
regulation of fibronectin matrix assembly, 187–188
RhoA regulation, 124–126, 212–213, 246
GTP loaded Ras, 122–123

H

H7, 250
Hematopoietic cells, 217, 219
Hirt supernatants, 240
H-Ras, 187–188, 236

I

IL-2 receptor, 4, 12
ILK, *see* Integrin-linked kinase
Image analysis, fluorescence microscopy
background correction, 297
misregistration, 297–298
pseudocolored images, 298
rapid image acquisition, 298
Immunofluorescence
cell spreading study, 13–14
in fixed cells, 57–58
fluorescent staining, 136, 189, 274–275
microinjection cell analysis, 253–254
in Wnt pathway studies, 331–333, 340
Immunogold labeling, 333–334
Immunohistochemical protocol, *see* Immunolocalization of proteins
Immunolocalization of proteins
blocking non-specific background, 268–269
buffer preparation, 275–276
β cytoplasmic domains, 272–274
embedding/sectioning, 266–268, 272–274
fixation considerations, 264–266
in situ study advantages, 264
permeablization, 268
primary antisera, 269–270
protocols, 272–275
secondary antisera/detection systems, 270–272
Immunoprecipitation, *see also* Coimmunoprecipitation
analysis of cyclin-cdk complexes, 138
of FAK, 51
antibody specific, 52–54
in chimeric receptor study, 5, 7
/immunoblotting, 54–55
lysis buffers used, 73
MAP kinase measurement, 118–119
myosin light chain phosphorylation analysis, 254–255
paxillin phosphorylation assay, 76–77
of Raf kinases, 120–122
tyrosine phosphorylation of Cas, 84
Western Blotting and, 76–78
Wnt molecular interactions, 339
Immuno-staining, 266
Indicators, fluorescent, 280; *see also* Dyes

Indo-1, 286, 287, 289, 292
Inhibitors
  actin-myosin interaction, 250
  contractility, 248
  dominant, function of, 14
  in signaling pathway studies, 218
  Wnt, 317
Inside-out signaling
  $\alpha_{IIb}\beta_3$
    affinity modulation, 204–207
    avidity modulation, 207–209
  contractility regulation by RhoA
    and focal adhesion assembly, 249–250
    microinjected cell analysis, 253–255
    microinjected protocols, 250–253
    nonmuscle and MLCK, 246–250
    role of, 256
    stress fiber attachment mediation, 246
  cytoplasmic binding partner role, 20
  expression cloning of proteins
    experimental considerations, 237–238
    protocols, 238–242
    use in protein identification, 235–236
  integrin-mediated adhesion
    experimental results and applications, 229–231
    flow cytometry protocol, 222–226
    GFP-fusion protein protocols, 219–226
    signaling pathway analysis approaches, 218–219
  platelet $\alpha_{IIb}\beta_3$, 203–204
    affinity modulation, 204–207
    avidity modulation, 207–209
    outside-in signaling, 209–212
Insulin, 155
Integrin β cytoplasmic domains, *see* β Cytoplasmic domains
Integrin function, 4, 102, 118–119, 151, 168
Integrin-linked kinase (ILK)
  identified as cytoplasmic partner, 34–35
  identified in library screen, 42
  $p59^{ILK}$ overexpression consequences, 44
  regulation of fibronectin matrix assembly, 186–187
Integrin-mediated Shc signaling, *see* Shc signaling pathway
Interaction plasmid rescue, 43–44
Interaction trap, 34, 35
Interleukin-2 (IL-2) receptor, 4
Intracellular [Ca^{2+}]
  calcium response experiments, 280
  cell migration relationship, 280
  microscopy experiment design
    dye loading and calibration, 291–292
    indicator selection, 286–290
    instrumentation, 292–294
    ratiometric vs. single-wavelength measurements, 286
    wide-field vs. confocal microscopy, 282–286
Isometric tension, 246, 249, 256

J

JNK pathway
  in vitro kinase assay, 90, 91–93
  in vivo assessment, 90
  luciferase trans-activator reporter system, 90, 94–95
  materials and buffers, 95–96
  pathway activation detection, 90–91

K

Keratin 18-cleavage assay, 163, 164–165
Keratinocyte differentiation, 157
Knockin embryos, 186

L

LEU2, 35, 37
LiAc (Lithium acetate), 37

Library screens, 35, 37, 40–42, 236
Library transformations, 37
LiCl treatment, 319
Ligand-mimetic monoclonal antibody, 204, 237, 238
LIM protein, 64, 142
Lithium acetate (LiAc), 37
Luciferase trans-activator reporter system, 90, 94–95

M

Magnetic sorting, 6, 7
MAP kinase cascades
  activation studies, 154
  Erk migration, 118
  FAK and paxillin regulating role, 151
  GTP loaded Ras measurement assay, 122–123
  integrin-mediated signaling and, 142, 152
  measurement of, 118–119
  mechanism for integrin regulation, 117–118
  morphology and signaling examination, 124–126
  pathway description and analysis, 118–120
  Raf tested for effects on MEK, 120–122
  Ras/Raf and Rap/Raf interactions assay, 123–124
MAPKs (mitogen-activated protein kinases), *see* MAP kinase cascades
Mapping, pins and paper, 29–30
MDCK cells, 162
MEK (MAPK/Erk kinases), 117, 121, 152
Mesoderm, 305–308
MG-63 cells, 10, 11, 13
Microinjection
  cell analysis
    cell labeling, 254–255
    immunofluorescence, 253–254
    myosin immunoprecipitation, 254–255
    silicone rubber substrata, 254
  protocols
    cell culture, 253
    procedures, 252–253
    purification of recombinant proteins, 250–252
  stress fiber formation, 28
Microspectrofluorometry, 281
Microtiter-well binding assays
  synthetic peptides, 25–26
Microtubule depolymerization, 250
Mitogen-activated protein kinases (MAPKs), *see* MAP kinase cascades
Mitogenic pathways
  differential activation of FAK and paxillin
    α5 and α6 role in, 151–152
    protocols, 152–154
  enhanced integrin signaling effects, 156
  flow cytometry used, 148–149
  muscle differentiation mechanism, 142–143, 156–157
  primary quail myoblast used to study
    benefits to using, 143–144
    cell culture protocols, 145
    cell cycle transition studied, 146–147
    expression vectors, 145–146
    isolation protocol, 144–146
    muscle differentiation mechanism, 142–143
    signaling research using, 156–157
  signal modulation by integrin subunits
    β1 cytoplasmic domain role in, 147
    protocols, 148–150
  terminal muscle differentiation in myoblasts, 146
    growth factor environment, 155
    protocols, 156
    regulation of onset, 154–155
Mitotic kinase, $p34^{cdc2}$, 246
MLCK, *see* Myosin light chain kinase
MLP (muscle LIM protein), 142
mRna
  concentration considerations, 345
  injection into *Xenopus* embryos
    alternate pathway tests, 326–327
    dorsalization, 322–323

purposes of, 321
specificity tests, 325
ventralization, 323–325
Muscle differentiation
mechanism for, 142–143, 156–157
primary quail myoblast used to study
benefits to using, 143–144
cell culture protocols, 145
cell cycle transition studied, 147
expression vectors, 145–146
isolation protocol, 144–145
muscle differentiation mechanism, 142–143
signaling research using, 156–157
terminal, 146
Muscle LIM protein (MLP), 142
Myoblast determination factor (MyoD), 142
Myoblast differentiation, *see* Primary quail myoblast
Myogenesis, 146
Myosin II, 246
Myosin light chain kinase (MLCK)
actin filament binding, 246
kinase and calmodulin binding, 247
phosphorylation analysis, 246, 247, 254–255

N

Nanogold labeling, 333–334
Neutrophils, 280
Nieuwkoop Center, 305
Non-affinity-purified polyclonal antisera, 269
Nonmuscle contractility, *see* Contractility Regulation by RhoA
NP-40 (Nonidet-40), 137, 330

O

Optical microscopy
[Ca^{2+}] study considerations, 280
advantages and disadvantages, 280–282
dye loading and calibration, 291–292
indicator selection, 286–290
instrumentation, 292–294
protocols
calibration, 296–297
cell loading with fura-2/AM, 295
image acquisition, 296
image analysis, 297–298
instrumentation, 295–296
sample chambers, 294
ratiometric vs. single-wavelength measurements, 286
scan time and illumination considerations, 285–286
wide-field vs. confocal microscopy
confocal advantages, 282–283, 285–286
wide-field use, 283–284
Outside-in signaling
$\alpha_{IIb}\beta_3$
CHO cell adhesion, 209–210
integrin-dependent morphological changes, 212–213
tyrosine phosphorylation/kinase activity, 210–212
cytoplasmic binding partner role, 20
expression cloning and, 235
platelet studies of, 204

P

p130^{Cas}, *see* Cas (Crk-associated substrate)
PAC1 (fibrinogen-mimetic monoclonal antibody), 204, 237, 238
PAGE, *see* SDS-polyacrylamide gel
Paraformaldehyde-lysine-periodate (PLP), 265
Paxillin
cell migration
materials, 75
possible role in, 74
protocols, 74–75
differential activation by α5 and α6, 152
FAK and phosphorylation, 64
-homologous proteins, 64
integrin link role, 151

LIM domains, 64
localization to focal adhesions, 51
cell adhesion effects, 66–67, 69–71
materials and buffers, 71–72
protocols, 67–71
multi-domain analysis using GST-fusion proteins
materials and buffers, 73–74
protocols, 72–73
purpose of, 72
phosphorylation analysis, 76–78
protein binding sites, 64
transfection assay
cDNA mutation used for, 64–65
materials, 66
protocols, 65–66
pBABE vector, 162
Peptides, *see also* Synthetic peptides
Peptides, soluble mitogen, 155
Pericellular matrix assembly, *see* Fibronectin matrix assembly
$p59^{ILK}$, *see* Integrin-linked kinase (ILK)
Phosphatidylinositol 3-kinase, 50
Phosphoinositide 3-OH kinase (PI 3-K), 218
Phosphorylation
cytoskeletal proteins and, 64
Erk analysis considerations, 133
FAK and, 64
myosin light chain, 246–249, 254–255
paxillin, 76–78
tyrosine, *see* Tyrosine phosphorylation
Phosphospecific antibodies, 119–120
PI 3-K (Phosphoinositide 3-OH kinase), 218
Pins and paper mapping technique, 29–30
Plasmids, bait
construction, interaction trap, 35–36
reporter, 38
rescue of interaction, 43–44
Platelet $\alpha_{IIb}\beta_3$
inside-out signaling, 203
affinity modulation, 204–207
avidity modulation, 207–209
outside-in signaling, 204
CHO cell adhesion, 209–210
integrin-dependent morphological changes, 212–213
tyrosine phosphorylation/kinase activity, 210–212
studies of, 203–204
PLP (paraformaldehyde-lysine-periodate), 265
Polymorphonuclear leukocytes, 280
Primary quail myoblast
benefits to using, 143–144
cell culture protocols, 145
cell cycle transition studied, 147
expression vectors, 145–146
isolation protocol, 144–145
muscle differentiation mechanism, 142–143
signaling research using, 156–157
Proline, 50–51
Protein-protein interactions
Cas studies
-crk signaling complex studies, 89–96
mediation role investigated, 86–89
multiple domains of, 82, 86, 89
tyrosine phosphorylation of, 83–85
chimeric receptor signaling studies
β cytoplasmic domain function, 4, 14–15
cell attachment inhibition, 9–12
cell spreading inhibition, 12–15
tyrosine phosphorylation activation, 5–9
focal adhesion kinase
interaction identification, 51–57
signal transducing complex, 49–51
subcellular localization, 57–58
paxillin studies
cell migration role, 74–75
effect on cell adhesion, 69–71
multi-domain analysis using GST-fusion proteins, 72–74
phosphorylation analysis protocols, 76–78
rate of localization to focal adhesion, 66–69
transfection assay, 64–66
specificity through Shc signaling pathway
coimmunoprecipitation of integrins, 104–106, 109, 111

experimental activation methods, 103–104
GST-fusion proteins assay, 113
integrin ligation, 106–108
tyrosine kinase assay, 110–111
tyrosine phosphorylation, 102, 112–113
synthetic peptides, cytoplasmic domain studies
affinity chromatography, 20–23
mapping, pins and paper, 29–30
microtiter-well binding assays, 25–26
resin pull-down experiments, 23–24
single-cell microinjection, 26–29
yeast two-hybrid screen
autotrophic selection importance, 47
bait plasmids preparation, 35–38
bait plasmids testing, 38–40
cDNA library screens, confirmed baits, 40–42
false positive isolation, 42, 47
interaction plasmid rescue, 44, 43
interaction trap, 34, 35
materials and reagents, 45–46
signal transduction mechanism, 34–35
Proteins, *see also* Amino acids
baits constructs, transcriptionally inert, 35
β cytoplasmic domains, *see* β Cytoplasmic domains
bound, identifying, 20–21, 22
Cas, *see* Cas (Crk-associated substrate)
caveolin-1
associated c-Fyn tyrosine kinase, 102, 110–111
coimmunoprecipitations, 104–106, 109
cyclin-cdk complex
cdk inhibitor analysis, 138
in vitro kinase activity analysis, 137
recombinant cyclin, 138–139
role of, 130
docking, 82
expression cloning
cDNA library parameters, 237
clone analysis, 241
experimental considerations, 237–238
flow cytometry techniques, 237–238
inside-out signaling for protein identification, 235–236
materials and equipment, 241–242
protein isolation screens, 236
protocols, 238–240
FAK binding, *see* Focal adhesion kinase
fibronectin, *see* Fibronectin matrix assembly
function determination, 264
fusion
in baits testing, 38, 39–40
construct importance, 35
GFP, *see* Green fluorescent protein
GST, *see* GST-fusion proteins
immunolocalization
blocking non-specific background, 268–269
buffer preparation, 275–276
embedding/sectioning, 266–268
fixation considerations, 264–266
in situ study advantages, 264
permeablization, 268
primary antisera, 269–270
protocols, 272–275
secondary antisera/detection systems, 270–272
kinase C
function determined with inhibitors, 218
H7 relationship to, 250
inhibitory phosphorylation, 246
stress fibers and, 187
microtiter-well binding assays, 25–26
myoblast determination factor, 142
paxillin, *see* Paxillin
-protein interactions, *see* Protein-protein interactions
recombinant, 91
preparation for cell cycle progression, 138–139
purification of, 250–252

serine/threonine kinase, 34
H7 activity, 250
ILK activity, 186–187
Ras effects on, 118
regulation by Wnt pathway, 309
structural, 246
verifying relevance of interaction, 22
pSH18-34, 38
Pull-down experiments
with GST-SH2-domains, 87–88
peptide-coated resin, 23–24

Q

Quin2, 289

R

Raf-1 kinase, 118, 187
Ras-Erk
GTP/GDP loading assay, 122–123
MAP kinase pathway, 118
and Raf interactions, 118, 123–124, 187
Shc activation of, 102–103
Ratiometric measurements, 286, 287–289
Recombinant proteins
JNK pathway and, 91
preparation for cell cycle progression, 138–139
purification of, 250–252
Reporters
domain, IL-2 receptor used as, 4
genes, 35, 91, 345
GFP in flow cytometry, 219
plasmids, 38
Resins, peptide-coated, 23–24
Reverse transcription, 342–344
RGD-containing integrin binding domain, 185
Rho, 187–188
RhoA
contractility and focal adhesion assembly
direct inhibition effects, 250
microtubule depolymerization and activation, 250
stress fiber regulation, 249
contractility and tension generating role, 246, 256
GTPases, 124–126, 212–213, 246
microinjected cell analysis
cell labeling, 254–255
immunofluorescence, 253–254
immunoprecipitation, 254–255
protocols, 250–253
silicone rubber substrata, 254
nonmuscle contractility regulation
myosin light chain kinases, 246–247
phosphorylation results, 247–250
R-Ras, 187–188

S

Scatchard plot, 25
SDS-polyacrylamide gel, *see also* Electrophoresis protocols
cell cycle progression analysis, 133–135
chimeric receptor studies, 54, 56
FAK study, 176
fibronectin matrix assembly, 193–194
paxillin studies, 76
used to assess kinase activity, 138
in Wnt study, 344
Ser-1/2, 246
Ser-19, 246, 247
Serine/threonine kinase, 34
H7 activity, 250
ILK activity, 186–187
Ras effects on, 118
regulation by Wnt pathway, 309
SH2, *see* Src Homology-2 (SH2)
SH3-domain, 82
Shc signaling pathway
activation of Ras, 102–103
caveolin-1 associated c-Fyn tyrosine kinase assay, 110–111
coimmunoprecipitation of integrins
with caveolin-1, 104–106, 109
Fyn with Shc, 111

experimental activation methods, 103–104
Grb-2 recruitment, 112
GST-fusion proteins assay, 113
integrin ligation
adhesion to anti-integrin antibody/ECM-coated dishes, 107–108
with antibody-coated polystyrene beads, 107
crosslinking with soluble antibodies, 106–107
reagents, 108
tyrosine phosphorylation of, 102, 112–113
Single-cell [Ca^{2+}], *see* Intracellular [Ca^{2+}]
Single-cell microinjection, synthetic peptides
to block FAK activation, 26
phenotypic changes, 28
protocols, 26–28
Single-wavelength measurements, 286
Sizzled, 307, 308
Sodium dodecyl sulfate polyacrylamide gel (PAGE), *see* SDS-polyacrylamide gel
Solid-phase binding assays
chromatography, 20–21
mapping, pins and paper, 29–30
microtiter-well, 25–26
resin pull-down experiments, 23–24
single-cell microinjection, 26–28
Soluble peptide mitogens, 155
Specificity, *see also* Shc signaling pathway
dorsal and ventral injections tests, 325
in fibronectin matrix assembly, 184
of identified protein interactions, 44
importance in dominant negative constructs, 329
inverse relationship with degree of use, 218
Spectrofluorometry, 280–281
Spemann Organizer, 306
Spot synthesis, 30
Spreading, cell, *see* Cell spreading inhibition
Src-family kinase
cell migration and association with FAK, 168
FAK combination with, 102
tyrosine and, 82, 83
Src Homology-2 (SH2)
binding motifs, 82
domains, 86, 118
GST- pull-down experiments, 87–88
tyrosine phosphorylation, 102
Stress fibers
fluorescent staining, 137
focal adhesion and, 249
formation analysis in microinjection assay, 28
myosin binding to, 246
protein kinase C inhibition and, 187
structural proteins and, 246
Study methods for integrin signaling
in situ immunolocalization
blocking non-specific background, 268–269
buffer preparation, 275–276
embedding/sectioning, 266–268
fixation considerations, 264–266
permeablization, 268
primary antisera, 269–270
protocols, 272–275
secondary antisera/detection systems, 270–272
optical microscopy studies
advantages and comparisons, 280–282
dye loading and calibration, 291–292
indicator selection, 286–290
instrumentation considerations, 292–294
protocols, 294–298
ratiometric vs. single-wavelength measurements, 286
wide-field vs. confocal microscopy, 282–286
Wnt pathway
axis formation manipulations, 318–320
β-catenin stability analysis, 341–342
dorsal and ventral injections, 321–327

embryo extract preparation, 330–331
endogenous components, interfering with, 327–329
expression/overexpression considerations, 345–347
gene expression detection, 342–345
GSK activity, 340
molecular interaction detection, 339–340
patterning during development, 305–308
properties of, 309–317
subcellular localization, 331–339
Xenopus embryo study, 304–308
Synthetic peptides
affinity chromatography study
cytoplasmic binding partners identified with, 20–21
potential for non-specific binding, 22
protocols, 21–22
mapping, pins and paper, 29–30
materials and instruments, 30
microtiter-well binding assays
binding interaction characterization, 25
protocols, 25–26
resin pull-down experiments
peptide-coated resin benefits and uses, 23
protocols, 24
single-cell microinjection
to block FAK activation, 26
phenotypic change considerations, 28
protocols, 26–28

T

Talin, 51, 102, 246, 254
T cell activation stimuli examples, 218
Titin, 142
Tumorigenesis, 184, 187
Tyrosine phosphorylation, *see also* Focal adhesion kinase
analysis with avidity modulation, 210–212
Cas relationship with, 82–86
chimeric receptors, activation by
increase in FAK observed, 5
materials and buffers, 8–9
protocols, 5–8
focal adhesion association, 50, 176
/kinase activity, 210–212
mitogenic pathways and, 152
outside-in signaling, 210–212
response to integrin activation, 168
of Shc, 102, 112–113

U

UV irradation, 318–319

V

Ventralization
by mRNA injection, 323–325
by UV irradation, 318–319
Vinculin, 72, 246, 254
Von Willebrand factor, 204, 209
VRhoA, 250; *see also* RhoA

W

Western Blotting
cyclin-cdk complex assays, 138
electrophoresis and, 133–135
and immunoprecipitation, 76–78
MAPK activation determination, 154
phosphotyrosine, 7
protein expression in yeast, 39
Whole mount immuno-staining, 266
Wide-field microscopy
vs. confocal, 282–286
excitation filters, 293

image acquisition times, 293–294
schematic diagrams, 292–294
Wnt5A pathway, 317, 326
Wnt pathway/*Xenopus* studies
axis formation manipulations
dorsalization, 319
dorsoanterior index, 320
secondary axes quantitation, 320
ventralization, 318–319
β-catenin stability analysis, 341–342
discussion, 304–308
dorsal and ventral injections
alternate pathway tests, 326–327
dorsalization, 322–323
purposes of, 321
specificity tests, 325
ventralization, 323–325
embryo extract preparation, 330–331
endogenous components, interfering with
dominant negative constructs, 328–329
engrailed repressor chimeras, 329
maternal component depletion, 327
zygotic component depletion, 328
expression/overexpression considerations, 345–347
gene expression detection
in situ hybridization, 344
reporter genes, 345
RT-PCR, 342–344
GSK activity detection, 340
molecular interaction detection
immunofluorescence, 340
immunoprecipitation, 339
mRna injection into *Xenopus* embryos
alternate pathway tests, 326–327
concentration considerations, 345
dorsalization, 322–323
purposes of, 321
specificity tests, 325
ventralization, 323–325
patterning during development, 305–308
properties of pathway
β-catenin considerations, 316–317, 327, 346–347
inhibitors, 317
table of components, 310–315
upstream activators, 309, 316
Wnt5A, 317
subcellular localization analysis
cell fractionation, 335–339
immunofluorescence, 331–333
immunogold labeling, 333–334
Wortmannin, 218
Wounding, 184
Wound method of motility assay, 171

X

Xenopus embryo studies, *see* Wnt pathway/*Xenopus* studies

Y

Yeast two-hybrid screen
autotrophic selection importance, 47
bait plasmids preparation
construct importance, 35
protocols, 35–37
yeast strain transformation methods, 37–38
bait plasmids testing
autoactivation protocol, 38–39
expression protocol, 39–40
cDNA library screens, confirmed baits, 40–42
false positive isolation, 42, 47
interaction plasmid rescue, 43–44
interaction trap, 34, 35
materials and reagents, 45–46
signal transduction mechanism, 34–35